CHEMICAL SAFETY DATA SHEETS

Volume 4b:

Toxic chemicals (m-z)

The Royal Society of Chemistry
Thomas Graham House
Science Park
Milton Road
Cambridge CB4 4WF
UK

Product team

Rebecca Allen (Editor)

Louise Catterick

Carol Fletcher

Jo Halsey

Michael Hannant

Gill Lyons

Colin Newham

Cover Design: John Tanton

Data Sheet Design: Iwan Thomas

Preface

Stuart Luxon

The information contained in this book has been compiled with the greatest care from sources believed to be reliable and to represent the best opinion on the subject as of late 1991. However, no warranty, guarantee, or representation is made by the Royal Society of Chemistry as to the correctness of any information herein, and the publisher assumes no responsibility in connection therewith: nor can it be assumed that all necessary warnings and precautionary measures are contained in this publication, or that other or additional information or measures may not be required or desirable because of particular or exceptional circumstances, or because of new or changed legislation.

©**Royal Society of Chemistry 1991**

ISBN 0-85186-321-3

Photocomposed by
Indispensable Publications Ltd.
Bugbrooke, Northamptonshire
Printed by Staples Printers Rochester Ltd.,
Kent

CONTENTS

	PREFACE	iv-vi
	INTRODUCTION	vii-xiv
80	Magnesium phosphide	1-3
81	Malononitrile	4-6
82	Mercury	7-11
83	Mercury(II) bromide	12-14
84	Mercury(II) chloride	15-18
85	Mercury(II) cyanide	19-21
86	Mercury(II) iodide	22-24
87	Mercury(II) nitrate	25-27
88	Mercury(II) oxide	28-31
89	Methacrylonitrile	32-34
90	*N*-Methylaniline	35-37
91	2-Methylaziridine	38-40
92	Methyl chloroformate	41-43
93	Methyl isocyanate	44-48
94	1-Methyl-3-nitro-1-nitrosoguanidine	49-51
95	Methyloxirane	52-55
96	Neopentyl glycol diacrylate	56-58
97	Nickel tetracarbonyl	59-62
98	Nicotine	63-66
99	5-Nitroacenaphthene	67-69
100	*m*-Nitroaniline	70-72
101	*o*-Nitroaniline	73-75
102	*p*-Nitroaniline	76-78
103	2-Nitro-*p*-anisidine	79-81
104	Nitrobenzene	82-85
105	Nitronaphthalene	86-88
106	2-Nitropropane	89-92
107	*o*-Nitrotoluene	93-95
108	*p*-Nitrotoluene	96-98
109	5-Nitro-*o*-toluidine	99-101
110	Osmium tetroxide	102-104
111	Pentachloroethane	105-108
112	Pentachlorophenol	109-112
113	*p*-Phenetidine	113-115
114	Phenol	116-121

115	*m*-Phenylenediamine	122-125
116	*p*-Phenylenediamine	126-129
117	Phenylhydrazine	130-132
118	Phenyl oxirane	133-136
119	Phosgene	137-140
120	Phosphorus, white (yellow)	141-144
121	4-Picoline	145-147
122	Potassium fluorosilicate	148-150
123	1,3-Propanesultone	151-153
124	3-Propanolide	154-156
125	Propargyl alcohol	157-159
126	Propyl chloroformate	160-162
127	Resorcinol	163-166
128	Selenium	167-170
129	Selenium dioxide	171-174
130	Selenium disulphide	175-177
131	Selenium hexafluoride	178-180
132	Selenium hydride	181-183
133	Selenium monochloride	184-186
134	Selenium monosulphide	187-189
135	Selenium tetrachloride	190-192
136	Strontium chromate	193-195
137	Sulphur dioxide	196-199
138	1,1,2,2-Tetrabromoethane	200-202
139	1,1,2,2-Tetrachloroethane	203-206
140	2,3,4,6-Tetrachlorophenol	207-209
141	Thioglycolic acid	210-212
142	*o*-Tolidine	213-216
143	Toluene-2,4-diisocyanate	217-220
144	*o*-Toluidine	221-224
145	*p*-Toluidine	225-227
146	Trichloroacetonitrile	228-230
147	Trichloroaniline	231-233
148	Trichloronitromethane	234-236
149	Triphenyltin acetate	237-239
150	Triphenyltin hydroxide	240-242
151	Uranium	243-246
152	Vinyl chloride	247-250
153	Vinyl cyclohexene diepoxide	251-253

154	Xylenols	254-256
155	Xylidines	257-260
156	Zinc chromate	261-264
157	Zinc phosphide	265-268
	Index of Names and Synonyms	269-280
	Index of CAS Registry Numbers	281-282

PREFACE

Over the past 25 years there have been a number of disastrous industrial incidents involving the escape of toxic chemicals. Unfortunately some of these have given rise to serious adverse effects not only inside the premises but also in the surrounding populated areas. The emotive reporting of these events has focussed public attention on the dangers associated with such substances and has given rise to concern regarding their safe use.

The incidents have also highlighted the fact that, either due to a lack of knowledge of the hazards or to a reckless disregard of available information, disastrous incidents can occur which can be prevented if relatively simple, well known anticipatory precautionary measures are taken. While the absence or incompleteness of underlying scientific knowledge to identify and quantify the hazards, particularly in the long term, is a risk with which we must live, ignoring available documented knowledge is a crime against all humanity, and brings everyone involved into disrepute. Indeed, to many members of the public, chemicals are seen as posing a threat to future living standards and it needs to be demonstrated that industry is taking a responsible attitude to the problems posed by their manufacture and use if such anxieties are to be alleviated.

To meet this worldwide concern detailed mandatory requirements have been promulgated in all developed countries which require an overall assessment of the risk any substance imposes in respect of workers, the public and the environment from the moment it is first produced until its final disposal. This requires a careful consideration of the properties of each substance, its physical form, method of use and ultimate disposal as waste.

This book sets out in a clear standardised form information in respect of some 78 toxic substances, from which an informed assessment of the hazards can be made and the necessary control measures devised.

Perhaps the most important and detailed legislative requirement is the Control of Substances Hazardous to Health Regulations which came into force in October 1989. These Regulations and the associated Codes of Practice require that, except when the risk is insignificant, a written assessment be made of the nature and degree of risk arising during the storage, use and disposal of every hazardous substance at every place at which persons are employed. They further require that adequate steps be taken to control any such risk that may be identified so as to protect not only employees but also any member of the public who may be affected by the work activity.

Risk to health is dependent on the inherent hazard of the substance and the degree of exposure of persons to that hazard. While the effects of exposure to chemicals are normally dose related and can therefore be quantified, many irritating substances may produce sensitisation in allergic individuals. After such sensitisation has occurred, exposure to very small amounts of certain chemicals may produce severe symptoms. It is important that such persons are identified and removed from all sources of exposure as soon sensitisation is suspected.

Using this book, the hazardous properties of the substance can be identified and quantified. Any hazard must then be eliminated or controlled by reducing personal exposure. Thus, if another substance can be used, presenting a lesser hazard, this should be the primary consideration. If, for process reasons, as is unfortunately often the case, this proves impossible, then suitable percautionary measures to adequately control exposure must be implemented.

It is important to realise that risk from toxic substances can occur in three ways. Firstly by external contact with the eyes, mucous membranes and skin, secondly if the substances are present in the inhaled air as gases, dusts or vapours, by inhalation and thirdly by ingestion. Assessment of the overall hazard must take account of all these routes of exposure. It will therefore be necessary to ascertain whether the substance can be inhaled, swallowed, absorbed through or irritate the skin or mucous membranes or can be injected through the skin by sharp objects. Good standards of personal hygiene normally render insignificant the dangers from ingestion.

In order to assess in more detail what control measures may be required, it is necessary to look at the work activity involving the use of the hazardous substance and the method and duration of worker exposure. The first step must be to minimise the numbers at risk and to reduce, as far as possible, the duration of exposure of those essential to the process under review.

Eye protection should be worn and where the danger exists of possible percutaneous absorption, impermeable gloves, gauntlets and protective clothing should be provided as appropriate. All equipment should be carefully selected to ensure that it is totally impermeable to all the chemicals to which the wearer may be exposed during normal work, maintenance and in dealing with spillages. Operatives should be given suitable instruction and training in its use and routine maintenance. Where the hazard is a serious one, cotton inner garments should be worn. These should be regularly laundered and will also serve as a check on any pin holes or leaks in the outer garments by indications of localised soiling.

If the presence of airborne contaminants in significant quantities is suspected it may be necessary to carry out personal monitoring to ascertain whether or not the limit values for airborne concentrations have been exceeded, and hence whether or not the control measures are adequate.

The process itself must be considered and suitable percautions devised to reduce the escape of gases, dusts or vapours into the working environment. A detailed consideration of this subject is outside the scope of this work but it must be appreciated that an accepted hierarchy of control measures exists.

   (a) Totally enclosed process and handling systems.
   (b) Partial enclosure and exhaust ventilation.
   (c) Local exhaust ventilation.
   (d) Good house keeping to remove dust deposits etc.
   (e) General ventilation.
   (f) Personal protection equipment.

Respiratory protective equipment should be considered only as a last resort except for specific tasks or very short duration where it would be uneconomic or impossible to provide alternative control measures. Where such equipment is used a rigorous training, selection and maintenance programme must be instituted.

The enclosures ventilating equipment and all other control measures must be regularly tested and maintained so as to ensure they are in efficient working order. If there is any doubt as to their adequacy, personal monitoring should be carried out as indicated above.

Training and instruction must play an important part in any system of work. Proper labelling will identify any hazards. The Classification, Packaging and Labelling Regulations (1984) set out the legislative requirements. The appropriate symbols, Risk and Safety phrases are set out in each data sheet where a legal requirement currently exists. It is recommended that a smiliar label should be used for every substance based on the parameters set out in the relevant CPL Regulations using the LD_{50} or LC_{50} set out in the Data Sheet.

To meet public concern in respect of environmental nature the Environmental Protection Act of 1990 consolidated and extended earlier enactments and provided a system of integrated pollution control. It introduced the concept of "Best Available Techniques not Entailing Excessive Costs" to control the escape of toxic chemicals into the environment. Detailed practical guidance has been published by the Department of the Environment entitled "Integrated Pollution Control. - A Practical Guide."

The Act also further regulates the disposal of solid wastes and requires a strict system of recording and licencing such activities.

I believe the book provides a concise systematic treatment of the subject that is clear and visually pleasing, which will enable the user to identify his overall responsibilities in respect of the listed chemicals. Where reliable information is not currently available or where exposure limits have not been published this is indicated. Current (1990-1991) relevant data derived from official sources in the UK, USA, France, Germany and Sweden together with information from the international Maritime Organisation, United Nations, World Health Organisation and International Labour Organisation etc. have been included so as to make the work as authoritative and up to date as possible.

Stuart G. Luxon FRSC CChem FIOH Dip. Occ. Hyg.

Health and Safety Consultant

Chair, RSC Health, Safety and Environment Committee

GUIDE TO THE USE OF THE DATA SHEETS

INTRODUCTION

So far as is possible these data sheets have been written to conform to the requirements of the Health & Safety Executive. In establishing a requirement to provide information on the hazards of chemicals used at work, a guidance booklet (*HS(G)27: Substances for use at work: the provision of information, HSE, 1985*) sets out minimum standards for data required to inform users of hazardous chemicals at work and the risks of chemicals for supply. In this booklet, it is stated that there is no single model format which can adequately and effectively cover all the categories of information required under Section 6 of the *Health & Safety at Work Act* and that the slavish use of a standard format could interfere with the clear and emphatic presentation of critical terms. As different formats of data sheets are required for different purposes, the purpose of this compilation of data sheets is to provide the essential data required for producing a data sheet, in terms of information on legislation, labelling and transportation data, agreed risk and safety phrases and other regulations pertaining to the safe use of chemicals.

In addition, it is intended that the data sheets will provide the opportunity for employers to fulfil their requirements under *COSHH (Control of Substances Hazardous to Health Regulations, 1989)* by providing additional information from the scientific and technical literature to enable risk assessment of handling and using chemicals. Every effort has been made to simplify the language used to describe physical and biological hazards. In referring to scientific literature, emphasis is placed on published literature which is freely available rather than manufacturers' literature which may not be as easily available. In the following paragraphs, details are given on the use of the various sections included within the data sheets. Details are also given on the references used to compile the information and the authority and status of this information.

RISK AND SAFETY PRECAUTIONS

Risk and safety precautions phrases used in connection with these data sheets are approved phrases for describing the risks involved in the use of hazardous chemicals and have validity in the United Kingdom and throughout the countries of the European Community. The approved texts have designated R (Risk) and S (Safety) numbers from which it is possible to provide translations for all approved languages adopted by the European Community. The risk and safety precautions phrases relate to the *Classification, Packaging and Labelling Regulations, 1984* and *The Road Traffic (Carriage of Dangerous Substances in Packages etc.) Regulations, 1986*. The information is provided in *The Authorised and Approved List. Information approved for the classification, packaging and labelling of dangerous substances for supply and conveyance by road (third edition) HSC, HMSO, 1990*. The risk and safety phrases should be used to describe the hazards of chemicals on data sheets for **use** and **supply**; for labelling of containers, storage drums, tanks etc., and for labelling of articles specified as dangerous for conveyance by road. The law specifies **supply** labels as those required to warn/inform the user of acute and chronic exposure hazards. They are to be placed on receptacles from which a dangerous substance is to be dispensed. The law also specifies **conveyance** labels used to inform the transporter, emergency services and the public, which must be placed on the outer packaging and which must take into account the acute risks posed by the substance. Further details on the use and compilation of labels may be found in the HSE publication *HS(R)22 A Guide to the Classification, Packaging and Labelling of Dangerous Substances Regulations, 1984*. In combination with the information provided under the **Road Transportation** section, it is possible to derive suitable **supply** and **conveyance** labels.

IDENTIFIERS

The identifiers section of the data sheet provides data on synonyms and approved codes used to define the precise chemical structure of the compound.

SYNONYMS

For common chemicals, several chemical names and numerous trade names may be applied to describe the chemical in question. In the synonyms field, many of these names are identified to aid users on the range of names which have been used to describe each substance. This information is also provided in the full index of alternative names available in the index.

CHEMICAL ABSTRACTS NUMBER

The Chemical Abstracts Registry Number is a number sequence adopted by the Chemical Abstracts Service (American Chemical Society, Columbus, Ohio, USA) to uniquely identify specific chemical substances. Where more than one Registry Number is included in this field, it indicates that a specific Registry Number is available for each specific chemical isomer of a substance and another is used where the precise isomer is not fully identified or the substance is a mixture of isomers. The Chemical Abstracts Registry Number has found many legislative uses, for example in the European Inventory of Existing Chemical Substances (EINECS) and the Toxic Substances Control Act (TSCA). The number contains no information relating to the chemical structure of a substance and is, in effect, a catalogue number relating to one of the 9 million or so unique chemical substances recorded in the Chemical Abstracts Registry: new numbers are assigned sequentially to each new compound identified by Chemical Abstracts Service.

NIOSH NUMBER

The NIOSH Number is a similar number sequence system used by the National Institute of Occupational Safety and Health, Cincinnati, USA. This system has been adopted by several organisations to list information on the toxicity and hazards of chemicals. Examples of these publications include *The Registry of Toxic Effects of Chemicals* and the *International Register of Potentially Toxic Chemicals United Nations Environment Programme*.

HAZCHEM CODE

The Hazchem Code is a code used by United Kingdom emergency services to classify hazards of chemicals transported by road. It is administered by the Chemical Industries Association.

UN NUMBER

The United Nations Number is a four-figure code used to identify hazardous chemicals and is used for identification of chemicals transported internationally by road, rail and by air.

THRESHOLD LIMIT VALUES

The airborne limits of permitted concentrations of hazardous chemicals represent conditions under which it is believed that nearly all workers may be repeatedly exposed day after day without adverse effect. These limits are subject to periodic revision and vary between different countries. The term *threshold limit* relates primarily to the USA, but equivalent terms are available in most industrialised countries. In these data sheets, comparable values are available for the USA, United Kingdom, West Germany, France and Sweden. The data relates to concentrations of substances expressed in *parts per million (ppm)* and *milligrams per cubic metre (mg/m^3)*. Additional assignations are given within the listings of limits, in the various countries, to indicate supplementary risks of chemicals, such as designation of carcinogenicity, sensitising effects and skin absorption hazards.

USA THRESHOLD LIMITS

The threshold limit values for the USA have been taken from the *Threshold Limit Values and Biological Exposure Indices, 1991-1992* produced by the American Conference of Governmental Industrial Hygienists, Cincinnati, USA. The limits relate to *Threshold Limit-Time Weighted Average, Threshold Limit-Short Term Exposure Limit* and *Threshold Limit-Ceiling Limit*. The Threshold Limit Value-Time Weighted Average (TLV-TWA) allows a time-weighted average concentration for a normal 8-hour working day and a 40-hour working week, to which nearly all workers may be repeatedly exposed day after day, without adverse effect. The Threshold Limit Value-Short Term Exposure Limit (TLV-STEL) is defined as a 15-minute, time-weighted average which should not be exceeded at any time during a workday, even if the 8-hour time-weighted average is within the TLV. It is designed to protect workers from chemicals which may cause irritancy, chronic or irreversible tissue damage, or narcosis likely to cause the likelihood of accidental injury. Many STELs have been deleted pending further toxicological assessment. With Threshold Limit-Ceiling Values (TLV-C) the concentration should not be exceeded during any part of the working day.

UK EXPOSURE LIMITS

The occupational limits relating to airborne substances hazardous to health are published by the Health and Safety Executive annually in *Guidance Note EH 40*. The values in the data sheets have been taken from the 1991 edition. Further changes to airborne occupational exposure limits are made by the Advisory Committee on Toxic Substances and advance notice of changes are reported in the Toxic Substances Bulletin, available from the Health and Safety Executive, St. Hugh's House, Stanley Precinct, Bootle, Merseyside L20 3QY.

In the United Kingdom, there are **Maximum Exposure Levels (MEL)** which are subject to regulation and which should not normally be exceeded. They derive from Regulations, Approved Codes of Practice, European Community Directives, or from the Health and Safety Commission. In addition, there are **Occupational Exposure Standards (OES)** which are considered to represent good practice and realistic criteria for the control of exposure. In an analogous fashion to the USA Threshold Limits, there are long-term limits, expressed as time-weighted average concentrations over an 8-hour working day, designed to protect workers against the effects of long-term exposure. The short-term exposure limit is for a time-weighted average of 10 minutes. For those substances for which no short-term limit is listed, it is recommended that a figure of three times the long-term exposure limit averaged over a 10-minute period be used as a guideline for controlling exposure to short-term excursions. In general, for most substances, the main route of entry is by inhalation but certain substances may penetrate the intact skin and become absorbed into the body. These substances are marked by the "Sk" notation. Other substances may cause allergic sensitisation such as a skin rash or asthma. Subsequent exposures of sensitised individuals may invoke a reaction at a level well below the exposure limit.

German MAK Values

Exposure limits in force in Germany are published by the Geschaftsstelle der Deutschen Forschungsgemeischaft, 40 Kennedyallee, D-5300 Bonn 2, in *Maximum Concentrations at the Workplace and Biological Tolerance Values for Working Materials, VCH Verlagsgesellschaft mbH, D-6940 Weinheim, West Germany*. The MAK value (Maximum Arbeitsplatz Konzentration) is defined as a maximum permissible concentration of a chemical compound present in the air within a working area, which according to current knowledge, does not impair the health of the employee or cause undue annoyance. Under these conditions, exposure can be repeated and of long duration over a daily period of eight hours, constituting an average working week of 40 hours. Within the MAK system for establishing airborne limit for exposure, there are special regulations adopted for carcinogenic substances and chemicals capable of producing embryotoxic and foetotoxic effects in pregnancy. In the case of carcinogenic substances there are 3 categories. Class A1 contains working materials which have been unequivocally proven carcinogenic by experience with humans and includes 2-naphthylamine, vinyl chloride and zinc chromate. In class A2 are compounds which, in the commission's opinion, have been proven to be unmistakably carcinogenic in animal experimentation. This class includes nickel carbonyl, 5-nitroacenaphthene, 2-nitronaphthalene, 2-nitropropane, pentachlorophenol, 1,3-propane sultone, β-propiolactone, propylene oxide, *o*-tolidine, *o*-toluidine and 4-vinyl-1-cyclohexane dioxide. For these substances, MAK values cannot be assigned as safe tolerance values cannot be established. In these cases a TRK value (Technische Richtkonzentrationen) is provided. In class B, results of cancer research suggest that further clarification is needed of the carcinogenic risk from exposure to the chemicals in question. MAK values for phenyl hydrazine, 1,1,2,2-tetrachloroethane, *p*-toluidine and 2,4-xylidine are tentatively retained and are subject to reevaluation and reassignment of carcinogenic potential. With regard to chemicals capable of causing hazards in pregnancy to the developing embryo or foetus, there are 3 groups of which group A contains chemicals from which an unequivocal risk of damage to the embryo or foetus has been established. In group B the risk of damage to the foetus or embryo must be considered to be probable and exposure of pregnant women may produce damage to the developing organism, even when MAK values are adhered to. Group C contains chemicals of which there is no reason to fear a risk of damage to the developing embryo or foetus when MAK values are adhered to. Group D contains compounds for which classification in groups A-C is not yet possible because although the data may indicate a trend, they are not sufficient for a final evaluation.

France

Exposure limits for France are published in *Valeurs Limites pour les Concentrations des Substances Dangereuses dans l'air des Lieux de Travail, INRS, ND 1653-129-87*, available from the Institut National de Recherche et de Securite, 30 Rue Olivier-Noyer, 75680 Paris cedex 14. The occupational exposure limits are *Valeurs des moyennes d'exposition (VME)* which is a time-weighted average value based on exposures to workers throughout the course of an 8-hour working day and the *Valeurs Limites d'exposition (VLE)* which is equivalent to a maximum concentration which not be exceeded in a 15-minute exposure.

Sweden

The Swedish Exposure Limits have been translated from Ordinance AFS 1987:12 issued by the National Swedish Board of Occupational Safety and Health and updated by the Newsletter of the National Board of the Occupational Safety and Health (Arbetarskyddsstyrelsen), S-17184 Solna, Sweden. The limit values in force in Sweden are the long-term limit (level limit value) covering exposure for a full working day, a ceiling limit value covering a maximum value of exposure during a 15-minute work period, and a short-term value, which is a time-weighted average value during 15 minutes. Certain carcinogenic substances may not be produced, used, or otherwise handled: these include β-naphthylamine. Permission is required from the Swedish Labour Inspectorate for handling 1,3-propane sultone, β-propiolactone and *o*-tolidine. Other carcinogenic substances are recognised by the Swedish authorities, and include nickel carbonyl, 2-nitropropane and vinyl chloride.

PHYSICAL PROPERTIES

Description

A simple description of the substance is given together with an indication of its odour and special properties such as hygroscopic nature or volatility.

Boiling point

The boiling point data are derived from various sources.

Melting point

The melting point data are derived from various sources.

Density

The density of each substance has been derived from a variety of sources. Wherever possible the data have been standardised.

Vapour density

The vapour density of each substance is derived from a variety of sources (water = 1.0).

Vapour pressure

The vapour pressure of each substance is expressed in millimetres of mercury at a particular temperature (usually around 20°C).

Flash point

The flash point is the lowest temperature at which the vapours of a volatile combustible substance will sustain combustion in air when exposed to a flame. The flash-point information is derived from various sources. Wherever possible the method of determination of the flash point is given.

Explosive limits

The lower explosive limit (LEL) is the minimum concentration (% by volume) of the vapour in air below which a flame is not propagated when an ignition source is present. The upper explosive limit (UEL) is the maximum concentration (% by volume) in air above which a flame is not propagated. The range becomes wider with increasing temperatures and in oxygen-rich atmospheres. The information has been derived from various sources, notably the National Fire Protection Association *Manual of Hazardous Reactions* and Sax, N. Irving *Dangerous Properties of Industrial Materials*.

Autoignition temperature

The autoignition temperature is the minimum temperature required to initiate or cause self-sustained combustion independent of the heat source. The information has been taken from various sources, notably Sax, N. Irving *Dangerous Properties of Industrial Materials*.

Solubility

Solubility information has been derived from various sources.

PACKAGING AND TRANSPORTATION

In this section, details on labelling of packages and containers, maximum quantities allowable for transportation, and transportation risk classification are presented for the carriage of dangerous substances by road, sea and air.

Road transportation

The information presented for the transportation of substances dangerous for conveyance by road is derived from *Information Approved for the Classification, Packaging and Labelling of Dangerous Substances for Supply and Conveyance by Road, HSC, 1990,* and *Approved Substance Identification Numbers, Emergency Action Codes and Classification for Dangerous Substances in Road Tankers and Tank Containers, HSC, 1989.* This gives details on the design of an appropriate hazard warning sign which must provide the United Nations approved Substance identification number, the hazard warning symbol, subsidiary risk symbols and the Hazchem Code.

Flammable Liquid   Toxic Substance

Example of Supply Label – Aniline

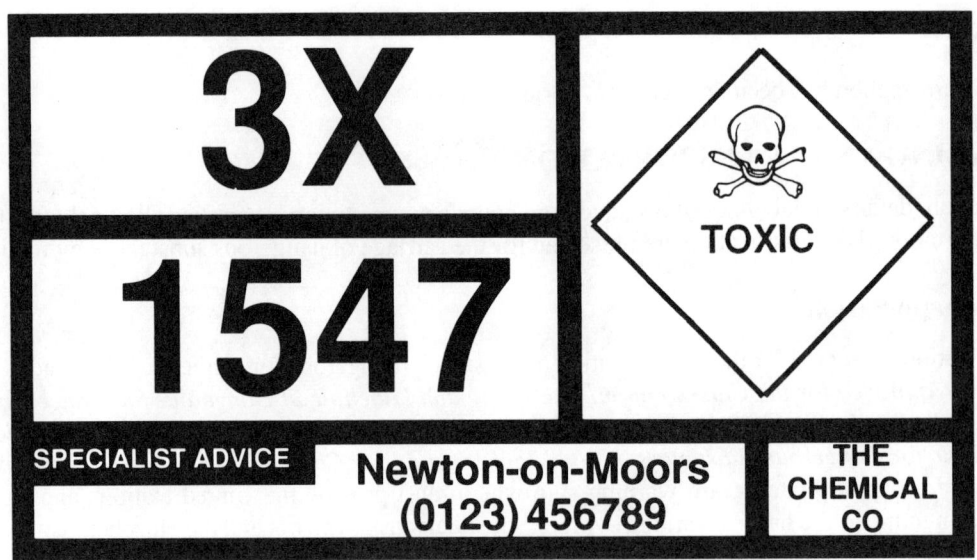

Example of Hazchem Tanker Sign (Vehicle Placard) - Aniline

Sea transportation

The transportation of hazardous cargo at sea is controlled by the International Convention for the Safety of Life at Sea. The International Maritime Dangerous Goods Code (IMDG Code) was introduced in 1965 in order to harmonise the practices and procedures followed by countries engaged in the carriage of dangerous goods by sea. The IMDG Code is administered by the International Maritime Organisation, 4 Albert Embankment, London SE1 7SR. The IMDG classification divides hazardous cargoes into 9 classes - Explosives, compressed gases, flammable liquids, flammable solids, substances liable to spontaneous combustion, oxidising substances, poisonous substances, radioactive materials, corrosives, and miscellaneous dangerous substances. Class 3 is subdivided by flash point. In this section, the IMDG Code, class, labelling information, and packaging group are given for each substance.

Air transportation

Details of regulations governing the transportation of dangerous goods by air are administered by the International Civil Aviation Organization (ICAO) and the International Air Transportation Agency (IATA). Dangerous goods are classified into the same 9 categories as the IMDG classification. Substances are identified using the United Nations Substance Identification Number. Details are given on the ICAO/IATA code, class, packaging group, packaging instructions, and maximum quantities allowed on cargo and passenger aircraft.

MANUFACTURE

Details of principal methods of manufacture of each substance are given. This information has been derived mainly from Kirk-Othmer *Encyclopedia of Chemical Technology*.

USES

Principal uses of the substances are given, with information on other significant uses in industrial processes.

CHEMICAL HAZARDS

Information on hazardous chemical reactions which have been reported in the scientific literature is included with special reference to information supplied in Bretherick's *Handbook of Reactive Chemical Hazards* (4th edition) and *National Fire Protection Association, Manual of Hazardous Reactions* (9th edition).

BIOLOGICAL HAZARDS

Details of the principal toxic effects from exposure, the metabolism, and specific legislation relating to the biological effects of handling the substances are covered in the following sections. The information is sub-divided into effects from vapour inhalation, contact with the eyes, skin contact, adverse effects resulting from absorption of the substance through the skin, and adverse effects resulting from swallowing the substance. Where possible, studies of human exposure are included, supplemented with toxicity studies conducted on animals.

CARCINOGENICITY

In view of the emotive response to the subject of carcinogenic substances, it is recommended that the information provided in this section is evaluated by the occupational hygienist. Information on the carcinogenicity of substances unequivocally proven to cause carcinogenicity in humans and laboratory animals, together with equivocal data from carcinogenicity assays in laboratory animals are covered where it has not been possible to establish an authoritative evaluation of the data. Where it has not been possible to find information on the carcinogenic potential of a particular substance, this is indicated

MUTAGENICITY

The potential of chemicals to induce changes in genetic materials at the gene or chromosome level has been extensively reviewed by reporting results of mutagenicity assays such as the Ames *Salmonella typhimurium* mutagenicity assay and other *in vitro* systems together with any evidence available from *in vivo* studies.

REPRODUCTIVE HAZARDS

Within this section, information on human experience and animal reproductive toxicity experiments has been extensively reviewed to provide an indication of the potential hazards from exposure to chemicals to pregnant workers, women of child-bearing age, and on male workers. In relation to fertility, sexual organ dysfunction, developmental changes to embryos and foetuses, teratogenicity, malformations, increases in spontaneous abortions or stillbirths, and other effects such as impotence, menstrual disorders or mental impairment in offspring. Where no information has been found to determine the potential for effects on reproduction, this is indicated.

FIRST AID

General instructions on first aid are given in relation to accidental exposure of the substance by eye contact, inhalation of vapours, ingestion of the liquid or contact of the substance with the skin. The first aid instructions are designed to be simple instructions, capable of being enacted at the site of the exposure, although, in certain cases, more advanced first aid instructions are given to aid the victim in cases where prompt treatment will effect more immediate relief from the symptoms of exposure.

HANDLING AND STORAGE

Information in this section includes choice of protective clothing materials, gloves, respiratory protection and breathing apparatus in response to particular hazards from exposure to the chemicals. Compatible materials for chemical plant and containers for the storage of materials are covered with general handling and storage instructions for the prevention of fire, explosions, or other accident. General instructions on flammability hazards and laboratory working procedures are included.

DISPOSAL

In this section, details of procedures for cleaning up spills of chemical and disposal of waste quantities of chemical are covered with special reference to the recommendations of the United Nations Environment Programme outlined in *Treatment and Disposal Methods for Waste Chemcials, International Register of Potentially Toxic Chemicals, IRPTC Data Profile Series No. 5* (UNEP, Geneva, Switzerland, 1985), and with information from *Dangerous Chemicals Emergency Spillage Guide* (Wolters Samson (UK) Ltd).

FIRE PRECAUTIONS

Within the fire precautions section, details on precautions to be adopted when tackling chemical fires, suitable fire extinguishing agents, unsuitable fire extinguishing agents, and hazards involving the generation of toxic and irritating fumes are given.

FURTHER READING

In the further reading section, significant monographs regarding the hazards and toxicity of the chemical substances are indicated. Full details of the major works are given below.

Dangerous Properties of Industrial Materials. Sax, N. Irving, Lewis, R. J. Van Nostrand Reinhold., 7th edition, 1989.

Documentation of the Threshold Limit Values and Biological Exposure Indices. American Conference of Governmental Industrial Hygienists Inc., 5th edition, 1986.

Kirk-Othmer Encyclopedia of Chemical Technology. 1-24. Wiley, 1979-1983.

Encyclopaedia of Occupational Safety & Health. International Labour Organisation, 3rd edition, 1983.

Hazards in the Chemical Laboratory. Bretherick, L. (Editor). Royal Society of Chemistry, 4th edition, 1986.

Handbook of Reactive Chemical Hazards. Bretherick, L. (Editor). Butterworths, 4th edition, 1990.

Patty's Industrial Hygiene and Toxicity. Volumes *1, 2A, 2B, 2C, 3*. Clayton, G. D.; Clayton, F. E. (Editors). Wiley, 1978-1982.

80. Magnesium Phosphide

MAGNESIUM PHOSPHIDE

Mg_3P_2

RISKS
Contact with water liberates toxic, highly flammable gas – Very toxic if swallowed (R15/29, R28)

SAFETY PRECAUTIONS
Keep locked up and out of reach of children – Do not breathe dust – In case of fire, use carbon dioxide, dry chemical powder, alcohol foam, or polymer foam. Never use water – In case of accident or if you feel unwell, seek medical advice immediately (show label where possible) (S1/2, S22, S43, S45)

IDENTIFIERS
SYNONYMS	none
CHEMICAL ABSTRACTS No.	12057-74-8
NIOSH No.	OM 4200000
HAZCHEM CODE	not available
UN No.	2011

THRESHOLD LIMIT VALUES
US TLV (TWA)	not available
US TLV (STEL)	not available

UK EXPOSURE LIMITS (OES)
Long-term (8 hr TWA value)	not available
Short-term (10 min TWA value)	not available

Germany
MAK	not available

France
VME	not available
VLE	not available

Sweden
Short-term limit	not available
Level limit	not available

PHYSICAL PROPERTIES
Description not available.
Boiling point	not available
Melting point	not available
Density	not available
Vapour density	not available
Vapour pressure	not available
Flash point	not available
Explosive limits	not available
Autoignition temperature	not available

Solubility not available.

PACKAGING AND TRANSPORTATION

Road transportation
hazard warning sign	2011 substance which in contact with water emits flammable gas; toxic
Hazchem code	not available

Sea transportation
IMDG page No.	4352
class	4.3
label	dangerous when wet; poison
packaging group	I

Air transportation
ICAO/IATA code (UN No.)	2011
class	4.3;6.1
label	danger if wet; poison
packaging group	I
packing instructions cargo	412
passenger	forbidden
passenger aircraft max. quantity	forbidden
cargo aircraft max. quantity	15 kilograms

80. Magnesium Phosphide

MANUFACTURE

Magnesium phosphide can be formed by direct combination of the elements, by reacting molten magnesium with liquid or gaseous phosphorus at temperatures above 650°C (1).

USES

Magnesium phosphide is used as a fumigant for the control of insect pests in stored food products (2).

CHEMICAL HAZARDS

Magnesium phosphide is flammable when exposed to heat, flame or oxidising materials. It reacts with water to liberate phosphine and may ignite. It ignites on heating in chlorine, bromine or in iodine vapours at higher temperatures. Reaction with nitric acid is incandescent. It emits toxic fumes of phosphorus oxides and phosphine when heated to decomposition.

BIOLOGICAL HAZARDS

Magnesium phosphide is moderately toxic if inhaled. Little toxicity data seems to be available.

Vapour Inhalation

Magnesium phosphide is moderately toxic by inhalation. The LC_{50} (1 hour) in rats is 580 ppm.

Eye Contact

No information is available concerning the effects on the eyes.

Skin Contact

No information is available concerning the effects on the skin.

Swallowing

No information is available on the effects of ingestion.

CARCINOGENICITY

No information is available concerning the carcinogenicity of magnesium phosphide.

MUTAGENICITY

No information is available concerning the mutagenicity of magnesium phosphide.

REPRODUCTIVE HAZARDS

No information is available concerning the reproductive hazards of magnesium phosphide.

FIRST AID

Eyes Wash the eye with flowing water for 10 minutes.

Lungs Remove casualty from area of exposure. If unconscious, do not give anything to drink, give artificial ventilation and chest compression or place in the recovery position as necessary. If conscious make the casualty lie or sit down quietly, give oxygen if available. Lung congestion may occur – a conscious casualty with breathing difficulties should be placed in a sitting position.

Mouth Do not make the casualty vomit. Treat unconscious casualties as for lungs but if conscious give 1 pint of water to drink; give repeated drinks of water (1 cupful every 10 minutes.) Lung congestion may occur – a conscious casualty with breathing difficulties should be placed in a sitting position.

Skin Remove contaminated clothing immediately, drench the affected area with running water for a least 10 minutes.

In all cases of exposure, the patient should be transferred to hospital as soon as possible.

HANDLING AND STORAGE

Magnesium phosphide should be handled wearing an approved respirator, chemical-resistant gloves, safety goggles and other protective clothing. It should be kept in a tightly closed container in a cool, dry place, away from heat and flame.

80. Magnesium Phosphide

DISPOSAL

Eliminate all sources of ignition and ventilate the area. Wearing a laboratory coat or overalls, safety glasses, gloves and for larger spills, approved self-contained breathing apparatus, absorb the spill onto sand, vermiculite or paper towels and scoop into plastic bags. Burn in an incinerator or arrange removal by a licenced contractor.

FIRE PRECAUTIONS

Fires involving magnesium phosphide should be extinguished using carbon dioxide, dry chemical powder, alcohol foam or polymer foam.

FURTHER READING

Bretherick, L. *Hazards in the Chemical Laboratory* (4th edition)

Bretherick, L. *Handbook of Reactive Chemical Hazards* (4th edition)

Sax, N. Irving *Dangerous Properties of Industrial Materials* (7th edition)

Encyclopaedia of Occupational Safety & Health

Patty's *Industrial Hygiene and Toxicology*

Kirk-Othmer *Encyclopedia of Chemical Technology*

National Fire Protection Association *Manual of Hazardous Reactions*

ACGIH Documentation of TLVs and BEIs (6th edition, 1986)

REFERENCES

1. Dorn, F. W. (Hoechst A. G.) Ger. Offen. 2,943,905 (Cl. C01B25/08) 1981
2. *European Directory of Agrochemcial Products volume 3: Insecticides* (4th edition) (RSC, London, 1990)

81. Malononitrile

MALONONITRILE
NCCH$_2$CN

RISKS
Toxic by inhalation, in contact with skin and if swallowed (R23/24/25)

SAFETY PRECAUTIONS
Do not breathe vapour – Take off immediately all contaminated clothing (S23, S27)

IDENTIFIERS

SYNONYMS malonic dinitrile; RCRA Waste No. U149; propanedinitrile; cyanoacetonitrile; dicyanomethane; malonic acid dinitrile; malonodinitrile; methylene cyanide; methylenedinitrile; USAF A-4600

CHEMICAL ABSTRACTS No.	109-77-3
NIOSH No.	OO 3150000
HAZCHEM CODE	2X
UN No.	2647

THRESHOLD LIMIT VALUES

US TLV (TWA)	not available
US TLV (STEL)	not available
UK EXPOSURE LIMITS (OES)	
Long-term (8 hr TWA value)	not available
Short-term (10 min TWA value)	not available
Germany	
MAK	not available
France	
VME	not available
VLE	not available
Sweden	
Short-term limit	not available
Level limit	not available

PHYSICAL PROPERTIES

Description	Colourless crystalline solid.
Boiling point	220°C
Melting point	30.5°C
Density	1.049 at 34°C
Vapour density	not available
Vapour pressure	not available
Flash point	130°C (open cup)
Explosive limits	not available
Autoignition temperature	not available
Solubility	Soluble in water, alcohol or ether.

PACKAGING AND TRANSPORTATION

Road transportation

hazard warning sign	2647 toxic substance
Hazchem code	2X

Sea transportation

IMDG page No.	6172
class	6.1
label	poison
packaging group	II

Air transportation

ICAO/IATA code (UN No.)	2647
class	6.1
label	poison
packaging group	II
packing instructions	
cargo	615
passenger	613
passenger aircraft max. quantity	25 kilograms
cargo aircraft max. quantity	100 kilograms

81. Malononitrile

MANUFACTURE

Malononitrile is produced commercially by the Lonza process from cyanogen chloride and acetonitrile. Both the reactants and products present handling and apparatus problems due to corrosiveness and toxicity.

USES

Malononitrile is used in leaching gold from ores, in the synthesis of vitamin B_1, and in the production of pesticides and merocyanine dyes (which are used in colour photography).

CHEMICAL HAZARDS

Malononitrile may polymerise violently on heating at 130°C, or in contact with strong bases at lower temperatures. It may decompose spontaneously above 100°C, particularly in the presence of impurities. A partially filled drum kept in an oven for 2 months at 70-80°C exploded violently (1). Hydrogen cyanide is released if malononitrile is exposed to air and ultra-violet radiation for a long time. Its properties and reactivity have been reviewed (2).

BIOLOGICAL HAZARDS

Workers with heart problems or who cannot smell malononitrile well should not work with this compound. Symptoms of poisoning are similar to cyanide and include methaemoglobinaemia, and may be delayed 10-20 minutes.

Vapour Inhalation

Malononitrile may cause methaemoglobinaemia.

Eye Contact

Malononitrile is a severe eye irritant.

Skin Contact

Malononitrile is a local irritant and may be absorbed through the skin.

Swallowing

The LD_{50} in mice is 18.6 mg/kg (3).

CARCINOGENICITY

No information is available concerning the carcinogenicity of malononitrile.

MUTAGENICITY

No information is available concerning the mutagenicity of malononitrile.

REPRODUCTIVE HAZARDS

No information is available concerning the reproductive hazards of malononitrile.

FIRST AID

Eyes Wash the eye with flowing water.

Lungs Remove casualty from area of exposure, wearing protective clothing and approved breathing apparatus if necessary. If unconscious, do not give anything to drink. If breathing has stopped DO NOT use mouth to mouth or mouth to nose ventilation but use a resuscitation bag and mask instead. If breathing, place in the recovery position. Break two amyl nitrate capsules open under the casualty's nose so that the vapour is inhaled. If Kelo-cyanor (dicobalt edetate) and personnel trained in its use are available and you are certain the casualty has been poisoned by cyanide, give it immediately. NOTE: KELO-CYANOR IS EXTREMELY DANGEROUS WHEN GIVEN TO ANYONE NOT SUFFERING FROM CYANIDE POISONING. If conscious make the casualty lie or sit down quietly. Oxygen may be beneficial.

Mouth Do not make the casualty vomit. Treat as for inhalation. If conscious give 1 pint of water immediately.

Skin Remove all contaminated clothing, wash affected areas with soap and copious amounts of water. Absorption through the skin may result in symptoms similar to those of ingestion and inhalation.

In all cases of exposure, the patient should be transferred to hospital as soon as possible.

81. Malononitrile

HANDLING AND STORAGE

Malononitrile should be handled wearing an approved respirator, rubber or PVC gloves, safety goggles and other protective clothing. It should only be used in a chemical fume hood. It should be kept in a tightly sealed container, in a cool dry place (4).

DISPOSAL

Eliminate all sources of ignition, ventilate the area and wearing self-contained breathing apparatus, a chemical resistant suit, gloves, and PVC safety boots, spread calcium hypochlorite over the (liquid) spill (or disperse excess calcium hypochlorite solution onto it). Mix and mop into polythene buckets, stand 24 hours and run to waste diluting x1000 with running cold water (see discharge limits from water authorities).

For solid spills either: a) sweep up, place in a large volume of water, add excess sodium hypochlorite, leave to stand 24 hours, then run to waste diluting x1000 with running water; or b) scoop into a large container, make alkaline with 2% sodium hydroxide solution and stir well. Add excess ferrous sulphate solution, mix well, stand for 2-3 hours and run to waste diluting x1000 with running water.

Large spills need specialist help – contact fire brigade.

FIRE PRECAUTIONS

Fires involving malononitrile should be extinguished using water spray, carbon dioxide, dry chemical powder, alcohol foam or polymer foam.

FURTHER READING

Sax, N. Irving *Dangerous Properties of Industrial Materials* (7th edition)

Encyclopedia of Occupational Safety & Health

Patty's *Industrial Hygiene and Toxicology*

Kirk-Othmer *Encyclopedia of Chemical Technology*

National Fire Protection Association *Manual of Hazardous Reactions*

ACGIH Documentation of TLVs and BEIs (6th edition, 1986)

REFERENCES

1. Anon. *CISHC Chem. Safety Summ.* 1978, **49**, 29
2. Freeman, F. *Chem. Rev.* 1969, (5), 591-624
3. Panov, Iv. K. *J. Eur. Toxicol.* 1969, **2**(6), 292-299
4. Lenga, R. E. *The Sigma-Aldrich Library of Chemical Safety Data* (2nd edition) (Sigma-Aldrich, Milwaukee, 1988)

MERCURY
Hg

RISKS
Toxic by inhalation – Danger of cumulative effects (R23, R33)

SAFETY PRECAUTIONS
Keep container tightly closed – If you feel unwell, seek medical advice (show label where possible) (S7, S44)

IDENTIFIERS

SYNONYMS quicksilver; NCI-C60399; RCRA Waste No. U151

CHEMICAL ABSTRACTS No.	7439-97-6
NIOSH No.	OV 4550000
HAZCHEM CODE	not available
UN No.	2809

THRESHOLD LIMIT VALUES

US TLV (TWA)	0.5 mg/m^3 (vapour)
US TLV (STEL)	not available

UK EXPOSURE LIMITS (OES)
Long-term (8 hr TWA value) .... 0.05 mg/m^3
Short-term (10 min TWA value) .... 0.15 mg/m^3

Germany
MAK .... 0.01 ppm (0.1 mg/m^3)

France
VME .... 0.05 mg/m^3 (vapour)
VLE .... not available

Sweden
Short-term limit .... not available
Level limit .... 0.05 mg/m^3 (vapour)

PHYSICAL PROPERTIES

Description Heavy silver mobile liquid.

Boiling point	356.9°C
Melting point	-38.89°C
Density	13.534 at 25°C
Vapour density	not available
Vapour pressure	2x10^{-3} mm at 25°C
Flash point	not available
Explosive limits	not available
Autoignition temperature	not available

Solubility Insoluble in water or organic solvents. Attacked by nitric acid or hot concentrated sulphuric acid.

PACKAGING AND TRANSPORTATION

Road transportation
hazard warning sign	not available
Hazchem code	not available

Sea transportation
IMDG page No.	8191
class	8
label	corrosive
packaging group	III

Air transportation
ICAO/IATA code (UN No.)	2809
class	8
label	corrosive
packaging group	I
packing instructions	
cargo	803
passenger	803
passenger aircraft max. quantity	35 kilograms
cargo aircraft max. quantity	35 kilograms

82. Mercury

82. Mercury

MANUFACTURE

Mercury occurs in several ores, but is obtained commercially by roasting cinnabar (the sulphide) in a rotary kiln or shaft furnace.

USES

Mercury is used in scientific and electrical equipment; in the manufacture of amalgams (eg. in dentistry); in the electrolytic production of chlorine and sodium hydroxide; and as a catalyst in polyurethane foam production.

CHEMICAL HAZARDS

Mercury reacts explosively with: dry tetracarbonylnickel (on shaking); peroxyformic acid (1); chlorine dioxide (on shaking); or 3-bromopropyne (2). Methyl silane explodes if shaken with mercury in oxygen (3). Concentrated solutions of silver perchlorate in 2-pentyne or 3-hexyne explode on contact with mercury (4). Fittings containing mercury should not be used for ethylene oxide service; traces of acetylene in the oxide could form metal acetylides able to detonate the vapour. Mercury ignites in a stream of chlorine at 200-300°C, and boron diiodophosphide ignites in contact with mercury vapour. Mercury reacts violently with dry bromine or sodium acetylide. It reacts violently and exothermically with sodium, rubidium, potassium, and amalgam formation with calcium at 390°C is violent. Mercury reacts similarly with lithium, but if large pieces of lithium are used the reaction may be explosive (5). A better method of forming lithium amalgam using p-cymene is available (6). Methyl azide is less stable in the presence of mercury (7).

BIOLOGICAL HAZARDS

Mercury metal or its vapour is a protoplasmic poison. Elemental mercury is particularly hazardous if spilled or heated. Many cases of subcutaneous or intravenous injection of mercury from broken thermometers or suicide attempts have been reported (8). The toxicity of inorganic mercury has been reviewed (9).

Vapour Inhalation

Inhalation of mercury vapour is the mose important route of uptake of elemental mercury (9). Elemental mercury vaporises at room temperature, and is readily absorbed by the lungs. It is then rapidly absorbed and distributed by the blood; about 1% is deposited in the mammalian brain where it is retained for a long time, and the rest is transported to the liver and kidneys where it is excreted through bile and urine.

Inhalation of very high concentrations causes acute pulmonary oedema and interstitial pneumonitis which may be fatal. In non-fatal cases dyspnoea and coughing may persist (8). Other effects include swollen mouth and gums, salivation, abdominal pain, chest pains, nausea, vomiting and diarrhoea (10). Kidney effects may occur at exposure levels lower than those causing central nervous system effects (9). Rabbits exposed to 28.8 mg/m^3 for 4 hours suffered severe brain, liver, kidney, heart and colon damage (10). The symptoms of chronic mercurialism are markedly different, affecting the central nervous system. Symptoms include tremors, erethism (changes in behaviour such as depression, despondency and irritability), and loss of appetite and weight (11). Erethism was more common in workers exposed to levels above 0.05 mg/m^3, but tremor was reported at both this and lower levels (12). Insomnia was reported in workers exposed to vapour concentrations between 0.05 and 0.1 mg/m^3 (13). After exposure ceases, the signs and symptoms of neurological impairment may regress slowly in milder cases, but may be exacerbated in more severe cases (9). A more uncommon result of subacute or chronic mercurialism is acrodynia, characterised by peeling skin on the feet and hands (8). Several reviews of human mercury poisoning are available (14,15), including a review of the neurotoxic effects of occupational exposure (16).

Eye Contact

It may cause eye irritation (17).

Skin Contact

Mercury is absorbed through intact skin causing symptoms outlined above. Contact dermatitis from mercury amalgam fillings (18) and mercury sensitivity in dental students have been reported (19). Mercury vapour may cause "Kawasaki" disease, which seems to be immunologically mediated and is similar to Pink disease (9). In a study on human volunteers, the rate of uptake of mercury vapour by the skin was calculated to be 2.2% that of the lungs, hence skin absorbtion poses a minor occupational hazard compared to that of inhalation, unless concentrations greatly exceed the TLV or if mercury droplets are trapped in clothes or boots (20).

Swallowing

In rats only small amounts of mercury are absorbed after oral administration of the liquid metal (8). Occasional incidental swallowing of metallic mercury is without harm.

CARCINOGENICITY

The excess risk of glioblastoma reported in Swedish dental workers is probably due to an occupational factor such as amalgam, chloroform or radiography (9). In a study involving intraperitoneal injection of pure metallic mercury only 12 of the 39 rats survived longer than 22 months; peritoneal sarcomas developed in 5 out of these 12 (21). A more recent review states "no positive report that mercury could be carcinogenic in man has appeared up to now and animal experiments have also provided negative results" (22). The WHO reported no evidence that inorganic mercury is carcinogenic (9).

MUTAGENICITY

The mutagenicity of organic and inorganic mercury compounds has been reviewed; organic mercury compounds were negative in the rec assay with *Bacillus subtilis*, but reportedly induced sex-linked recessive lethals in *Drosophila melanogaster*; in plant material, organic and inorganic mercury compounds inactivated the spindle-fibre mechanism at cell division causing aneuploidy and/or polyploidy (22). A study of chloralkali plant workers exposed to metallic mercury did not find significant increases in chromosome aberrations in peripheral lymphocytes (this test is not considered a good indicator of *in vivo* chemical damage) (22). Both positive and negative results have been reported in tests for chromosome aberrations in workers exposed to metallic mercury vapour (9).

REPRODUCTIVE HAZARDS

That mercury crosses the placenta has been demonstrated in the foetus of a monkey that had been exposed to mercury vapour (23). Complications in pregnancy and childbirth, low birth weight (24), menstrual disturbances, spontaneous abortion, premature deliveries, and an increased incidence of malformations (9,25) have been reported in women occupationally exposed to mercury. There was a significant, positive association between total mercury levels (TMLs) in the hair of occupationally exposed (dentistry) women and the occurrence of reproductive failures in their history. The relation between TMLs in the scalp hair and the prevalence of menstrual cycle disorders was statistically significant (26). The death of a 3 month old baby with signs of mercury poisoning born to a woman working in the manufacture of mercury soap was reported (27). Embryotoxicity and teratogenicity of organic mercury compounds have been reported in many test systems (22). In pregnant rats, acute exposure to elemental mercury vapour caused increased resorptions, while chronic exposure caused foetuses with cranial defects (9). The standard of epidemiological studies is considered inadequate to decide whether mercury vapour adversely affects fertility or foetal development in the absence of the well-known signs of mercury poisoning (9).

FIRST AID

Eyes Wash the eye with flowing water for 10 minutes.

Lungs Remove casualty from area of exposure. If unconscious, do not give anything to drink. Give artificial ventilation and chest compression or place in the recovery position as necessary. If conscious make the casualty lie or sit down quietly, give oxygen if available. Convulsions may occur and may cause unconsciousness.

Mouth Do not make the casualty vomit. Treat unconscious casualties as for lungs but if conscious give 1 pint of water to drink.

Skin Remove all contaminated clothing, wash affected areas with soap and copious amounts of water. Absorption through the skin may result in symptoms similar to those of ingestion or inhalation.

In all cases of exposure, the patient should be transferred to hospital as soon as possible.

HANDLING AND STORAGE

A high standard of cleanliness is needed, and wooden floors should be avoided, as well as cracks and crevices where mercury could collect. Containers should be kept in trays to catch spills. Processes involving mercury should be fully enclosed or exhaust ventilation provided. Protective clothing should be worn to prevent skin contact, and respirators when vapour exposure is possible. Processes involving heating mercury should be carried out in a fume cupboard with local exhaust ventilation (28). Safe working practices have been reviewed (29). Mercury should be kept in containers covered by a layer of water, in a fireproof building, away from acetylene, ammonia and azides (17).

DISPOSAL

Ventilate the area, wear a laboratory coat, rubber gloves, self-contained breathing apparatus or an approved canister respirator for mercury vapour. For small spills of metallic mercury, remove by suction into a plastic bottle. For small spills of metallic mercury compounds, EITHER treat all affected areas with a 50:50 mixture of flowers of sulphur and calcium hydroxide in water; OR cover with zinc powder. Leave for 12 hours and then remove dried yellow mixture with water and reclean surfaces. For other mercury compounds, cover with sand, using 10-20 times by weight. Then arrange for

82. Mercury

removal by a licenced contractor. If the compound is soluble in water, small amounts may be dissolved in a bucket of water, diluted and the solution run to waste, if trade effluent regulations permit this.

For large spills contact the fire-brigade and police before taking action.

FIRE PRECAUTIONS

Mercury is not combustible, so an extinguisher suitable for surrounding fire conditions should be used.

FURTHER READING

Bretherick, L. *Hazards in the Chemical Laboratory* (4th edition)

Bretherick, L. *Handbook of Reactive Chemical Hazards* (4th edition)

Sax, N. Irving *Dangerous Properties of Industrial Materials* (7th edition)

Encyclopaedia of Occupational Safety & Health

Patty's *Industrial Hygiene and Toxicology*

Kirk-Othmer *Encyclopedia of Chemical Technology*

National Fire Protection Association *Manual of Hazardous Reactions*

ACGIH Documentation of TLVs and BEIs (6th edition, 1986)

Dangerous Prop. Ind. Mater. Rep. 1981, **1**(3), 70-72

Mercury and its hazards *Cah. Suiss. Secur. Trav.* 1988, (145), 1-45

REFERENCES

1. D'Ans, J.; et al. *Ber.* 1915, **48**, 1136
2. *Dangerous Substances: Guidance on Fires and Spillages* (Section 1: Inflammable Liquids) (HMSO, London, 1972) p. 27
3. Stock, A.; et al. *Ber.* 1919, **52**, 706
4. Comyns, A. E.; et al. *J. Am. Chem. Soc.* 1957, **79**, 4324
5. Smith, G. McP.; et al. *J. Am. Chem. Soc.* 1909, **31**, 799
6. Alexander, J.; et al. *J. Chem. Educ.* 1970, **47**, 277
7. Currie, C. L.; et al. *Can. J. Chem.* 1963, **41**, 1048
8. Gosselin, R. E.; et al. *Clinical Toxicology of Commercial Products* (5th edition) (Williams & Wilkins, Baltimore, 1984)
9. *Environmental Health Criteria 118: Inorganic Mercury* (WHO, Geneva, 1991)
10. Ashe, W. F.; et al. *Arch. Ind. Hyg. Occup. Med.* 1953, **7**, 19
11. Smith, R. G.; et al. *Am. Ind. Hyg. Assoc. J.* 1970, **31**, 687
12. Turrian, H.; et al. *Schweiz. Med. Wochenschr.* 1956, **86**, 109
13. Danziger, S. J.; Possick, P. A. *J. Occup. Med.* 1973, **15**, 15
14. Gerstner, H. B.; et al. NTIS *Report* 1977, ORNL/TIRC 77/1
15. Burt, S. D. *AAOHN. J.* 1986, **34**(11), 543-546
16. Foa, V. *Drug. Chem. Toxicol.* 1985, **3**(Neurotoxicology), 323-343
17. *International Chemical Safety Cards: Mercury* (CEC, Luxembourg, 1990)
18. Feuerman, E. J.; et al. *Int. J. Dermatol.* 1975, **14**, 657
19. White, R. R.; et al. *J. Am. Dent. Assoc.* 1976, **92**, 1204
20. Hursh, J. B.; et al. *Arch. Environ. Health.* 1989, **44**(2), 120-127
21. Druckrey, H.; et al. *Zeitschrift fur Krebsforschung* 1957, **61**, 511-519
22. Leonard, A.; et al. *Mutat. Res.* 1983, **114**(1), 1-18
23. Smith, R. G.; et al. *Unpublished results of work supported by the Chlorine institute, Wayne State University, Michigan,* 1970

24. Mishonova, V.; et al. *Gig. Tr. Prof. Zabol.* 1980, (2), 21-23
25. Marinova, G.; et al. *Probl. Akush Ginekol.* 1973, **1**, 75-77
26. Sikorski. R.; et al. *Int. Arch. Occup. Environ. Health* 1987, **59**(6), 551-557
27. Fogg, E.; Jennings, A. *New. Sci.* 1985, **106**(1456), 9
28. *Lab. Hazards Bull.* May 1986, 16-19
29. Moore, D. J.; Timbs, A. E. *Chem. Br.* 1984, **20**(7), 622-624

83. Mercury(II) Bromide

MERCURY(II) BROMIDE

$HgBr_2$

RISKS
Very toxic by inhalation, in contact with skin and if swallowed – Danger of cumulative effects (R26/27/28, R33)

SAFETY PRECAUTIONS
Keep locked up and out of reach of children – Keep away from food, drink and animal feeding stuffs – After contact with skin, wash immediately with plenty of water – In case of accident or if you feel unwell, seek medical advice immediately (show label where possible) (S1/2, S13, S28, S45)

IDENTIFIERS

SYNONYMS mercury bromide ($HgBr_2$); dibromomercury; mercuric bromide; mercuric dibromide; mercury dibromide

CHEMICAL ABSTRACTS No.	7789-47-1
NIOSH No.	OV 7415000
HAZCHEM CODE	not available
UN No.	1634

THRESHOLD LIMIT VALUES

US TLV (TWA)	0.1 mg/m^3
US TLV (STEL)	not available

UK EXPOSURE LIMITS (OES)
Long-term (8 hr TWA value) ... 0.05 mg/m^3 (as Hg)
Short-term (10 min TWA value) .. 0.15 mg/m^3 (as Hg)

Germany
MAK .... not available

France
VME .... 0.1 mg/m^3
VLE .... not available

Sweden
Short-term limit .... not available
Level limit .... 0.05 mg/m^3 (as Hg)

PHYSICAL PROPERTIES

Description White crystals or crystalline powder. Sensitive to light.

Boiling point	sublimes at 322°C
Melting point	237°C
Density	6.109 at 25°C
Vapour density	not available
Vapour pressure	1 mm Hg at 136.5°C
Flash point	not available
Explosive limits	not available
Autoignition temperature	not available

Solubility Very soluble in hot alcohol, methanol, hydrogen chloride, hydrogen bromide, alkali bromide solutions. Slightly soluble in chloroform.

PACKAGING AND TRANSPORTATION

Road transportation
hazard warning sign	1634 toxic substance
Hazchem code	not available

Sea transportation
IMDG page No.	6179
class	6.1
label	marine pollutant; poison
packaging group	II

Air transportation
ICAO/IATA code (UN No.)	1634
class	6.1
label	poison
packaging group	II
packing instructions	
cargo	615
passenger	613
passenger aircraft max. quantity	25 kilograms
cargo aircraft max. quantity	100 kilograms

83. Mercury(II) Bromide

MANUFACTURE

Mercury(II) bromide is produced by precipitation, using mercury(II) nitrate and sodium bromide solution. The washed compound is dried at below 75°C.

USES

Mercury(II) bromide has been used for a few medicinal purposes.

CHEMICAL HAZARDS

Mercury(II) bromide reacts vigorously with indium at 350°C (1,2). It is incompatible with sodium or potassium (3).

BIOLOGICAL HAZARDS

Acute exposure to mercury compounds causes gastrointestinal symptoms, anuria and uraemia, but is uncommon in industry. The signs and symptoms of chronic mercurialism include tremors and neuropsychiatric disturbances.

Vapour Inhalation

Inhalation is the main route of entry into the body for both elemental mercury vapour and mercury compound dusts. Acute exposure to high concentrations can cause gingivitis, abdominal cramps and diarrhoea, cough, fever, restlessness, bronchitis and inflammation of the lungs (4).

Eye Contact

Mercury(II) bromide may cause irritation (5).

Skin Contact

The dermal LD_{50} in rats is 100 mg/kg (5). Inorganic mercury compounds cause dermatitis (6).

Swallowing

The oral LD_{50} in rats and mice is 40 mg/kg and 35 mg/kg respectively (4). Acute mercury toxicity is usually characterised by shock, cardiovascular collapse, kidney failure and severe gastrointestinal damage (6).

CARCINOGENICITY

The WHO reported no evidence that inorganic mercury was carcinogenic, and is classified by the U.S. EPA in group O (not classifiable as to human carcinogenicity) (6).

MUTAGENICITY

No information specific to mercury(II) bromide is available, but other inorganic mercury compounds are mutagenic in bacterial and mammalian assays (6).

REPRODUCTIVE HAZARDS

No information is available concerning the reproductive hazards of mercury(II) bromide, but other inorganic mercury compounds are teratogenic in animal studies and there have been reports suggesting inorganic mercury compounds cause spontaneous abortion in women (6).

FIRST AID

Eyes Wash the eye with flowing water for 10 minutes.

Lungs Remove casualty from area of exposure. If unconscious, do not give anything to drink. Give artificial ventilation and chest compression or place in the recovery position as necessary. If conscious make the casualty lie or sit down quietly, give oxygen if available. Convulsions may occur and may cause unconsciousness.

Mouth Do not make the casualty vomit. Treat unconscious casualties as for lungs but if conscious give 1 pint of water to drink.

Skin Remove all contaminated clothing, wash affected areas with soap and copious amounts of water. Absorption through the skin may result in symptoms similar to those of ingestion or inhalation.

In all cases of exposure, the patient should be transferred to hospital as soon as possible.

83. Mercury(II) Bromide

HANDLING AND STORAGE

Mercury(II) bromide should be handled wearing an approved respirator, chemical-resistant gloves, safety goggles and other protective clothing. It should only be used in a chemical fume hood. In plants producing mercurials, disposable uniforms and disposable mercury vapour-absorbing masks should be used. There should be adequate ventilation, good housekeeping, and steel trays should be used in order to catch spills. Floors should be non-porous and washed regularly with calcium sulphide solution. It should be kept in a tightly closed container, in a cool dry place (5). It should be protected from light.

DISPOSAL

Ventilate the area, wear a laboratory coat, rubber gloves, self-contained breathing apparatus or an approved canister respirator for mercury vapour. For small spills of metallic mercury, remove by suction into a plastic bottle. For small spills of metallic mercury compounds, EITHER treat all affected areas with a 50:50 mixture of flowers of sulphur and calcium hydroxide in water; OR cover with zinc powder. Leave for 12 hours and then remove dried yellow mixture with water and reclean surfaces. For other mercury compounds, cover with sand, using 10-20 times by weight. Then arrange for removal by a licensed contractor. If the compound is soluble in water, small amounts may be dissolved in a bucket of water, diluted and the solution run to waste, if trade effluent regulations permit this.
For large spills contact the fire-brigade and police before taking action.

FIRE PRECAUTIONS

Mercury(II) bromide is not combustible, and fire extinguishers appropriate to surrounding fire conditions should be used.

FURTHER READING

Bretherick, L. *Hazards in the Chemical Laboratory* (4th edition)

Bretherick, L. *Handbook of Reactive Chemical Hazards* (4th edition)

Sax, N. Irving *Dangerous Properties of Industrial Materials* (7th edition)

Encyclopaedia of Occupational Safety & Health

Patty's *Industrial Hygiene and Toxicology*

Kirk-Othmer *Encyclopedia of Chemical Technology*

National Fire Protection Association *Manual of Hazardous Reactions*

ACGIH Documentation of TLVs and BEIs (6th edition, 1986)

REFERENCES

1. *Handling Chemicals Safely* (Dutch Assoc. Saf. Experts, Dutch Chem. Ind. Assoc., Dutch Saf. Inst., 1980) p. 614
2. Leleu, J. *Cah. Notes Doc.* 1980, **98**, 131-133
3. Leleu, J. *Cah. Notes Doc.* 1975, **79**, 265-270
4. *Laboratory Hazards Data Sheet No. 47: Mercury and mercury compounds* (RSC, London, 1986)
5. Lenga, R. E. *The Sigma-Aldrich Library of Chemical Safety Data* (2nd edition) (Sigma-Aldrich, Milwaukee, 1988)
6. *Environmental Health Criteria 118: Inorganic Mercury* (WHO, Geneva, 1991)

84. Mercury(II) Chloride

MERCURY(II) CHLORIDE
HgCl$_2$

RISKS
Very toxic by inhalation, in contact with skin and if swallowed – Danger of cumulative effects (R26/27/28, R33)

SAFETY PRECAUTIONS
Keep locked up and out of reach of children – Keep away from food, drink and animal feeding stuffs – After contact with skin, wash immediately with plenty of water – In case of accident or if you feel unwell, seek medical advice immediately (show label where possible) (S1/2, S13, S28, S45)

IDENTIFIERS

SYNONYMS mercury chloride (HgCl$_2$); bichloride of mercury; Calochlor; corrosive sublimate; dichloromercury; Emisan 6; mercuric bichloride; mercuric chloride; mercury bichloride; mercury dichloride; mercury perchloride; sublimate; Sulem

CHEMICAL ABSTRACTS No.	7487-94-7
NIOSH No.	OV 9100000
HAZCHEM CODE	2X
UN No.	1624

THRESHOLD LIMIT VALUES

US TLV (TWA)	0.1 mg/m^3
US TLV (STEL)	not available

UK EXPOSURE LIMITS (OES)
Long-term (8 hr TWA value) ... 0.05 mg/m^3 (as Hg)
Short-term (10 min TWA value) .. 0.15 mg/m^3 (as Hg)

Germany
MAK .... not available

France
VME .... 0.1 mg/m^3
VLE .... not available

Sweden
Short-term limit .... not available
Level limit .... 0.05 mg/m^3 (as Hg)

PHYSICAL PROPERTIES

Description	Crystals or white granules or powder.
Boiling point	302°C
Melting point	277°C
Density	5.44 at 25°C
Vapour density	not available
Vapour pressure	1 mm Hg at 136°C
Flash point	not available
Explosive limits	not available
Autoignition temperature	not available

Solubility Soluble in water, alcohol, benzene, ether, glycerol, acetic acid, methanol, acetone, carbon disulphide or pyridine.

PACKAGING AND TRANSPORTATION

Road transportation
hazard warning sign	1624 toxic substance
Hazchem code	2X

Sea transportation
IMDG page No.	6175
class	6.1
label	marine pollutant; poison
packaging group	II

Air transportation
ICAO/IATA code (UN No.)	1624
class	6.1
label	poison
packaging group	II
packing instructions	
cargo	615
passenger	613
passenger aircraft max. quantity	25 kilograms
cargo aircraft max. quantity	100 kilograms

84. Mercury(II) Chloride

MANUFACTURE

Mercury(II) chloride is produced by direct oxidation of mercury with excess chlorine.

USES

Mercury(II) chloride is used as a catalyst in the production of other mercury salts, as an intermediate in organic synthesis, and in analytical chemistry. It is also used in fungicides; with sodium chloride in photography; in batteries; in preserving wood and anatomical specimens; in tanning leather; for etching and browning steel and iron; and in electroplating aluminium.

CHEMICAL HAZARDS

Mercury(II) chloride interacts with sodium *aci*-nitromethanide forming mercury nitromethanide which is converted by acids to mercury fulminate, which is endothermic and used as a detonator (1).

BIOLOGICAL HAZARDS

Mercury(II) chloride is a severe eye and skin irritant, and is poisonous if swallowed. It is particularly hazardous due to its water solubility and high vapour pressure. Acute exposure to mercury compounds causes gastrointestinal symptoms, anuria and uremia, but is uncommon in industry. The signs and symptoms of chronic mercurialism include tremors and neuropsychiatric disturbances. Its toxicity to the kidney has been reviewed (2). Mutagenic and reproductive effects have been reported.

Vapour Inhalation

Mercury(II) chloride is highly destructive to the upper respiratory tract and absorbtion via the lungs causes systemic effects which are detailed above (3).

Eye Contact

Mercury(II) chloride is a severe eye irritant. It may cause corneal ulceration (4).

Skin Contact

Mercury(II) chloride is a severe skin irritant. Patch testing of health care workers revealed that dentists and surgeons were commonly allergic to mercury(II) chloride (5). It is absorbed through the skin, and the dermal LD_{50} in rats is 41 mg/kg (3). "Pink disease" (irritability, insomnia, sweating, photophobia, rash and cold, painful and swollen extremities) has occurred in children before mercury(II) chloride in teething powder was withdrawn (6).

Swallowing

Ingestion causes nausea, vomiting, reduced urine volume or anuria and respiratory obstruction. In a case of attempted suicide by ingestion of about 2 g of mercury(II) chloride, symptoms included a choking feeling in the throat, and intense burning in the retrosternal and epigastric regions within minutes, followed by vomiting and haematemesis after an hour. The patient suffered acute kidney failure with rhabdomyolysis (7). Several cases of human poisoning, some fatal, have been reported. Gastrointestinal and kidney lesions are the commonest autopsy findings (6). The lowest lethal dose in humans has been reported as 29 mg/kg. The oral LD_{50} in rats is 1 mg/kg (3).

CARCINOGENICITY

The WHO reported no evidence that inorganic mercury was carcinogenic, and is classified by the U.S. EPA in group O (not classifiable as to human carcinogenicity). Mercury(II) chloride in the drinking water was not significantly tumourigenic to mice in lifetime studies (6).

MUTAGENICITY

Mercury(II) chloride is mutagenic in *Escherichia coli* (8). Positive results were reported in the rec assay with *Bacillus subtilis* (9). In an investigation of *in vitro* clastogenic capacity of mercury(II) chloride in human peripheral blood lymphocytes, a "rather important" clastogenic effect and dissociation of the acrocentrics was reported (10). The frequency of sister chromatid exchanges in cultured human lymphocytes was close to that for untreated controls (11). Mercury(II) chloride was not a potent inducer of dominant-lethal mutations in female mice (12). Subcutaneous injection of up to 12.8 mg Hg/kg to female golden hamsters caused a slight increase in incidence of chromosome aberrations in bone marrow cells (13). It also induced gene mutations in mouse lymphoma cells, DNA damage in mouse and rat fibroblasts, single stranded DNA breaks in cultured mouse embryo cells, and effects on DNA repair in mammalian cells (6). DNA damage has also been reported in cultured Chinese hamster ovary cells (14).

84. Mercury(II) Chloride

REPRODUCTIVE HAZARDS

Mercury(II) chloride was not a potent inducer of dominant-lethal mutations in female mice. Single intraperitoneal injection caused slight reduction in the number of implants and living embryos, and in long term reproductive performance in female mice (12). Foetuses in pregnant mice exposed to mercury(II) chloride fumes at levels of 2 and 0.2 mg/kg showed significant disorders. At 2 mg/kg 32% of foetuses died, and the survivors were seriously weight retarded, 68% showed serious bone growth retardation, and 38% showed liver chromosome aberrations (15). Oral adminstration of mercury(II) chloride to pregnant rats caused adverse foetal effects only at maternally toxic doses (16). In inhalation tests on pregnant rats, the no embryotoxic effect level was considered to be 0.000276 mg/m^3 (17). Abortion has been reported in a case of acute mercury intoxication (following ingestion of 2.5 g mercury(II) chloride). The abortion was attributed to foetal and placental mercury poisoning (18). Injection of mercury(II) chloride caused reduced fertility and alterations in testicular tissue of male mice and rats respectively (6).

FIRST AID

Eyes Wash the eye with flowing water for 10 minutes.

Lungs Remove casualty from area of exposure. If unconscious, do not give anything to drink. Give artificial ventilation and chest compression or place in the recovery position as necessary. If conscious make the casualty lie or sit down quietly, give oxygen if available. Convulsions may occur and may cause unconsciousness.

Mouth Do not make the casualty vomit. Treat unconscious casualties as for lungs but if conscious give 1 pint of water to drink.

Skin Remove all contaminated clothing, wash affected areas with soap and copious amounts of water. Absorption through the skin may result in symptoms similar to those of ingestion or inhalation.

In all cases of exposure, the patient should be transferred to hospital as soon as possible.

HANDLING AND STORAGE

Mercury(II) chloride should be handled wearing an approved respirator, chemical-resistant gloves, safety goggles and other protective clothing. It should only be used in a chemical fume hood (3). In plants producing mercurials, disposable uniforms and disposable mercury vapour-absorbing masks should be used. There should be adequate ventilation, good housekeeping, and steel trays should be used in order to catch spills. Floors should be non-porous and washed regularly with calcium sulphide solution. It should be kept in a tightly closed container, in a cool dry place (3) or refrigerator (4).

DISPOSAL

Ventilate the area, wear a laboratory coat, rubber gloves, self-contained breathing apparatus or an approved canister respirator for mercury vapour. For small spills of metallic mercury, remove by suction into a plastic bottle. For small spills of metallic mercury compounds, EITHER treat all affected areas with a 50:50 mixture of flowers of sulphur and calcium hydroxide in water; OR cover with zinc powder. Leave for 12 hours and then remove dried yellow mixture with water and reclean surfaces. For other mercury compounds, cover with sand, using 10-20 times by weight. Then arrange for removal by a licensed contractor. If the compound is soluble in water, small amounts may be dissolved in a bucket of water, diluted and the solution run to waste, if trade effluent regulations permit this.
For large spills contact the fire-brigade and police before taking action.

FIRE PRECAUTIONS

Mercury(II) chloride is not combustible, and fire extinguishers appropriate to surrounding fire conditions should be used.

FURTHER READING

Bretherick, L. *Hazards in the Chemical Laboratory* (4th edition)

Bretherick, L. *Handbook of Reactive Chemical Hazards* (4th edition)

Sax, N. Irving *Dangerous Properties of Industrial Materials* (7th edition)

Encyclopaedia of Occupational Safety & Health

Patty's *Industrial Hygiene and Toxicology*

Kirk-Othmer *Encyclopedia of Chemical Technology*

National Fire Protection Association *Manual of Hazardous Reactions*

ACGIH Documentation of TLVs and BEIs (6th edition, 1986)

84. Mercury(II) Chloride

REFERENCES

1. Nef, J. U. *Ann.* 1894, **280**, 263, 305
2. Oken, D. E. *Prog. Biochem. Pharmacol.* 1972, **7**, 219-247
3. Lenga, R. E. *The Sigma-Aldrich Library of Chemical Safety Data* (2nd edition) (Sigma-Aldrich, Milwaukee, 1988)
4. Keith, L. H.; Walters, D. B. (editors) *Compendium of Safety Data Sheets for Research and Industrial Chemicals* (VCH, Deerfield Park, 1987)
5. Rudzki, E.; et al. *Contact Dermatitis* 1989, **20**(4), 247-250
6. *Environmental Health Criteria 118: Inorganic Mercury* (WHO, Geneva, 1991)
7. Chugh, K. S.; et al. *Med. J. Aust.* 1978, **2**, 125-126
8. Partington, C. R. *Univ. Microfilms Int., Order No. 7809422* 1977
9. Kanematsu, N.; et al. *Mutat. Res.* 1980, **77**(2), 109-116
10. Verschaeve, L.; et al. *Mutat. Res.* 1985, **157**(2-3), 221-226
11. Ogawa, H. *Kyoto-furitsu Ika Daigaku Zassi* 1979, **88**(4), 505-539
12. Suter, K. E. *Mutat. Res.* 1975, **30**(3), 365-374
13. Watanabe, T.; et al. *Teratology* 1982, **25**, 381-384
14. Cantoni, O.; Costa, M. *Proc. Am. Assoc. Cancer Res.* 1983, **24**, 74
15. Selypes, A.; et al. *Munkavedelem* 1983, **29**(7-9), 152-154
16. McAnulty, P. A.; et al. *Teratology* 1982, **25**(1), 26A
17. Grins, N.; Govorunova, N. N. *Gig. Sanit.* 1981, (5), 67-68
18. Afonso, J. F.; de Alvarez, R. R. *Am. J. Obstet. Gynecol.* 1960, **80**, 145-154

85. Mercury(II) Cyanide

MERCURY(II) CYANIDE

$Hg(CN)_2$

RISKS
Very toxic by inhalation, in contact with skin and if swallowed – Danger of cumulative effects (R26/27/28, R33)

SAFETY PRECAUTIONS
Keep locked up and out of reach of children – Keep away from food, drink and animal feeding stuffs – After contact with skin, wash immediately with plenty of water – In case of accident or if you feel unwell, seek medical advice immediately (show label where possible) (S1/2, S13, S28, S45)

IDENTIFIERS

SYNONYMS   mercury cyanide (Hg(CN)$_2$); Cianurina; dicyanomercury; mercuric cyanide; mercury dicyanide

CHEMICAL ABSTRACTS No.	592-04-1
NIOSH No.	OW 1515000
HAZCHEM CODE	not available
UN No.	1636

THRESHOLD LIMIT VALUES

US TLV (TWA)	0.1 mg/m^3
US TLV (STEL)	not available

UK EXPOSURE LIMITS (OES)
Long-term (8 hr TWA value)   ...   0.05 mg/m^3 (as Hg)
Short-term (10 min TWA value)   ..   0.15 mg/m^3 (as Hg)

Germany
MAK   ....   not available

France
VME   ....   0.1 mg/m^3
VLE   ....   not available

Sweden
Short-term limit   ....   not available
Level limit   ....   0.05 mg/m^3 (as Hg)

PHYSICAL PROPERTIES

Description   Colourless odourless prisms, darkened by light.

Boiling point	decomposes at 320°C
Melting point	not available
Density	3.996
Vapour density	not available
Vapour pressure	not available
Flash point	not available
Explosive limits	not available
Autoignition temperature	not available

Solubility   Slightly soluble in ether. Soluble in water, alcohol or methanol.

PACKAGING AND TRANSPORTATION

Road transportation

hazard warning sign	1636 toxic substance
Hazchem code	not available

Sea transportation

IMDG page No.	6182
class	6.1
label	marine pollutant; poison
packaging group	II

Air transportation

ICAO/IATA code (UN No.)	1636
class	6.1
label	poison
packaging group	II
packing instructions	
cargo	615
passenger	613
passenger aircraft max. quantity	25 kilograms
cargo aircraft max. quantity	100 kilograms

85. Mercury(II) Cyanide

MANUFACTURE

Mercury(II) cyanide is produced by reaction of an aqueous slurry of yellow mercury(II) oxide with excess cyanide. The mixture is heated to 95°C, filtered, crystallised, isolated and dried.

USES

Mercury(II) cyanide can be used in photography, in the manufacture of cyanogen gas, and as an antiseptic.

CHEMICAL HAZARDS

Mercury(II) cyanide is a friction-and impact-sensitive explosive, and can initiate detonation of liquid hydrogen cyanide (1). It reacts explosively with sodium nitrite on heating (2), and the cyanogen released by thermal decomposition of mercury(II) cyanide reacts explosively with magnesium. It ignites in fluorine on gentle warming.

BIOLOGICAL HAZARDS

Mercury(II) cyanide may cause cyanosis (3). Acute exposure to mercury compounds causes gastrointestinal symptoms, anuria and uraemia, but is uncommon in industry. The signs and symptoms of chronic mercurialism include tremors and neuropsychiatric disturbances.

Vapour Inhalation

Mercury(II) cyanide may be fatal if inhaled (3). Symptoms include cough, choking, diarrhoea, breathing difficulties and kidney damage. Signs of cyanosis include dizziness, rapid breathing, headache, tiredness and unconsciousness (4).

Eye Contact

Mercury(II) cyanide may cause eye irritation (3,4).

Skin Contact

Mercury(II) cyanide may cause irritation (3). Skin absorption may cause diarrhoea, breathing difficulties and kidney damage (4).

Swallowing

The oral LD_{50} in rats and mice is 26 mg/kg and 33 mg/kg respectively (5). The lowest toxic dose in humans has been reported as 10 mg/kg, and toxic effects included nausea or vomiting, diarrhoea, kidney changes, tiredness, and hypermotility. Ingestion causes symptoms of both mercury and cyanide poisoning as outlined above; in dogs the effects of mercury appear more prominent, while in humans signs of cyanide poisoning appear first (4).

CARCINOGENICITY

The WHO reported no evidence that inorganic mercury was carcinogenic, and is classified by the U.S. EPA in group O (not classifiable as to human carcinogenicity) (6).

MUTAGENICITY

No information specific to mercury(II) cyanide is available, but other inorganic mercury compounds are mutagenic in bacterial and mammalian assays (6).

REPRODUCTIVE HAZARDS

No information is available concerning the reproductive hazards of mercury(II) cyanide, but other inorganic mercury compounds are teratogenic in animal studies and there have been reports suggesting inorganic mercury compounds cause spontaneous abortion in women (6).

FIRST AID

Eyes Wash the eye with flowing water.

Lungs Remove casualty from area of exposure, wearing protective clothing and approved breathing apparatus if necessary. If unconscious, do not give anything to drink. If breathing has stopped DO NOT use mouth to mouth or mouth to nose ventilation but use a resuscitation bag and mask instead. If breathing, place in the recovery position. Break two amyl nitrate capsules open under the casualty's nose so that the vapour is inhaled. If Kelo-cyanor (dicobalt edetate) and personnel trained in its use are available and you are certain the casualty has been poisoned by cyanide, give it immediately. NOTE: KELO-CYANOR IS EXTREMELY DANGEROUS WHEN GIVEN TO ANYONE NOT SUFFERING

85. Mercury(II) Cyanide

FROM CYANIDE POISONING. If conscious make the casualty lie or sit down quietly. Oxygen may be beneficial.

Mouth   Do not make the casualty vomit. Treat as for inhalation. If conscious give 1 pint of water immediately.

Skin   Remove all contaminated clothing, wash affected areas with soap and copious amounts of water. Absorption through the skin may result in symptoms similar to those of ingestion and inhalation.

In all cases of exposure, the patient should be transferred to hospital as soon as possible.

HANDLING AND STORAGE

Mercury(II) cyanide should be handled wearing an approved respirator, chemical-resistant gloves, safety goggles and other protective clothing. It should only be used in a chemical fume hood. In plants producing mercurials, disposable uniforms and disposable mercury vapour-absorbing masks should be used. There should be adequate ventilation, good housekeeping, and steel trays should be used in order to catch spills. Floors should be non-porous and washed regularly with calcium sulphide solution. It should be kept in a tightly closed container, in a cool, dry, well- ventilated place, away from light (3) and sources of ignition, and protected against physical damage. Outside or detached storage is preferred.

DISPOSAL

Eliminate all sources of ignition, ventilate the area and wearing self-contained breathing apparatus, a chemical resistant suit, gloves, and PVC safety boots, spread calcium hypochlorite over the (liquid) spill (or disperse excess calcium hypochlorite solution onto it). Mix and mop into polythene buckets, stand 24 hours and run to waste diluting x1000 with running cold water (see discharge limits from water authorities).

For solid spills either: a) sweep up, place in a large volume of water, add excess sodium hypochlorite, leave to stand 24 hours, then run to waste diluting x1000 with running water; or b) scoop into a large container, make alkaline with 2% sodium hydroxide solution and stir well. Add excess ferrous sulphate solution, mix well, stand for 2-3 hours and run to waste diluting x1000 with running water.

Large spills need specialist help – contact fire brigade.

FIRE PRECAUTIONS

Mercury(II) cyanide is not combustible and an extinguisher suitable to the surrounding fire conditions should be used (3).

FURTHER READING

Bretherick, L. *Hazards in the Chemical Laboratory* (4th edition)

Bretherick, L. *Handbook of Reactive Chemical Hazards* (4th edition)

Sax, N. Irving *Dangerous Properties of Industrial Materials* (7th edition)

Encyclopaedia of Occupational Safety & Health

Patty's *Industrial Hygiene and Toxicology*

Kirk-Othmer *Encyclopedia of Chemical Technology*

National Fire Protection Association *Manual of Hazardous Reactions*

ACGIH *Documentation of TLVs and BEIs* (6th edition, 1986)

REFERENCES

1. Wohler, L.; et al. *Chem. Ztg.* 1926, **50**, 761
2. Eiter, K.; et al. Austrian Pat. 176,784 (1953)
3. Lenga, R. E. *The Sigma-Aldrich Library of Chemical Safety Data* (2nd edition) (Sigma-Aldrich, Milwaukee, 1988)
4. *Dangerous Prop. Ind. Mater. Rep.* 1986, **6**(1), 68-72
5. Vernot, E. H.; et al. *Toxicol. Appl. Pharmacol.* 1977, **42**(2), 417-423
6. *Environmental Health Criteria 118: Inorganic Mercury* (WHO, Geneva, 1991)

86. Mercury(II) Iodide

MERCURY(II) IODIDE

HgI_2

RISKS
Very toxic by inhalation, in contact with skin and if swallowed – Danger of cumulative effects (R26/27/28, R33)

SAFETY PRECAUTIONS
Keep locked up and out of reach of children – Keep away from food, drink and animal feeding stuffs – After contact with skin, wash immediately with plenty of water – In case of accident or if you feel unwell, seek medical advice immediately (show label where possible) (S1/2, S13, S28, S45)

IDENTIFIERS

SYNONYMS mercury iodide (HgI_2); mercuric diiodide; mercuric iodide; mercury biiodide; mercury diiodide; red mercuric iodide

CHEMICAL ABSTRACTS No.	7774-29-0
NIOSH No.	OW 5250000
HAZCHEM CODE	not available
UN No.	1638

THRESHOLD LIMIT VALUES

US TLV (TWA)	0.1 mg/m^3
US TLV (STEL)	not available

UK EXPOSURE LIMITS (OES)
Long-term (8 hr TWA value) ... 0.05 mg/m^3 (as Hg)
Short-term (10 min TWA value) . 0.15 mg/m^3 (as Hg)

Germany
MAK ... not available

France
VME ... 0.1 mg/m^3
VLE ... not available

Sweden
Short-term limit ... not available
Level limit ... 0.05 mg/m^3 (as Hg)

PHYSICAL PROPERTIES

Description Scarlet red, heavy odourless, almost tasteless powder. Sensitive to light. At 130°C becomes yellow, and red again on cooling.

Boiling point	sublimes at approx. 350°C
Melting point	259°C
Density	6.28
Vapour density	not available
Vapour pressure	not available
Flash point	not available
Explosive limits	not available
Autoignition temperature	not available

Solubility Very soluble in alkali iodides, mercury chloride or sodium thiosulphate.

PACKAGING AND TRANSPORTATION

Road transportation

hazard warning sign	1638 toxic substance
Hazchem code	not available

Sea transportation

IMDG page No.	6183
class	6.1
label	marine pollutant; poison
packaging group	II

Air transportation

ICAO/IATA code (UN No.)	1638
class	6.1
label	poison
packaging group	II

packing instructions
cargo
liquid ... 612
solid ... 615
passenger
liquid ... 610
solid ... 613

passenger aircraft max. quantity
liquid ... 5 litres
solid ... 25 kilograms

86. Mercury(II) Iodide

MANUFACTURE

Mercury(II) iodide is prepared by precipitation from a solution of mercury(II) chloride with potassium iodide.

USES

Mercury(II) iodide is used as an analytical reagent, in photography and in the treatment of skin diseases. Mercury(II) iodide has been used in skin-lightening soaps and creams. Their distribution is banned in the EEC and several African countries, but they have been found to have been reimported illegally from certain African countries.

CHEMICAL HAZARDS

Mercury(II) iodide reacts violently, ignition often occurring, with chlorine tetrachloride at ambient or slightly elevated temperature.

BIOLOGICAL HAZARDS

Acute exposure to mercury compounds causes gastrointestinal symptoms, anuria and uraemia, but is uncommon in industry. The signs and symptoms of chronic mercurialism include tremors and neuropsychiatric disturbances (memory loss, insomnia, irritability and depression).

Vapour Inhalation

High concentrations are very destructive to the upper respiratory tract (1).

Eye Contact

High concentrations are very destructive to the eyes (1).

Skin Contact

High concentrations are very destructive to the skin (1). The dermal LD_{50} in rats is 75 mg/kg.

Swallowing

The lowest lethal dose in humans is 357 mg/kg. The oral LD_{50} in rats is 18 mg/kg.

CARCINOGENICITY

The WHO reported no evidence that inorganic mercury was carcinogenic, and is classified by the U.S. EPA in group O (not classifiable as to human carcinogenicity) (2).

MUTAGENICITY

No information specific to mercury(II) iodide is available, but other inorganic mercury compounds are mutagenic in bacterial and mammalian assays (2).

REPRODUCTIVE HAZARDS

Mercury(II) iodide is an experimental teratogen. Inhalation of 0.00487 mg/m^3 and 0.0248 mg/m^3 caused decreased fertility, foetal weight, anomalous growth and increased foetal mortality (3).

FIRST AID

Eyes Wash the eye with flowing water for 10 minutes.

Lungs Remove casualty from area of exposure. If unconscious, do not give anything to drink. Give artificial ventilation and chest compression or place in the recovery position as necessary. If conscious make the casualty lie or sit down quietly, give oxygen if available. Convulsions may occur and may cause unconsciousness.

Mouth Do not make the casualty vomit. Treat unconscious casualties as for lungs but if conscious give 1 pint of water to drink.

Skin Remove all contaminated clothing, wash affected areas with soap and copious amounts of water. Absorption through the skin may result in symptoms similar to those of ingestion or inhalation.

In all cases of exposure, the patient should be transferred to hospital as soon as possible.

86. Mercury(II) Iodide

HANDLING AND STORAGE

Mercury(II) iodide should be handled wearing an approved respirator, chemical-resistant gloves, safety goggles and other protective clothing. It should only be used in a chemical fume hood. In plants producing mercurials, disposable uniforms and disposable mercury vapour-absorbing masks should be used. There should be adequate ventilation, good housekeeping, and steel trays should be used in order to catch spills. Floors should be non-porous and washed regularly with calcium sulphide solution. It should be kept in a tightly closed container, in a cool, dry place and protected from light (1).

DISPOSAL

Ventilate the area, wear a laboratory coat, rubber gloves, self-contained breathing apparatus or an approved canister respirator for mercury vapour. For small spills of metallic mercury, remove by suction into a plastic bottle. For small spills of metallic mercury compounds, EITHER treat all affected areas with a 50:50 mixture of flowers of sulphur and calcium hydroxide in water; OR cover with zinc powder. Leave for 12 hours and then remove dried yellow mixture with water and reclean surfaces. For other mercury compounds, cover with sand, using 10-20 times by weight. Then arrange for removal by a licensed contractor. If the compound is soluble in water, small amounts may be dissolved in a bucket of water, diluted and the solution run to waste, if trade effluent regulations permit this.
For large spills contact the fire-brigade and police before taking action.

FIRE PRECAUTIONS

Mercury(II) iodide is not combustible and an extinguisher suitable to the surrounding fire conditions should be used (1).

FURTHER READING

Bretherick, L. *Hazards in the Chemical Laboratory* (4th edition)

Bretherick, L. *Handbook of Reactive Chemical Hazards* (4th edition)

Sax, N. Irving *Dangerous Properties of Industrial Materials* (7th edition)

Encyclopaedia of Occupational Safety & Health

Patty's *Industrial Hygiene and Toxicology*

Kirk-Othmer *Encyclopedia of Chemical Technology*

National Fire Protection Association *Manual of Hazardous Reactions*

ACGIH *Documentation of TLVs and BEIs* (6th edition, 1986)

REFERENCES

1. Lenga, R. E. *The Sigma-Aldrich Library of Chemical Safety Data* (2nd edition) (Sigma-Aldrich, Milwaukee, 1988)

2. *Environmental Health Criteria 118: Inorganic Mercury* (WHO, Geneva, 1991)

3. Govorunova, N. N.; et al. *Gig. Sanit.* 1981, (5), 73-74

MERCURY(II) NITRATE
Hg(NO$_3$)$_2$

87. Mercury(II) Nitrate

RISKS
Very toxic by inhalation, in contact with skin and if swallowed – Danger of cumulative effects (R26/27/28, R33)

SAFETY PRECAUTIONS
Keep locked up and out of reach of children – Keep away from food, drink and animal feeding stuffs – After contact with skin, wash immediately with plenty of water – In case of accident or if you feel unwell, seek medical advice immediately (show label where possible) (S1/2, S13, S28, S45)

IDENTIFIERS

SYNONYMS nitric acid, mercury(2+) salt; Citrine ointment; mercuric nitrate; mercury dinitrate; mercury(2+) nitrate; mercury nitrate (Hg(NO)$_3$)

CHEMICAL ABSTRACTS No.	10045-94-0
NIOSH No.	OW 8225000
HAZCHEM CODE	not available
UN No.	1625

THRESHOLD LIMIT VALUES

US TLV (TWA)	0.1 mg/m^3
US TLV (STEL)	not available

UK EXPOSURE LIMITS (OES)
Long-term (8 hr TWA value) ... 0.05 mg/m^3 (as Hg)
Short-term (10 min TWA value) ... 0.15 mg/m^3 (as Hg)

Germany
MAK .... not available

France
VME .... 0.1 mg/m^3
VLE .... not available

Sweden
Short-term limit .... not available
Level limit .... 0.05 mg/m^3 (as Hg)

PHYSICAL PROPERTIES

Description White-yellowish deliquescent powder. Odour of nitric acid.

Boiling point	decomposes
Melting point	79°C
Density	4.39
Vapour density	not available
Vapour pressure	not available
Flash point	not available
Explosive limits	not available
Autoignition temperature	not available

Solubility Soluble in a small amount of water, and in dilute acids.

PACKAGING AND TRANSPORTATION

Road transportation

hazard warning sign	1625 toxic substance
Hazchem code	not available

Sea transportation

IMDG page No.	6175
class	6.1
label	marine pollutant; poison
packaging group	II

Air transportation

ICAO/IATA code (UN No.)	1625
class	6.1
label	poison
packaging group	II
packing instructions	
cargo	615
passenger	613
passenger aircraft max. quantity	25 kilograms
cargo aircraft max. quantity	100 kilograms

87. Mercury(II) Nitrate

MANUFACTURE

Mercury(II) nitrate is prepared by exothermic dissolution of mercury in hot, concentrated, nitric acid. The reaction is complete when a cloud of mercury(I) chloride is not formed when treated with sodium chloride solution. The product crystallises on cooling.

USES

Mercury(II) nitrate is used in felt and mercury fulminate manufacture, as a radioactive iodine scrubber, and to destroy phylloxera.

CHEMICAL HAZARDS

Mercury(II) nitrate forms explosive compounds with acetylene; ethanol (1); ferrocene (2); isobutene; phosphine; and mixtures with potassium cyanide explode on heating if contained in narrow ignition tubes. It reacts violently with phosphinic acid; gas oil or cracked naphtha (3-5).

BIOLOGICAL HAZARDS

Mercury(II) nitrate is poisonous intraperitoneally, subcutaneously and if swallowed. Acute exposure to mercury compounds causes gastrointestinal symptoms, anuria and uraemia, but is uncommon in industry. The signs and symptoms of chronic mercurialism include tremors and neuropsychiatric disturbances (memory loss, insomnia, irritability and depression) (6).

Vapour Inhalation

Inhalation may cause tightness and pain in the chest, coughing and dyspnoea (6).

Eye Contact

Mercury(II) nitrate may cause eye irritation, conjunctival and corneal ulceration (6).

Skin Contact

Mercury(II) nitrate may cause skin irritation and dermatitis (6). The dermal LD_{50} in rats and mice is 26 mg/kg and 25 mg/kg respectively.

Swallowing

The oral LD_{50} in rats is 26 mg/kg. Oral administration caused severe diarrhoea in mice and rats (7).

CARCINOGENICITY

The WHO reported no evidence that inorganic mercury was carcinogenic, and is classified by the U.S. EPA in group O (not classifiable as to human carcinogenicity) (8).

MUTAGENICITY

No information specific to mercury(II) nitrate is available, but other inorganic mercury compounds are mutagenic in bacterial and mammalian assays (8).

REPRODUCTIVE HAZARDS

Chronic mercury poisoning in female mice following adminstration of mercury(II) nitrate caused disturbances of the oestrus cycle (9).

FIRST AID

Eyes Wash the eye with flowing water for 10 minutes.

Lungs Remove casualty from area of exposure. If unconscious, do not give anything to drink. Give artificial ventilation and chest compression or place in the recovery position as necessary. If conscious make the casualty lie or sit down quietly, give oxygen if available. Convulsions may occur and may cause unconsciousness.

Mouth Do not make the casualty vomit. Treat unconscious casualties as for lungs but if conscious give 1 pint of water to drink.

Skin Remove all contaminated clothing, wash affected areas with soap and copious amounts of water. Absorption through the skin may result in symptoms similar to those of ingestion or inhalation.

87. Mercury(II) Nitrate

In all cases of exposure, the patient should be transferred to hospital as soon as possible.

HANDLING AND STORAGE

Mercury(II) nitrate should be handled wearing an approved dust mask, chemical-resistant gloves, safety goggles and other protective clothing. In plants producing mercurials, disposable uniforms and disposable mercury vapour-absorbing masks should be used. There should be adequate ventilation, good housekeeping, and steel trays should be used in order to catch spills. Floors should be non-porous and washed regularly with calcium sulphide solution. It should be kept in a tightly closed container and protected from light (6).

DISPOSAL

Ventilate the area, wear a laboratory coat, rubber gloves, self-contained breathing apparatus or an approved canister respirator for mercury vapour. For small spills of metallic mercury, remove by suction into a plastic bottle. For small spills of metallic mercury compounds, EITHER treat all affected areas with a 50:50 mixture of flowers of sulphur and calcium hydroxide in water; OR cover with zinc powder. Leave for 12 hours and then remove dried yellow mixture with water and reclean surfaces. For other mercury compounds, cover with sand, using 10-20 times by weight. Then arrange for removal by a licensed contractor. If the compound is soluble in water, small amounts may be dissolved in a bucket of water, diluted and the solution run to waste, if trade effluent regulations permit this.
For large spills contact the fire-brigade and police before taking action.

FIRE PRECAUTIONS

Mercury(II) nitrate is not flammable and an extinguisher suitable for surrounding fire conditions should be used (6).

FURTHER READING

Bretherick, L. *Hazards in the Chemical Laboratory* (4th edition)

Bretherick, L. *Handbook of Reactive Chemical Hazards* (4th edition)

Sax, N. Irving *Dangerous Properties of Industrial Materials* (7th edition)

Encyclopaedia of Occupational Safety & Health

Patty's *Industrial Hygiene and Toxicology*

Kirk-Othmer *Encyclopedia of Chemical Technology*

National Fire Protection Association *Manual of Hazardous Reactions*

ACGIH Documentation of TLVs and BEIs (6th edition, 1986)

REFERENCES

1. Leleu, M. J. *Cah. Notes Doc.* 1976, **82**, 121-125
2. Sallot, G. P.; et al. *Proc. Int. Pyrotech. Semin.* 1984, 589-602
3. Mixer, R. Y. *Chem. Eng. News* 1948, **26**, 2434
4. Ball, J. *Chem. Eng. News* 1948, **26**, 3300
5. Leleu, M. J. *Cah. Notes Doc.* 1979, **96**, 451-460
6. *Dangerous Prop. Ind. Mater. Rep.* 1988, **8**(4), 42-49
7. Grins, N.; et al. *Gig. Sanit.* 1981, (8), 12-14
8. *Environmental Health Criteria 118: Inorganic Mercury* (WHO, Geneva, 1991)
9. Lach, H.; Srebryo, Z. *Acta Biol. Cracov., Ser. Zool.* 1972, **15**(1), 121-130

88. Mercury(II) Oxide

MERCURY(II) OXIDE
HgO

RISKS
Very toxic by inhalation, in contact with skin and if swallowed – Danger of cumulative effects (R26/27/28, R33)

SAFETY PRECAUTIONS
Keep locked up and out of reach of children – Keep away from food, drink and animal feeding stuffs – After contact with skin, wash immediately with plenty of water – In case of accident or if you feel unwell, seek medical advice immediately (show label where possible) (S1/2, S13, S28, S45)

IDENTIFIERS

SYNONYMS mercury oxide (HgO); C.I. 77760; mercuric oxide; mercuric oxide (HgO); mercury monoxide; mercury(2+) oxide; red mercuric oxide; Santar; Santar M; yellow mercury oxide

CHEMICAL ABSTRACTS No.	21908-53-2
NIOSH No.	OW 8750000
HAZCHEM CODE	not available
UN No.	1641

THRESHOLD LIMIT VALUES

US TLV (TWA)	0.1 mg/m^3
US TLV (STEL)	not available

UK EXPOSURE LIMITS (OES)
Long-term (8 hr TWA value) ... 0.05 mg/m^3 (as Hg)
Short-term (10 min TWA value) .. 0.15 mg/m^3 (as Hg)

Germany
MAK .... not available

France
VME .... 0.1 mg/m^3
VLE .... not available

Sweden
Short-term limit .... not available
Level limit .... 0.05 mg/m^3 (as Hg)

PHYSICAL PROPERTIES

Description Bright red or orange-red, heavy, odourless, crystalline powder or scales, orthorhomubic structure. Decomposes on exposure to light into mercury and oxygen. OR Yellow or orange-yellow, heavy odourless powder, orthorhombic structure, becomes red on heating, yellow again on cooling.

Boiling point	not available
Melting point	decomposes at 300°C
Density	11.14
Vapour density	not available
Vapour pressure	not available
Flash point	not available
Explosive limits	not available
Autoignition temperature	not available

Solubility Practically insoluble in water. Soluble in dilute hydrochloric and nitric acids.

PACKAGING AND TRANSPORTATION

Road transportation

hazard warning sign	1641 toxic substance
Hazchem code	not available

Sea transportation

IMDG page No.	6184
class	6.1
label	marine pollutant; poison
packaging group	II

Air transportation

ICAO/IATA code (UN No.)	1641
class	6.1
label	poison
packaging group	II
packing instructions	
cargo	615
passenger	613
passenger aircraft max. quantity	25 kilograms
cargo aircraft max. quantity	100 kilograms

88. Mercury(II) Oxide

MANUFACTURE

Yellow mercury(II) oxide (particles under 5 μm) is produced by precipitation from solutions of most water soluble mercury(II) salts by addition of alkali. The most economical salts are the dioxide and the nitrate. Red mercury(II) oxide (particles over 8 μm) is produced by heat-induced decomposition of mercury(I) nitrate or by hot precipitation. Careful reaction conditions are needed in both cases.

USES

Red mercury(II) oxide is used in marine bottom paints, porcelain pigments, as a depolariser in dry batteries, and in Kjeldahl determinations. Yellow mercury(II) oxide is also used in preparation of organomercury compounds, and as an analytical reagent.

CHEMICAL HAZARDS

Mercury(II) oxide can act as a powerful catalyst and/or oxidant under appropriate conditions. It reacts explosively with phosphorus (on impact or boiling with water); sulphur (on heating); hydrazine hydrate; phosphonic acid; sodium-potassium alloys (on impact) (1); magnesium or potassium (on heating) (2). A violent explosion occurred during preparation of 2-ethoxy-1-iodo-3-butene from butadiene, ethanol, iodine and mercury(II) oxide (3). Interaction of chlorine and ethylene is explosive in the presence of mercury(II) oxide. Red mercury(II) oxide reacts explosively with acetyl nitrate (4). It forms an explosive compound with chlorine (5). Chlorine and methane react explosively over yellow mercury(II) oxide and mixtures with over 20 vol. % of chlorine are explosive (6). Red mercury(II) oxide decomposes hydrogen peroxide vigorously, but traces of nitric acid inhibit this and promote formation of red mercury(II) peroxide which explodes on impact even if wet (if the mercury(II) oxide was finely divided). Oxides of mercury are very active and the parent metal and alloys must be excluded from peroxide handling systems (7). It reacts incandescently with phospham. It reacts violently with methanediol (without a diluent) and contact with hydrogen trisulphide causes violent decomposition and ignition. Mixtures of diboron tetrafluoride with mercury(II) oxide prepared at -80°C ignited at 20°C (8). Interaction with disulphur dichloride is rapid and violently exothermic. A case of exposure in a school laboratory is reported when mercury(II) oxide was used instead of silver oxide in an oxidation-reduction experiment (9,10).

BIOLOGICAL HAZARDS

Acute exposure to mercury compounds causes gastrointestinal symptoms, anuria and uraemia, but is uncommon in industry. The signs and symptoms of chronic mercurialism include tremors and neuropsychiatric disturbances (memory loss, irritability, depression, insomnia) (11).

Vapour Inhalation

Mercury(II) oxide may cause irritation, and if absorbed via the lungs may cause systemic effects described above (12).

Eye Contact

Mercury(II) oxide may cause irritation (12).

Skin Contact

The dermal LD_{50} in rats is 315 mg/kg (12). It is a strong skin irritant, and an allergen (11).

Swallowing

The oral LD_{50} in rats and mice is 18 mg/kg and 22 mg/kg respectively (13).

CARCINOGENICITY

The WHO reported no evidence that inorganic mercury was carcinogenic, and is classified by the U.S. EPA in group O (not classifiable as to human carcinogenicity) (14).

MUTAGENICITY

No information specific to mercury(II) oxide is available, but other inorganic mercury compounds are mutagenic in bacterial and mammalian assays (14).

REPRODUCTIVE HAZARDS

Oral administration of mercury(II) oxide (equivalent to 2 mg Hg) in peanut oil to rats on days 5, 12 or 19 of pregnancy caused retarded foetal growth and inhibition of eye formation (14).

88. Mercury(II) Oxide

FIRST AID

Eyes Wash the eye with flowing water for 10 minutes.

Lungs Remove casualty from area of exposure. If unconscious, do not give anything to drink. Give artificial ventilation and chest compression or place in the recovery position as necessary. If conscious make the casualty lie or sit down quietly, give oxygen if available. Convulsions may occur and may cause unconsciousness.

Mouth Do not make the casualty vomit. Treat unconscious casualties as for lungs but if conscious give 1 pint of water to drink.

Skin Remove all contaminated clothing, wash affected areas with soap and copious amounts of water. Absorption through the skin may result in symptoms similar to those of ingestion or inhalation.

In all cases of exposure, the patient should be transferred to hospital as soon as possible.

HANDLING AND STORAGE

Mercury(II) oxide should be handled wearing an approved respirator, chemical-resistant gloves, safety goggles and other protective clothing. It should only be used in a chemical fume hood. In plants producing mercurials, disposable uniforms and disposable mercury vapour-absorbing masks should be used. There should be adequate ventilation, good housekeeping, and steel trays should be used in order to catch spills. Floors should be non-porous and washed regularly with calcium sulphide solution. It should be kept in a tightly closed container, in a cool dry place (12).

DISPOSAL

Ventilate the area, wear a laboratory coat, rubber gloves, self-contained breathing apparatus or an approved canister respirator for mercury vapour. For small spills of metallic mercury, remove by suction into a plastic bottle. For small spills of metallic mercury compounds, EITHER treat all affected areas with a 50:50 mixture of flowers of sulphur and calcium hydroxide in water; OR cover with zinc powder. Leave for 12 hours and then remove dried yellow mixture with water and reclean surfaces. For other mercury compounds, cover with sand, using 10-20 times by weight. Then arrange for removal by a licensed contractor. If the compound is soluble in water, small amounts may be dissolved in a bucket of water, diluted and the solution run to waste, if trade effluent regulations permit this.
For large spills contact the fire-brigade and police before taking action.

FIRE PRECAUTIONS

Mercury(II) oxide is not combustible, and fire extinguishers appropriate to surrounding fire conditions should be used.

FURTHER READING

Bretherick, L. *Hazards in the Chemical Laboratory* (4th edition)

Bretherick, L. *Handbook of Reactive Chemical Hazards* (4th edition)

Sax, N. Irving *Dangerous Properties of Industrial Materials* (7th edition)

Encyclopaedia of Occupational Safety & Health

Patty's *Industrial Hygiene and Toxicology*

Kirk-Othmer *Encyclopedia of Chemical Technology*

National Fire Protection Association *Manual of Hazardous Reactions*

ACGIH Documentation of TLVs and BEIs (6th edition, 1986)

REFERENCES

1. Leleu, J. *Cah. Notes Doc.* 1975, **79**, 265-270
2. Leleu, J. *Cah. Notes Doc.* 1975, **80**, 397-399
3. Trent, J.;et al. *Chem. Eng. News* 1966, **44**(33), 7
4. Chertien, A.; et al. *Compt. Rend.* 1945, **220**, 823
5. Tabata, Y.; et al. *J. Haz. Mat.* 1987, **17**, 55
6. Eisenlohr, D. H. U.S. Pat. 2,989,571 (1961)
7. *Hydrogen peroxide data manual* (Laporte Chems. Ltd., Luton, 1960)

88. Mercury(II) Oxide

8. Holliday, A. K.; et al. *J. Chem. Soc.* 1964, 2732
9. *Appl. Ind. Hyg.* 1988, **3**(9), R4-R5
10. Shelnitz, M.; et al. *Morbid. Mortal. Weekly Rep.* 1988, **37**(10), 153-155
11. *Dangerous Prop. Ind. Mater. Rep.* 1989, **9**(5), 49-57
12. Lenga, R. E. *The Sigma-Aldrich Library of Chemical Safety Data* (2nd edition) (Sigma-Aldrich, Milwaukee, 1988)
13. Vernot, E. H.; et al. *Toxicol. Appl. Pharmacol.* 1977, **42**(2), 417-423
14. *Environmental Health Criteria 118: Inorganic Mercury* (WHO, Geneva, 1991)

89. Methacrylonitrile

METHACRYLO-NITRILE

$H_2C=C(CH_3)CN$

RISKS
Highly flammable – Toxic by inhalation, in contact with skin and if swallowed – May cause sensitisation by skin contact (R11, R23/24/25, R43)

SAFETY PRECAUTIONS
Keep container in a well ventilated place – Keep away from sources of ignition – No Smoking – Handle and open container with care – Do not empty into drains – In case of accident or if you feel unwell, seek medical advice immediately (show label where possible) (S9, S16, S18, S29, S45)

IDENTIFIERS

SYNONYMS 2-propenenitrile, 2-methyl-; methacrylonitrile; 2-cyanopropene; isopropene cyanide; isopropenylnitrile; methacrylnitrile; α-methacrylonitrile; α-methylacrylonitrile; 2- methylacrylonitrile; 2-methylpropenenitrile

CHEMICAL ABSTRACTS No.	126-98-7
NIOSH No.	UD 1400000
HAZCHEM CODE	3WE
UN No.	3079

THRESHOLD LIMIT VALUES

US TLV (TWA)	not available
US TLV (STEL)	not available

UK EXPOSURE LIMITS (OES)
Long-term (8 hr TWA value) .... 1 ppm (3 mg/m^3)
Short-term (10 min TWA value) .... not available

Germany
MAK .... not available

France
VME .... 1 ppm (3 mg/m^3)
VLE .... not available

Sweden
Short-term limit .... not available
Level limit .... not available

PHYSICAL PROPERTIES

Description Clear colourless liquid.

Boiling point	90.3°C
Melting point	-35.8°C
Density	0.8 at 20°C
Vapour density	2.31
Vapour pressure	60 mm Hg at 21.5°C
Flash point	13°C (open cup)
Explosive limits	2%-6.8%
Autoignition temperature	not available

Solubility Slightly soluble in water. Miscible with acetone, octane and toluene at 20-25°C.

PACKAGING AND TRANSPORTATION

Road transportation
hazard warning sign ....
.... 3079 flammable liquid; toxic substance
Hazchem code 3WE

Sea transportation
IMDG page No.	3250
class	3.2
label	flammable liquid; poison; marine pollutant
packaging group	I

Air transportation
ICAO/IATA code (UN No.)	3079
class	3;6.1
label	liquid flammable; poison
packaging group	I
packing instructions	
cargo	303
passenger	forbidden
passenger aircraft max. quantity	forbidden
cargo aircraft max. quantity	30 litres

89. Methacrylonitrile

MANUFACTURE

Methacrylonitrile is prepared by dehydration of methacrylamide, or from isopropylene oxide and ammonia.

USES

Methacrylonitrile is used as an intermediate in preparation of acids, amides, amines, esters and nitriles, and in the preparation of homopolymers and copolymers.

CHEMICAL HAZARDS

Methacrylonitrile may autopolymerise. It is incompatible with strong acids, strong bases, strong oxidizing or reducing agents (1).

BIOLOGICAL HAZARDS

Its acute toxicity is comparable to acrylonitrile. Its hazards have been reviewed (2) and summarised (3).

Vapour Inhalation

Methacrylonitrile is highly toxic by inhalation, and the vapour has very poor warning properties (4). Acute inhalation toxicity in mice is 800 ppm (30 minute exposure) and 230 ppm (8 hour exposure) (5). Central nervous system effects, microscopic brain lesions and diarrhoea were reported in chronic exposure studies on dogs.

Eye Contact

Methacrylonitrile causes mild eye irritation and lacrimation.

Skin Contact

Methacrylonitrile is a local skin irritant, and is readily absorbed through the skin.

Swallowing

The oral LD_{50} in rats is 0.25 ml/kg (4,6).

CARCINOGENICITY

No information is available concerning the carcinogenicity of methacrylonitrile.

MUTAGENICITY

No information is available concerning the mutagenicity of methacrylonitrile.

REPRODUCTIVE HAZARDS

No information is available concerning the reproductive hazards of methacrylonitrile.

FIRST AID

Eyes Wash the eye with flowing water.

Lungs Remove casualty from area of exposure, wearing protective clothing and approved breathing apparatus if necessary. If unconscious, do not give anything to drink. If breathing has stopped DO NOT use mouth to mouth or mouth to nose ventilation but use a resuscitation bag and mask instead. If breathing, place in the recovery position. Break two amyl nitrate capsules open under the casualty's nose so that the vapour is inhaled. If Kelo-cyanor (dicobalt edetate) and personnel trained in its use are available and you are certain the casualty has been poisoned by cyanide, give it immediately. NOTE: KELO-CYANOR IS EXTREMELY DANGEROUS WHEN GIVEN TO ANYONE NOT SUFFERING FROM CYANIDE POISONING. If conscious make the casualty lie or sit down quietly. Oxygen may be beneficial.

Mouth Do not make the casualty vomit. Treat as for inhalation. If conscious give 1 pint of water immediately.

Skin Remove all contaminated clothing, wash affected areas with soap and copious amounts of water. Absorption through the skin may result in symptoms similar to those of ingestion and inhalation.

In all cases of exposure, the patient should be transferred to hospital as soon as possible.

89. Methacrylonitrile

HANDLING AND STORAGE

Methacrylonitrile should be handled wearing an approved respirator, butyl rubber gloves (7), safety goggles and other protective clothing. It should only be used in a chemical fume hood (1). Methacrylonitrile is stabilised with 50 ppm hydroquinone monomethyl ether. It should be kept in a tightly sealed container, in an explosion proof refrigerator (8). The vapour may travel a considerable distance to a source of ignition and flash back.

DISPOSAL

Eliminate all sources of ignition, ventilate the area and wearing self-contained breathing apparatus, a chemical resistant suit, gloves, and PVC safety boots, spread calcium hypochlorite over the (liquid) spill (or disperse excess calcium hypochlorite solution onto it). Mix and mop into polythene buckets, stand 24 hours and run to waste diluting x1000 with running cold water (see discharge limits from water authorities).

For solid spills either: a) sweep up, place in a large volume of water, add excess sodium hypochlorite, leave to stand 24 hours, then run to waste diluting x1000 with running water; or b) scoop into a large container, make alkaline with 2% sodium hydroxide solution and stir well. Add excess ferrous sulphate solution, mix well, stand for 2-3 hours and run to waste diluting x1000 with running water.

Large spills need specialist help – contact fire brigade.

FIRE PRECAUTIONS

Fires involving methacrylonitrile should be extinguished using water spray, carbon dioxide, dry chemical powder, alcohol foam or polymer foam. Container explosion may occur under fire conditions (1).

FURTHER READING

Sax, N. Irving *Dangerous Properties of Industrial Materials* (7th edition)

Encyclopedia of Occupational Safety & Health

Patty's *Industrial Hygiene and Toxicology*

Kirk-Othmer *Encyclopedia of Chemical Technology*

National Fire Protection Association *Manual of Hazardous Reactions*

ACGIH *Documentation of TLVs and BEIs* (6th edition, 1986)

Dangerous Prop. Ind. Mater. Rep. 1986, **6**(1), 76-81

REFERENCES

1. Lenga, R. E. *The Sigma-Aldrich Library of Chemical Safety Data* (2nd edition) (Sigma-Aldrich, Milwaukee, 1988)
2. Ball, L. E.; Greene, J. L. *Encycl. Polym. Sci. Technol.* 1971, **15**, 319-353
3. *Toxic. Subst. Bull.* 1987, (7), 8-11
4. Pozzani, U. C.; et al. *Am. Ind. Hyg. Assoc. J.* 1968, **29**(3), 202-210
5. McOmie, W. A. *J. Ind. Hyg. Toxicol.* 1949, **31**, 113-116
6. Smyth, H. F.; et al. *Am. Ind. Hyg. Assoc. J.* 1962, **23**, 95
7. Forsberg, K.; Mansdorf, S. Z. *Quick selection guide to chemical protective clothing* (Van Nostrand Reinhold, New York, 1989)
8. Keith, L. H.; Walters, D. B. (editors) *Compendium of Safety Data Sheets for Research and Industrial Chemicals* (VCH, Deerfield Park, 1987)

N-METHYLANILINE

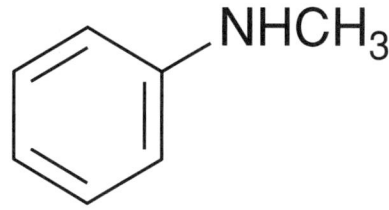

RISKS
Toxic by inhalation, in contact with skin and if swallowed – Danger of cumulative effects (R23/24/25, R33)

SAFETY PRECAUTIONS
After contact with skin, wash immediately with plenty of water – Wear suitable gloves – If you feel unwell, seek medical advice (show label where possible) (S28, S37, S44)

IDENTIFIERS
SYNONYMS benzenamine, *N*-methyl-; aniline, *N*-methyl-; anilinomethane; (methylamino)benzene; methylaniline; *N*- methylbenzenamine; methylphenylamine; *N*-methylphenylamine; *N*-monomethylaniline; *N*-phenylmethylamine

CHEMICAL ABSTRACTS No.	100-61-8
NIOSH No.	BY 4550000
HAZCHEM CODE	3X
UN No.	2294

THRESHOLD LIMIT VALUES

US TLV (TWA)	0.5 ppm (2.2 mg/m^3)
US TLV (STEL)	not available

UK EXPOSURE LIMITS (OES)
Long-term (8 hr TWA value) .... 0.5 ppm (2 mg/m^3)
Short-term (10 min TWA value) .... not available

Germany
MAK .... 0.5 ppm (2 mg/m^3)

France
VME .... 0.5 ppm (2 mg/m^3)
VLE .... not available

Sweden
Short-term limit .... not available
Level limit .... not available

PHYSICAL PROPERTIES
Description Colourless or slightly yellow liquid, becoming brown on standing.

Boiling point	194-196°C
Melting point	-57°C
Density	0.989 at 20°C
Vapour density	3.7
Vapour pressure	0.3 mm Hg at 20°C
Flash point	79.44°C (closed cup)
Explosive limits	not available
Autoignition temperature	not available

Solubility Soluble in alcohol or ether. Slightly soluble in water.

PACKAGING AND TRANSPORTATION

Road transportation
hazard warning sign	2294 harmful substance
Hazchem code	3X

Sea transportation
IMDG page No.	6188
class	6.1
label	harmful stow away from foodstuffs
packaging group	III

Air transportation
ICAO/IATA code (UN No.)	2294
class	6.1
label	keep away from food
packaging group	III
packing instructions	
cargo	618
passenger	611
passenger aircraft max. quantity	60 litres
cargo aircraft max. quantity	220 litres

90. N-Methylaniline

MANUFACTURE
N-Methylaniline is prepared by heating aniline chloride and methyl alcohol under pressure.

USES
N-Methylaniline is used as an acid acceptor, solvent, and in organic synthesis.

CHEMICAL HAZARDS
N-Methylaniline is incompatible with acids, acid chlorides, acid anhydrides, strong oxidising agents or carbon dioxide. It emits nitrogen oxides when heated to decomposition (1).

BIOLOGICAL HAZARDS
N-Methylaniline is toxic if inhaled, swallowed or absorbed through the skin. Effects may include headache, fatigue, dizziness, confusion, loss of appetite, cyanosis and convulsions (2). Effects on the blood, including methaemoglobinaemia and Heinz body formation, have been reported in animals following inhalation exposure.

Vapour Inhalation
N-Methylaniline may be irritating to the respiratory tract, causing coughing, wheezing, sore throat, dyspnoea, headache, nausea, vomiting and methaemoblobinemia (1).

Eye Contact
N-Methylaniline may cause irritation to the eyes.

Skin Contact
N-Methylaniline may cause skin irritation and is readily absorbed through the skin.

Swallowing
The oral LD_{50} in rabbits is 280 mg/kg (3).

CARCINOGENICITY
N-Methylaniline induced cancer of the oesophagus in rats and mice when fed together with sodium nitrite (4,5).

MUTAGENICITY
N-Methylaniline was not mutagenic in *Salmonella typhimurium* TA98 (6).

REPRODUCTIVE HAZARDS
No information is available concerning the reproductive hazards of N-methylaniline.

FIRST AID
Eyes Wash the eye with flowing water for 10 minutes.

Lungs Remove casualty from area of exposure. If unconscious, do not give anything to drink. Give artificial ventilation and chest compression or place in the recovery position as necessary. If conscious make the casualty lie or sit down quietly, give oxygen if available. Convulsions may occur and may cause unconsciousness. Shock may result – if so do not give any drinks, and if conscious, lie casualty flat with legs raised.

Mouth Do not make the casualty vomit. Treat unconscious casualties as for lungs, but if conscious give 1 pint of water to drink.

Skin Remove contaminated clothing immediately, wash the affected area with soap and copious amounts of water. Absorption through the skin may cause symptoms similar to those of inhalation.

In all cases of exposure, the patient should be transferred to hospital as soon as possible.

HANDLING AND STORAGE
N-Methylaniline should be handled wearing an approved respirator, chemical-resistant gloves, safety goggles and other protective clothing. It should only be used in a chemical fume hood (1). It should be kept in a tightly closed container in a cool, dry place, away from heat and flame.

90. *N*-Methylaniline

DISPOSAL

Wear a laboratory coat, safety spectacles, butyl rubber gloves and suitable safety shoes, and have an approved self-contained breathing apparatus or canister respirator available. Absorb small liquid spills onto paper towels, evaporate in an iron pan in a fume cupboard, add crumpled paper and burn carefully. Brush small solid spills onto paper, put in an iron pan, cover with crumpled paper and burn carefully in a safe place outside. For large spills, EITHER cover with sand and mix carefully. Shovel into containers, disperse in an excess solution of dilute hydrochloric acid, mix well and leave to stand for 24 hours stirring occasionally. Carefully decant acid extract into the drains, diluting with a large volume of cold tap water. Wash the sand thoroughly with cold water. OR cover with a sand- soda ash mix (90-10), mix well, shovel into cardboard boxes, pack with crumpled paper and incinerate.

FIRE PRECAUTIONS

Fires involving *N*-methylaniline should be extinguished using water spray, carbon dioxide, dry chemical powder, alcohol foam or polymer foam (1).

FURTHER READING

Sax, N. Irving *Dangerous Properties of Industrial Materials* (7th edition)

Encyclopedia of Occupational Safety & Health

Patty's *Industrial Hygiene and Toxicology*

Kirk-Othmer *Encyclopedia of Chemical Technology*

National Fire Protection Association *Manual of Hazardous Reactions*

ACGIH *Documentation of TLVs and BEIs* (6th edition, 1986)

REFERENCES

1. Lenga, R. E. *The Sigma-Aldrich Library of Chemical Safety Data* (2nd edition) (Sigma-Aldrich, Milwaukee, 1988)
2. Keith, L. H.; Walters, D. B. (editors) *Compendium of Safety Data Sheets for Research and Industrial Chemicals* (VCH, Deerfield Park, 1987)
3. Treon, J. F.; et al. *J. Ind. Hyg. Tox.* 1949, **31**, 1
4. Mirvish, S. S. *Top. Chem. Carcinog., Proc. Int. Symp., 2nd* 1971, 279-295
5. Sander, J.; Schweinsberg, F. **N**-Nitroso Compounds Anal. Form., Proc. Work, Conf. 1971, 97-103
6. Ho, C. H.; et al. *Mutat. Res.* 1981, **85**(5), 335-345

91. 2-Methylaziridine

2-METHYLAZIRIDINE

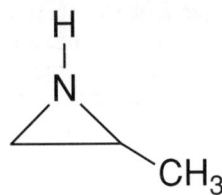

RISKS

May cause cancer – Highly flammable – Very toxic by inhalation, in contact with skin and if swallowed – Risk of serious damage to eyes (R45, R11, R26/27/28, R41)

SAFETY PRECAUTIONS

Avoid exposure-obtain special instructions before use – In case of contact with eyes, rinse immediately with plenty of water and seek medical advice – In case of accident or if you feel unwell, seek medical advice immediately (show label where possible) (S53, S26, S45)

IDENTIFIERS

SYNONYMS aziridine, 2-methy-; 2-methylethylenimine; propylenimine; propyleneimine; 1,2-propylenimine; 2-methylazacyclopropane; RCRA Waste No. P067

CHEMICAL ABSTRACTS No.	75-55-8
NIOSH No.	CM 8050000
HAZCHEM CODE	not available
UN No.	1921

THRESHOLD LIMIT VALUES

US TLV (TWA)	2 ppm (4.7 mg/m^3)
US TLV (STEL)	not available

UK EXPOSURE LIMITS (OES)
Long-term (8 hr TWA value)	not available
Short-term (10 min TWA value)	not available

Germany
MAK	not available

France
VME	not available
VLE	not available

Sweden
Short-term limit	not available
Level limit	not available

PHYSICAL PROPERTIES

Description Colourless oily liquid, with a strong ammonia-like smell.

Boiling point	66°C
Melting point	-65°C
Density	0.802 at 25°C
Vapour density	2.0
Vapour pressure	112 mm Hg at 20°C
Flash point	-10°C
Explosive limits	not available
Autoignition temperature	not available

Solubility Miscible with water. Soluble in most organic solvents.

PACKAGING AND TRANSPORTATION

Road transportation
hazard warning sign	not available
Hazchem code	not available

Sea transportation
IMDG page No.	3274
class	3.2
label	flammable liquid; poison
packaging group	1

Air transportation
ICAO/IATA code (UN No.)	1921
class	3
label	not available
packaging group	1
packing instructions	
cargo	304
passenger	306
passenger aircraft max. quantity	1 litre
cargo aircraft max. quantity	30 litres

91. 2-Methylaziridine

MANUFACTURE

2-Methylaziridine is manufactured by combining ammonia and 1,2-dichloropropane at high temperature. It can also be prepared by addition of hydrogen chloride to 1-amino-2-propanol, followed by treatment with sodium hydroxide.

USES

2-Methylaziridine derivatives have been used in several industries including adhesives, paper, textiles, pharmaceuticals, rubber and agrochemicals (1). It has also been used in oil refining and oil additives.

CHEMICAL HAZARDS

2-Methylaziridine is a very dangerous fire hazard if exposed to heat or flame. It may polymerise explosively on contact with acids or acidic fumes, and must always be stored over solid alkali (2). It emits toxic fumes of nitrogen oxides when heated to decomposition.

BIOLOGICAL HAZARDS

Acute effects are like those of ethylenimine, but it is less toxic to rats and guinea pigs.

Vapour Inhalation

2-Methylaziridine is irritating to the nose and throat and may cause nausea, vomiting and breathing difficulties (1).

Eye Contact

2-Methylaziridine is a severe eye irritant (1).

Skin Contact

2-Methylaziridine is irritating to the skin and causes burns. It is poisonous by skin contact (1).

Swallowing

2-Methylaziridine is poisonous if swallowed, and causes mouth and stomach irritation (1). The oral LD_{50} in rats is 19 mg/kg.

CARCINOGENICITY

2-Methylaziridine is a potent carcinogen when given orally to rats (3,4). The IARC evaluation is that there is "limited evidence" for carcinogenicity in animal studies (5).

MUTAGENICITY

Increased frequency of sister chromatid exchanges is reported in phytohaemagglutinin-stimulated human whole blood cultures (6). 2-Methylaziridine is mutagenic in a standard *Salmonella typhimurium* assay using TA1535 and TA1538 and gave positive results in *Escherichia coli* (7). Mutagenicity is also reported in *S. typhimurium* TA100 (8,9).

REPRODUCTIVE HAZARDS

No information is available concerning the reproductive hazards of 2-methylaziridine.

FIRST AID

Eyes Wash the eye with flowing water for 10 minutes.

Lungs Remove casualty from area of exposure. If unconscious, do not give anything to drink, give artificial ventilation and chest compression or place in the recovery position as necessary. If conscious make the casualty lie or sit down quietly, give oxygen if available. Lung congestion may occur – a conscious casualty with breathing difficulties should be placed in a sitting position. Convulsions may occur and may cause unconsciousness. Shock may result – do not give any drinks, and if conscious lie casualty flat with legs raised.

Mouth Do not make the casualty vomit. Treat unconscious casualties as for lungs but if conscious give 1 pint of water to drink immediately; give repeated drinks of water (1 cupful every 10 minutes). Convulsions may occur and may cause unconsciousness.

Skin Remove contaminated clothing immediately, drench the affected area with running water for at least 10 minutes.

In all cases of exposure, the patient should be transferred to hospital as soon as possible.

91. 2-Methylaziridine

HANDLING AND STORAGE

2-Methylaziridine should be handled wearing an approved respirator, chemical-resistant gloves, safety goggles and other protective clothing. It should only be used in a chemical fume hood (10). It should be kept in a tightly closed container, refrigerated and away from heat, sparks or open flame. The vapour may travel a considerable distance to a source of ignition and flash back. Mild steel, stainless steel or glass materials are suitable for storage and handling of aziridines. Storage tanks should have a water-spray system, high temperature alarm, and relief valves or rupture disks vented to an acid scrubber. Storage tanks should be padded with pure nitrogen at 136 kPa.

DISPOSAL

Eliminate all sources of ignition, ventilate the area and wear a laboratory coat or overalls, butyl rubber gloves, approved self-contained breathing apparatus or all purpose canister respirator. Cover the spill with sodium bisulphite and dilute greatly with water, OR neutralise with dilute sulphuric acid. Transfer outside in buckets and wash down the drain with a large excess of running water. Large spills may also be absorbed onto vermiculite, mixed well with sodium carbonate and calcium hydroxide mixture, wrapped in paper and burned outside in an incinerator.

FIRE PRECAUTIONS

Fires involving 2-methylaziridine should be extinguished using water spray, carbon dioxide, dry chemical powder, alcohol foam or polymer foam. Container explosion may occur under fire conditions.

FURTHER READING

Bretherick, L. *Hazards in the Chemical Laboratory* (4th edition)

Bretherick, L. *Handbook of Reactive Chemical Hazards* (4th edition)

Sax, N. Irving *Dangerous Properties of Industrial Materials* (7th edition) *Encyclopedia of Occupational Safety & Health*

Patty's *Industrial Hygiene and Toxicology*

Kirk-Othmer *Encyclopedia of Chemical Technology*

National Fire Protection Association *Manual of Hazardous Reactions*

ACGIH *Documentation of TLVs and BEIs* (6th edition, 1986)

REFERENCES

1. Keith, L. H.; Walters, D. B. (editors) *Compendium of Safety Data Sheets for Research and Industrial Chemicals* (VCH, Deerfield Park, 1987)
2. Inlow, R. O.; et al. *J. Inorg. Nucl. Chem.* 1975, **37**, 2353
3. Ulland, B.; et al. *Nature (London)* 1971, **230**, 460-461
4. Weisburger, E. K.; et al. *JNCI, J. Natl. Cancer Inst.* 1981, **67**(1), 75-88
5. IARC *Monographs on the evaluation of the carcinogenic risk of chemicals to humans* 1975, **9**, 61-65
6. Tucker, J. D.; et al. *Environ. Mol. Mutagen.* 1985, **7**(suppl. 3), 48
7. Rosencranz, H. S.; et al. *JNCI, J. Natl. Cancer Inst.* 1979, **62**(4), 873-892
8. Simmon, V. F.; et al. *JNCI, J. Natl. Cancer Inst.* 1979, **62**(4), 893-899
9. Dunkel, V. C.; et al. *Environ. Mutagen.* 1984, **6**(suppl. 2), 1
10. Lenga, R. E. *The Sigma-Aldrich Library of Chemical Safety Data* (2nd edition) (Sigma-Aldrich, Milwaukee, 1988)

92. Methyl Chloroformate

METHYL CHLOROFORMATE
ClCO$_2$CH$_3$

RISKS
Highly flammable – Toxic by inhalation – Irritating to eyes, respiratory system and skin (R11, R23, R36/37/38)

SAFETY PRECAUTIONS
Keep container in a well ventilated place – Keep away from sources of ignition – No Smoking – Take precautionary measures against static discharges – If you feel unwell, seek medical advice (show label where possible) (S9, S16, S33, S44)

IDENTIFIERS

SYNONYMS   carbonochloridic acid, methyl ester; formic acid, chloro-, methyl ester; chlorocarbonic acid, methyl ester; chloroformic acid, methyl ester; methoxycarbonyl chloride; methyl chlorocarbonate

CHEMICAL ABSTRACTS No.	79-22-1
NIOSH No.	FG 3675000
HAZCHEM CODE	3WE
UN No.	1238

THRESHOLD LIMIT VALUES

US TLV (TWA)	not available
US TLV (STEL)	not available
UK EXPOSURE LIMITS (OES)	
Long-term (8 hr TWA value)	not available
Short-term (10 min TWA value)	not available
Germany	
MAK	not available
France	
VME	not available
VLE	not available
Sweden	
Short-term limit	not available
Level limit	not available

PHYSICAL PROPERTIES

Description	Colourless liquid.
Boiling point	71.4°C
Melting point	not available
Density	1.223 at 20°C
Vapour density	3.26
Vapour pressure	127 mm Hg at 20°C
Flash point	24.4°C (open cup)
Explosive limits	not available
Autoignition temperature	504°C

Solubility   Slightly soluble in water with gradual decomposition. Miscible with alcohol, benzene, chloroform or ether.

PACKAGING AND TRANSPORTATION

Road transportation

hazard warning sign	1238 toxic substance; flammable liquid; corrosive substance
Hazchem code	3WE

Sea transportation

IMDG page No.	6193
class	6.1
label	flammable liquid; corrosive; poison
packaging group	I

Air transportation

ICAO/IATA code (UN No.)	1238
class	6.1;3;8
label	not available
packaging group	not available
packing instructions	
cargo	forbidden
passenger	forbidden
passenger aircraft max. quantity	forbidden

92. Methyl Chloroformate

MANUFACTURE

Methyl chloroformate is prepared from methyl alcohol and phosgene.

USES

Methyl chloroformate is used as an intermediate for various products including mixed or symmetrical carbonates for use as unsaturated monomer polymerisation initiators. Carbonates derived from chloroformates are used to manufacture pharmaceuticals.

CHEMICAL HAZARDS

Methyl chloroformate is incompatible with acids, amines, strong bases, and alcohols. It may decompose on exposure to moisture or water. It emits hydrogen chloride, carbon dioxide and phosgene when heated to decomposition (1). It is a dangerous fire hazard on exposure to heat, sources of ignition or oxidisers. It produces corrosive and toxic fumes on contact with water or steam. Acyl halides tend to react violently with protic organic solvents, water, dimethylformamide and dimethyl sulphoxide. They may also react hazardously with ethers.

BIOLOGICAL HAZARDS

Methyl chloroformate is corrosive, and is highly toxic by inhalation and if swallowed.

Vapour Inhalation

Chloroformates are highly toxic in subacute inhalation studies with rats (2). The LC_{50} (1 hour) in rats is 88 ppm. Lower chloroformate vapours cause a pneumonia-like condition in animal studies. Chronic exposure causes changes in neuromuscular excitability, body temperature, respiration rate and organ weights (3).

Eye Contact

Methyl chloroformate is corrosive and causes lacrimation and conjunctival irritation.

Skin Contact

Methyl chloroformate was corrosive to rabbits' skin in experiments based on OECD guidelines (4). The dermal LD_{50} in rabbits and mice is 7120 mg/kg and 1750 mg/kg respectively. Chloroformates are vesicants, with effects like those of hydrochloric acid.

Swallowing

The oral LD_{50} in rats is 60 mg/kg.

CARCINOGENICITY

No information is available concerning the carcinogenicity of methyl chloroformate.

MUTAGENICITY

No information is available concerning the mutagenicity of methyl chloroformate.

REPRODUCTIVE HAZARDS

No information is available concerning the reproductive hazards of methyl chloroformate.

FIRST AID

Eyes Wash the eye with flowing water for 10 minutes.

Lungs Remove casualty from area of exposure. If unconscious, do not give anything to drink. Give artificial ventilation and chest compression or place in the recovery position as necessary. If conscious make the casualty lie or sit down quietly, give oxygen if available. Lung congestion may occur – a conscious casualty with breathing difficulties should be placed in a sitting position.

Mouth Do not make the casualty vomit. Treat unconscious casualties as for lungs but if conscious give 1 pint of water to drink immediately give repeated drinks of water (1 cupful every 10 minutes).

Skin Remove all contaminated clothing, wash affected areas with soap and copious amounts of water. Absorption through the skin may result in symptoms similar to those of ingestion and inhalation.

92. Methyl Chloroformate

In all cases of exposure, the patient should be transferred to hospital as soon as possible.

HANDLING AND STORAGE

Methyl chloroformate should be handled wearing an approved respirator, chemical-resistant gloves, safety goggles and other protective clothing. It should only be used in a chemical fume hood (1). It should be kept in a tightly closed container, away from heat and sources of ignition. The vapour may travel a considerable distance to a source of ignition and flash back. It should be refrigerated on arrival, vented from time to time, and opened carefully. Chloroformates should be transferred through closed systems. Equipment should be of stainless steel or nickel with glass pumps, lines and valves.

DISPOSAL

Eliminate all sources of ignition, ventilate the area and wear a laboratory coat or overalls, rubber gloves, approved compressed air breathing apparatus and safety boots. Absorb small spills onto paper towels, remove to a safe open air site and allow to evaporate in a metal tray. Wash spillage site with detergent. For large spills, absorb onto vermiculite-sodium carbonate mixture (90-10) or sand-soda ash (90-10) and mix carefully. EITHER transport in dry buckets to a safe open air area for atmospheric evaporation OR shovel into paper boxes and incinerate.

FIRE PRECAUTIONS

Fires involving methyl chloroformate should be extinguished using carbon dioxide, dry chemical powder, alcohol foam or polymer foam. DO NOT use water. Container explosion may occur under fire conditions.

FURTHER READING

Bretherick, L. *Hazards in the Chemical Laboratory* (4th edition)

Bretherick, L. *Handbook of Reactive Chemical Hazards* (4th edition)

Sax, N. Irving *Dangerous Properties of Industrial Materials* (7th edition)

Encyclopedia of Occupational Safety & Health

Patty's *Industrial Hygiene and Toxicology*

Kirk-Othmer *Encyclopedia of Chemical Technology*

National Fire Protection Association *Manual of Hazardous Reactions*

ACGIH *Documentation of TLVs and BEIs* (6th edition, 1986)

REFERENCES

1. Lenga, R. E. *The Sigma-Aldrich Library of Chemical Safety Data* (2nd edition) (Sigma-Aldrich, Milwaukee, 1988)
2. Gage, J. C. *Br. J. Ind. Med.* 1970, **27**(1), 1-18
3. Gurova, A. L.; et al. *Gig. Sanit.* 1977, (5), 97-99
4. Grundler, O. J.; et al. *Food Chem. Toxicol.* 1985, **23**(6), 615-617

93. Methyl Isocyanate

METHYL ISOCYANATE
CH3NCO

RISKS
Extremely flammable – Toxic by inhalation, in contact with skin and if swallowed – Irritating to eyes, respiratory system and skin (R12, R23/24/25, R36/37/38)

SAFETY PRECAUTIONS
Keep container in a well ventilated place – Never add water to this product – In case of fire, use carbon dioxide or dry chemical powder. Do not use water – If you feel unwell, seek medical advice (show label where possible) (S9, S30, S43, S44)

IDENTIFIERS

SYNONYMS   methane, isocyanato-; isocyanic acid, methyl ester; isocyanatomethane; MIC; RCRA Waste No. P064; TL 1450

CHEMICAL ABSTRACTS No.	624-83-9
NIOSH No.	NQ 9450000
HAZCHEM CODE	not available
UN No.	2480

THRESHOLD LIMIT VALUES

US TLV (TWA)	0.02 ppm (0.047 mg/m^3)
US TLV (STEL)	not available

UK EXPOSURE LIMITS (MEL)
Long-term (8 hr TWA value)   0.02 mg/m^3 (as -NCO)
Short-term (10 min TWA value)   0.07 mg/m^3 (as -NCO)

Germany
MAK   0.01 ppm (0.025 mg/m^3)

France
VME   0.02 ppm (0.5 mg/m^3)
VLE   not available

Sweden
Short-term limit   not available
Level limit   not available

PHYSICAL PROPERTIES

Description	Colourless liquid.
Boiling point	39°C
Melting point	-17°C
Density	0.96 at 20°C
Vapour density	1.97
Vapour pressure	348 torr at 20°C
Flash point	-7°C (open cup)
Explosive limits	5.3%-26%
Autoignition temperature	534°C

Solubility   Decomposes in water. Dissolves without decomposition in hydrocarbons, halogenated hydrocarbons and dimethylsulphoxide.

PACKAGING AND TRANSPORTATION

Road transportation
hazard warning sign	2480 toxic substance; flammable liquid
Hazchem code	not available

Sea transportation
IMDG page No.	6197
class	6.1
label	poison; flammable liquid
packaging group	II

Air transportation
ICAO/IATA code (UN No.)	2480
class	6.1;3
label	not available
packaging group	not available
packing instructions	
cargo	forbidden
passenger	forbidden
passenger aircraft max. quantity	forbidden
cargo aircraft max. quantity	forbidden

93. Methyl Isocyanate

MANUFACTURE

Methyl isocyanate is manufactured by reaction of methylamine with phosgene.

USES

Methyl isocyanate is used as an intermediate in the manufacture of carbamate pesticides, and in the production of polyurethane foams and plastics.

CHEMICAL HAZARDS

Methyl isocyanate is a very dangerous fire hazard when exposed to heat, flame or oxidisers. It reacts exothermically with water. Much has been published on the likely causes of the Bhopal disaster (1-5). The accident occurred when an exothermic reaction in a methyl isocyanate storage vessel raised the internal temperature above the boiling point, and the cooling system could not cope with the continuing exotherm. The partially inoperative scrubbing system could not deal with the rapid vapour release and the flare system did not ignite. The cause of the exothermic reaction could have been contamination of the storage vessel by moisture, methylamine, acidic or metallic impurities. Accidental formation of methyl isocyanate in laboratories and factories has been reviewed (6). It may polymerise rapidly in the presence of zinc, iron, tin, their salts and other catalysts. It has exploded violently when dropped onto a hot surface, and the dried-out material may explode on exposure to heat or shock.

BIOLOGICAL HAZARDS

Recent data on the effects on human health of methyl isocyanate has come primarily from communities in Bhopal, India. In 1984 an accident at the Union Carbide carbaryl plant resulted in a release of methyl isocyanate which killed 5,000 to 8,000 people and an estimated 200,000 were exposed to the vapour (5). There have been a series of follow-up studies on survivors (7). Prior to this there had been limited investigation of the health effects, particularly long-term, of methyl isocyanate due to experimental difficulties, relatively small industrial use, and low potential for human exposure when used in closed production processes (8,9). The major effect on Bhopal survivors is lung damage. There are a few reports of neurological symptoms and minor chromosome aberrations. No significant changes in human placenta have been observed, and suggestions of male sterility have been discounted (10).

Vapour Inhalation

The lungs are the major target organ. Inhalation of methyl isocyanate vapour causes vomiting, coughing (11), and lung damage (oedema, permanent fibrosis, emphysema and bronchitis). A follow up study 3 years after Bhopal found breathlessness to be twice as common in those exposed (12). Lingering respiratory illness appears to be the main long-term effect of exposure (13,14). Survivors of Bhopal have also reported neuromuscular dysfunction, an effect which seems to be supported by findings in cultured rat brain cells (15). In a series of experiments on human volunteers, exposure to concentrations of 1.75 ppm caused eye, nose and throat irritation. Effects disappeared in most volunteers in under 10 minutes, and there was no consistent relationship between odour detection and concentration. Exposure to 21 ppm was unbearable (16-18). The LC_{50} (6 hours) ranges from 5.4 ppm (guinea pigs) to 12.2 ppm (mice) (19).

Eye Contact

Methyl isocyanate causes eye irritation and lacrimation, and exposure can result in permanent eye damage (20). Animal studies suggest that exposure to concentrations causing eye injury would be fatal as a result of lung damage (8). A follow up study 3 years after Bhopal found an excess of eye irritation, eyelid infection, cataract and decreased visual acuity (12). There is no evidence of irreversible eye damage or cases of blindness in Bhopal survivors in two studies (11,21). Burning and watering of the eyes, photophobia and corneal erosion were reported (11).

Skin Contact

Methyl isocyanate causes skin irritation and sensitisation. It can be absorbed through the skin. The dermal LD_{50} in rabbits for undiluted methyl isocyanate is 0.22 ml/kg, resulting in haemorrhage and oedema of the skin, and definite reactions occurred in interdermal sensitisation tests on guinea pigs (22).

Swallowing

The oral LD_{50} in mice and rats is 120 mg/kg and 69 mg/kg respectively.

CARCINOGENICITY

The carcinogenicity prediction and battery selection (CPBS) method predicts that methyl isocyanate has a significant potential for inducing cancer in rodents (23).

93. Methyl Isocyanate

MUTAGENICITY

Methyl isocyanate appears to be weakly genotoxic (8). Following the Bhopal accident, almost all seriously dyspnoeic patients developed at least two categories of chromosomal aberration (24). Methyl isocyanate is not mutagenic in *Salmonella* or *Drosophila*, but caused sister chromatid exchanges (SCE) and chromosome aberrations in Chinese hamster ovary cells either with or without metabolic activation (25). A small significant increase in SCE frequency was reported in cultured lung cells of mice exposed to methyl isocyanate (1-6 ppm, 6 hours/day for 4 consecutive days) but not in peripheral blood lymphocytes (26).

REPRODUCTIVE HAZARDS

There is little evidence that methyl isocyanate affects fertility, but exposure during pregnancy seems to increase the risk of spontaneous abortions and neonatal death (8). Exposure of mice to 9 ppm for 3 hours on day 8 of pregnancy caused loss of all embryos in 70% of animals. The authors suggest that foetal toxicity was partly independent of maternal lung damage (27). In another study on mice, inhalation exposure (0.1 or 3 ppm for 6 hours/day on days 14-17 of pregnancy) had no or minimal effects on fertility and reproduction (28).

FIRST AID

Eyes Wash the eye with flowing water for 10 minutes.

Lungs Remove casualty from area of exposure. If unconscious, do not give anything to drink, give artificial ventilation and chest compression or place in the recovery position as necessary. If conscious make the casualty lie or sit down quietly, give oxygen if available. Lung congestion may occur – a conscious casualty with breathing difficulties should be placed in a sitting position. Convulsions may occur and may cause unconsciousness. Allergic asthma (wheezy breathing) may occur and immediate treatment is needed – Ventolin may be useful.

Mouth Do not make the casualty vomit. Treat unconscious casualties as for lungs but if conscious give 1 pint of water to drink immediately. Convulsions may occur and may cause unconsciousness.

Skin Remove contaminated clothing immediately, drench the affected area with running water for at least 10 minutes.

In all cases of exposure, the patient should be transferred to hospital as soon as possible.

HANDLING AND STORAGE

Methyl isocyanate should be handled wearing an approved respirator, polyvinyl alcohol gloves (29), safety goggles and other protective clothing. It should only be used in chemical fume hood (30). Containers should be stainless steel or glass-lined materials, and transfer hoses should be flexible stainless steel. It must be protected by a 'blanket' of dry nitrogen, and bulk storage should be cooled to 0°C. Drums may be stored at ambient temperature but not in direct sunlight (3). It should be kept away from moisture, acids, bases or metal catalysts. It should be kept cool, away from sparks and flame, in areas separate from oxidisers. Pressure may develop in storage, and containers must be oversized to allow for expansion. The vapour may travel a considerable distance to source of ignition and flash back. Caps have disintegrated in storage, and improved types are now used for methyl isocyanate (31). Reducing the risks associated with storage by development of *in situ* methyl isocyanate generation processes has been discussed (32).

DISPOSAL

Eliminate all sources of ignition, ventilate the area and wearing self-contained breathing apparatus, a chemical resistant suit, gloves, and PVC safety boots, spread calcium hypochlorite over the (liquid) spill (or disperse excess calcium hypochlorite solution onto it). Mix and mop into polythene buckets, stand 24 hours and run to waste diluting x1000 with running cold water (see discharge limits from water authorities).

For solid spills either: a) sweep up, place in a large volume of water, add excess sodium hypochlorite, leave to stand 24 hours, then run to waste diluting x1000 with running water; or b) scoop into a large container, make alkaline with 2% sodium hydroxide solution and stir well. Add excess ferrous sulphate solution, mix well, stand for 2-3 hours and run to waste diluting x1000 with running water.

Large spills need specialist help – contact fire brigade.

FIRE PRECAUTIONS

Fires involving methyl isocyanate should be extinguished using carbon dioxide or dry chemical powder. Toxic and irritating methyl isocyanate vapours and hydrogen cyanide will be produced in a fire. Protective clothing for emergency response personnel has been evaluated. None of the materials provided adequate protection for at least an hour, but of the 22 chemical protective total encapsulating suit materials tested, Chem Fab™ Teflon-Nomex™ laminate gave the best protection (33). Total encapsulating suits with the respirator inside the suit should be worn during emergency response operations (33).

FURTHER READING

Bretherick, L. *Hazards in the Chemical Laboratory* (4th edition)

Bretherick, L. *Handbook of Reactive Chemical Hazards* (4th edition)

Sax, N. Irving *Dangerous Properties of Industrial Materials* (7th edition)

Encyclopedia of Occupational Safety & Health

Patty's *Industrial Hygiene and Toxicology*

Kirk-Othmer *Encyclopedia of Chemical Technology*

National Fire Protection Association *Manual of Hazardous Reactions*

ACGIH Documentation of TLVs and BEIs (6th edition, 1986)

Morel, C.; et al. Toxicology card No. 162 Methyl isocyanate *Cah. Notes Doc.* 1981, **104**, 451-454

Recommendations for the handling of aromatic isocyanates (Int. Isocyanate Inst., Inc., New York, 1976)

Dangerous Prop. Ind. Mater. Rep. 1989, **9**(3), 68-74

Dangerous Prop. Ind. Mater. Rep. 1985, **5**(2), 68-70

REFERENCES

1. Anon. *Chem. & Ind. (London)* 1985, 202
2. Anon. *Loss Prev. Bull.* 1985, (063), 1-8
3. Worthy, W. *Chem. Eng. News* 1985, **63**(6), 27-33
4. D'Silva, T. D. J. *J. Org. Chem.* 1986, **51**, 3781-3788
5. Bowander, B.; et al. *J. Haz. Mater.* 1988, **19**, 237-269
6. Lohs, L.; et al. *Z. Chem.* 1985, **25**(6), 197-206
7. *Indian J. Exp. Biol.* 1988, **26**, 149-176
8. Bucher, J. R. *Fundam. Appl. Toxicol.* 1987, **9**(3), 367-379
9. Dagani, R. *Chem. Eng. News* 1985, **63**(6), 37-40
10. Jayaraman, K. S. *Nature (London)* 1987, **329**(6142), 752
11. Anderson, N.; et al. *Br. J. Ind. Med.* 1988, **45**(7), 469-475
12. Anderson, N.; et al. *Br. J. Ind. Med.* 1990, **47**(8), 553-558
13. *Chem. Eng. News* 1986, **64**(9), 4
14. *Occup. Saf. Health.* 1986, **16**(4), 4
15. Anderson, D.; et al. *Br. J. Ind. Med.* 1990, **47**(9), 596-601
16. Kimmerle, G.; Eben, A. *Arch. Toxikol.* 1964, **20** 235
17. Mellon Inst. *Special Report 26-75* to Union Carbide Corp., Chem. & Plastics Operations Divn. 1970
18. Mellon Inst. *Special Report 26-23* 1963
19. *Project Report 45-62* (Union Carbide, Bushy Run Res. Centre, 1982
20. Rye, W. A. *JOM, J. Occup. Med.* 1973, **15**, 306
21. Anderson, N.; et al. *Lancet* 1984, **II 8417/8**, 1481
22. Mellon Inst. *Special Report 26-75* to Union Carbide Chem. Co. 1963
23. Ennever, F. K.; Rosenkranz, H. S. *Toxicol. Appl. Pharmacol.* 1987, **91**, 502-505
24. Goswami, H. K.; et al. *Hum. Genet.* 1990, **84**(2), 172-176
25. Mason, J. M.; et al. *Environ. Mutagen.* 1987, **9**(1), 19-28
26. Kligerman, A. D.; et al. *Environ. Mutagen.* 1987, **9**(1), 29-36
27. Varma, D. R.; et al. *J. Toxicol. Environ. Health* 1990, **30**(1), 1-14
28. McConnell, E. E.; et al. *Environ. Sci. Technol.* 1987, **21**(2), 188-193

93. Methyl Isocyanate

29. Forsberg, K.; Mansdorf, S. Z. *Quick selection guide to chemical protective clothing* (Van Nostrand Reinhold, New York, 1989)
30. Lenga, R. E. *The Sigma-Aldrich Library of Chemical Safety Data* (2nd edition) (Sigma-Aldrich, Milwaukee, 1988)
31. Anon. *CISHC Chem. Safety Summ.* 1979, **50**, 93
32. Hendershot, D. C. *Environ. Prog.* 1988, **7**(3), 180-184
33. Berardeinelli, S. P.; Moyer, E. S. *Am. Ind. Hyg. Assoc. J.* 1987, **48**(4), 324-329

1-METHYL-3-NITRO-1-NITROSO-GUANIDINE

CH3N(NO)C(=NH)NHNO2

RISKS
May cause cancer – Harmful by inhalation – Irritating to eyes and skin (R45, R20, R36/38)

SAFETY PRECAUTIONS
Avoid exposure-obtain special instructions before use – If you feel unwell, seek medical advice (show label where possible) (S53, S44)

IDENTIFIERS

SYNONYMS guanidine, N-methyl-N'-nitro-N-nitroso-; guanidine, 1-methyl-3-nitro-1-nitroso-; MNNG; nitrosoguanidine; methylnitronitrosoguanidine; methylnitrosoguanidine; 1-methyl-1-nitroso-3-nitro-guanidine; N-methyl-N-nitroso-N'-nitro-guanidine; MNG; N'- nitro-N-nitroso-N-methyl-guanidine; N-nitroso-N-methylnitro- guanidine; NSC 9369; RCRA Waste No. U163

CHEMICAL ABSTRACTS No.	70-25-7
NIOSH No.	MF 4200000
HAZCHEM CODE	not available
UN No.	not available

THRESHOLD LIMIT VALUES

US TLV (TWA)	not available
US TLV (STEL)	not available
UK EXPOSURE LIMITS (OES)	
Long-term (8 hr TWA value)	not available
Short-term (10 min TWA value)	not available
Germany	
MAK	not available
France	
VME	not available
VLE	not available
Sweden	
Short-term limit	not available
Level limit	not available

PHYSICAL PROPERTIES

Description Pale yellow, orange or pink crystals.

Boiling point	not available
Melting point	decmposes at 118-123.5°C
Density	not available
Vapour density	not available
Vapour pressure	not available
Flash point	not available
Explosive limits	not available
Autoignition temperature	not available

Solubility Slightly soluble in water. Soluble in polar organic solvents.

PACKAGING AND TRANSPORTATION

Road transportation

hazard warning sign	not available
Hazchem code	not available

Sea transportation

IMDG page No.	not available
class	not available
label	not available
packaging group	not available

Air transportation

ICAO/IATA code	not available
class	not available
label	not available
packaging group	not available
packing instructions	
cargo	not available
passenger	not available
passenger aircraft max. quantity	not available
cargo aircraft max. quantity	not available

94. 1-Methyl-3-Nitro-1-Nitrosoguanidine

MANUFACTURE
1-Methyl-3-nitro-1-nitrosoguanidine is prepared by nitrosation of *N*-methyl-*N*-nitroguanidine.

USES
1-Methyl-3-nitro-1-nitrosoguanidine is used in research as a mutagen, and as a tumour inducer in laboratory animals. It was used in the 1940s and 1950s as a diazomethane precursor.

CHEMICAL HAZARDS
1-Methyl-3-nitro-1-nitrosoguanidine detonates under high impact. A sample exploded when melted in a sealed capillary tube (1). The crude product from the aqueous nitrosation is pyrophoric, but recrystallised material is stable (2). It may explode on heating, and alkaline hydroysis gives the toxic, irritating, flammable and explosive diazomethane. It is incompatible with acids, bases, oxidising or reducing agents and moisture, and may decompose if exposed to light (3).

BIOLOGICAL HAZARDS
1-Methyl-3-nitro-1-nitrosoguanidine is a severe irritant, carcinogen and mutagen.

Vapour Inhalation
1-Methyl-3-nitro-1-nitrosoguanidine is a severe irritant and may cause coughing, sore throat, shortness of breath, headache, nausea and vomiting (3).

Eye Contact
1-Methyl-3-nitro-1-nitrosoguanidine is a severe irritant (3).

Skin Contact
1-Methyl-3-nitro-1-nitrosoguanidine is a severe irritant and may cause dermatitis. It can be absorbed through the skin (3).

Swallowing
The oral LD_{50} in rats is 90 mg/kg.

CARCINOGENICITY
In the studies reviewed by the IARC they concluded that 1-methyl-3-nitro-1- nitrosoguanidine is carcinogenic in rats, mice, hamsters, rabbits and dogs, having a mainly local carcinogenic effect following oral administration and other routes of exposure (4). Single doses caused primarily local tumours in laboratory animals orally and by other routes of exposure (5).

MUTAGENICITY
1-Methyl-3-nitro-1-nitrosoguanidine is a direct acting mutagen in *Salmonella typhimurium* TA1538, TA1535 (6), TA100 (7), TA98 and TA1537, and in *Escherichia coli* (8). It also gave positive results in the *umu* test (9), and in cultured human cells (5). Its mutagenicity has been reviewed (10).

REPRODUCTIVE HAZARDS
1-Methyl-3-nitro-1-nitrosoguanidine causes malformations of the brain, face, vertebrae, ribs and limbs of the foetus following intraperitoneal injection of doses ranging from 40-80 mg/kg to pregnant mice (11,12).

FIRST AID
Eyes Wash the eye with flowing water for at least 15 minutes.

Lungs Remove casualty from area of exposure. If unconscious, do not give anything to drink. Give artificial ventilation and chest compression or place in the recovery position as necessary. If conscious make the casualty lie or sit down quietly, give oxygen if available. Lung congestion may occur – a conscious casualty with breathing difficulties should be placed in a sitting position. Allergic asthma (wheezy breathing) may occur and immediate treatment is needed – Ventolin may be useful.

Mouth Do not make the casualty vomit. Treat unconscious casualties as for lungs but if conscious give 1 pint of water to drink.

Skin Remove contaminated clothing immediately, wash affected area with soap and copious amounts of water.

94. 1-Methyl-3-Nitro-1-Nitrosoguanidine

Absorption through the skin may result in symptoms similar to those of inhalation or ingestion.

In all cases of exposure, the patient should be transferred to hospital as soon as possible.

HANDLING AND STORAGE

1-Methyl-3-nitro-1-nitrosoguanidine should be handled wearing an approved respirator, chemical-resistant gloves, safety goggles and other protective clothing. It should only be used in a chemical fume hood (3). It should be kept refrigerated in a tightly closed container, away from heat, sources of ignition, or light. It is capable of causing a dust explosion.

DISPOSAL

Wearing approved self-contained breathing apparatus, rubber boots, heavy rubber gloves and disposable overalls, shut off all sources of ignition and ventilate the area. Sweep the spill into a bag and hold for waste disposal. Use non-sparking tools. Wash the spill site thoroughly.

Spills of 1-methyl-3-nitro-1-nitrosoguanidine can be decontaminated with a solution of sulphamic acid in 2M hydrochloric acid (12,13).

FIRE PRECAUTIONS

Fires involving 1-methyl-3-nitro-1-nitrosoguanidine should be extinguished using carbon dioxide, dry chemical powder, alcohol foam or polymer foam. Air-supplied respirators and full-face masks should be worn.

FURTHER READING

Bretherick, L. *Hazards in the Chemical Laboratory* (4th edition)

Bretherick, L. *Handbook of Reactive Chemical Hazards* (4th edition)

Sax, N. Irving *Dangerous Properties of Industrial Materials* (7th edition)

Encyclopedia of Occupational Safety & Health

Patty's *Industrial Hygiene and Toxicology*

Kirk-Othmer *Encyclopedia of Chemical Technology*

National Fire Protection Association *Manual of Hazardous Reactions*

ACGIH Documentation of TLVs and BEIs (6th edition, 1986)

REFERENCES

1. Eisendrath, J. N. *Chem. Eng. News* 1953, **31**, 3016
2. Aldrich advert. *J. Org. Chem.* 1974, **39**(7), cover iv
3. Lenga, R. E. *The Sigma-Aldrich Library of Chemical Safety Data* (2nd edition) (Sigma-Aldrich, Milwaukee, 1988)
4. IARC *Monographs on the evaluation of the carcinogenic risk of chemicals to humans* 1974, **4**, 183-195
5. *Dangerous Prop. Ind. Mater. Rep.* 1985, **5**(4), 59-65
6. Rosenkranz, H. S.; Poirer, L. A. *JNCI, J. Natl. Cancer Inst.* 1979, **62**(4), 873-892
7. Simmon, V. F. *JNCI, J. Natl. Cancer Inst.* 1979, **62**(4), 893-899
8. Dunkel, V. C.; et al. *Environ. Mutagen.* 1984, **6**(suppl. 2), 1
9. Nakamura, S.; et al. *Mutat. Res.* 1987, **192**, 239-246
10. Gichner, T.; Veleminsky, J. *Mutat. Res.* 1982, **99**(2), 129-242
11. Manson, J. M.; Miller, M. L. *Teratol. Carcinogen. Mutagen.* 1983, **3**, 335
12. Inouye, M.; Murakami. U. *Teratology* 1978, **18**, 263-268
13. Lunn, G.; Sansone, E. B. *Food Chem. Toxicol.* 1988, **26**(5), 481-484
14. Castegnaro, M.; et al. *Laboratory decontamination and destruction of carcinogens in laboratory wastes: some N-nitrosamides* (OUP, Oxford, 1984)

95. Methyloxirane

METHYLOXIRANE

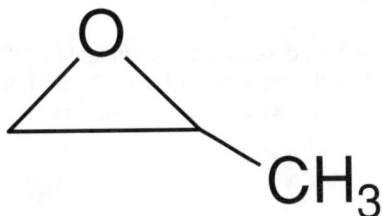

RISKS
May cause cancer – Extremely flammable – Harmful by inhalation, in contact with skin and if swallowed – Irritating to eyes, respiratory system and skin (R45, R12, R20/21/22, R36/37/38)

SAFETY PRECAUTIONS
Avoid exposure-obtain special instructions before use – Keep container tightly closed, in a cool well ventilated place – Keep away from sources of ignition – No Smoking – Take precautionary measures against static discharges – If you feel unwell, seek medical advice (show label where possible) (S53, S3/7/9, S16, S33, S44)

IDENTIFIERS

SYNONYMS oxirane, methyl-; propylene oxide; propane, 1,2-epoxy-; AD 6; AD 6 (suspending agent); epoxypropane; 1,2-epoxypropane; propene oxide; propylene epoxide; 1,2-propylene oxide; methyl ethylene oxide; NCI-C50099

CHEMICAL ABSTRACTS No.	75-56-9
NIOSH No.	TZ 2975000
HAZCHEM CODE	2PE
UN No.	1280

THRESHOLD LIMIT VALUES

US TLV (TWA)	20 ppm (48 mg/m^3)
US TLV (STEL)	not available

UK EXPOSURE LIMITS (OES)
Long-term (8 hr TWA value) ... 20 ppm (50 mg/m^3)
Short-term (10 min TWA value) .. 100 ppm (240 mg/m^3)

Germany
MAK ... not available

France
VME ... 20 ppm (50 mg/m^3)
VLE ... not available

Sweden
Short-term limit ... 10 ppm (25 mg/m^3)
Level limit ... 5 ppm (12 mg/m^3)

PHYSICAL PROPERTIES

Description	Colourless liquid.
Boiling point	34°C
Melting point	-112°C
Density	0.8304 at 20°C
Vapour density	2.0
Vapour pressure	445 mm Hg at 20°C
Flash point	-37°C (open cup)
Explosive limits	2.1%-37%
Autoignition temperature	449°C
Solubility	Soluble in water, ether or ethanol.

PACKAGING AND TRANSPORTATION

Road transportation

hazard warning sign	1280 flammable liquid
Hazchem code	2PE

Sea transportation

IMDG page No.	3143
class	3.1
label	flammable liquid
packaging group	I

Air transportation

ICAO/IATA code (UN No.)	1280
class	3
label	liquid flammable
packaging group	I
packing instructions	
cargo	304
passenger	306
passenger aircraft max. quantity	1 litre
cargo aircraft max. quantity	30 litres

95. Methyloxirane

MANUFACTURE

Methyl oxirane is produced by the chlorohydrin process in which 1-chloro-2-propanol and 2-chloro-1-propanol react with calcium oxide or potassium hydroxide. It is also produced by an indirect oxidation process involving propene and an organic hydroperoxide.

USES

Propylene oxide is used as an intermediate in the production of a range of chemicals such as polyether polyols and propylene glycol. It is also used in the preparation of lubricants, surfactants and oil demulsifiers. Small amounts are used for sterilisation of medical equipment or fumigation of soil and foodstuffs.

CHEMICAL HAZARDS

Propylene oxide forms explosive mixtures with air, and is a very dangerous fire and explosion hazard if exposed to heat or flame. It reacts explosively with epoxy resin (1) or sodium hydroxide (2). It reacts with ethylene oxide and polyhydric alcohol to form thermally unstable polyether alcohol. It is incompatible with ammonium hydroxide, chlorosulphonic acid, hydrogen chloride, hydrogen fluoride, nitric acid, oleum or sulphuric acid. Safe handling of propylene oxide on a laboratory scale has been discussed (3).

BIOLOGICAL HAZARDS

Propylene oxide is a severe irritant and is moderately toxic. It should be regarded as a carcinogenic risk to humans. It is mutagenic in microbial and mammalian assays, and experimental reproductive effects have been reported. Its properties and health hazards have been reviewed (4).

Vapour Inhalation

Propylene oxide is irritating to the upper respiratory tract, and may cause drowsiness and weakness. A case of poisoning was reported in Russia after exposure to 1500 ppm for 10 minutes; symptoms included lung and eye irritation, headache, diarrhoea, weakness and collapse. Recovery was complete the following day after treatment with oxygen and antihistamines (5). It caused severe nose irritation, lacrimation, salivation, dyspnoea, central nervous system depression in animals at 4100 or 9500 mg/m^3. Chronic exposure to 457 ppm caused lung damage in rats and guinea pigs (6).

Eye Contact

Corneal burns have been reported in humans (7). Gaseous propylene oxide is irritating to the eyes, and the liquid caused severe injury to rabbits' eyes (8).

Skin Contact

Propylene oxide causes burns, blisters and oedema in six minutes in close contact with the skin (6). Skin injury is not likely, however, if it can evaporate freely. Cases of allergic contact dermatitis were reported in 2 laboratory assistants from a pre-injection swab containing 1% propylene oxide (9).

Swallowing

The oral LD_{50} in rats has been reported as 520 to 1140 mg/kg, with damage to the liver and stomach lining occurring at these levels (10).

CARCINOGENICITY

In inhalation studies the NTP found clear evidence of carcinogenicity in mice and some evidence in rats. Propylene oxide produced local sarcomas in rats or mice by oral administration, inhalation and subcutaneous administration. The IARC evaluation is that there is "sufficient evidence" for carcinogenicity in animals, but "inadequate evidence" in humans. Therefore for practical purposes propylene oxide should be treated as if it represented a carcinogenic risk to humans (11).

MUTAGENICITY

Propylene oxide is mutagenic to microorganisms including *Salmonella typhimurium* (12) and insects (13). Mutations, DNA damage and chromosomal effects have been reported in mammalian cells *in vitro*. It induced chromosome aberrations and sister chromatid exchanges in Chinese hamster ovary cells *in vitro* (14). *In vivo* studies found increased micronuclei in mice after intraperitoneal injection, but no chromosome effects or dominant lethal effects in monkeys and mice following inhalation of propylene oxide (10). Exposure of RL1 rat liver cell lines caused a significant dose-related increase in chromatid aberrations (15).

95. Methyloxirane

REPRODUCTIVE HAZARDS

An increased rate of resorption was reported in rats exposed to 1190 mg/m^3, 7 hours/day on days 7-16 of pregnancy. Similar exposure of rats on days 1-16 of pregnancy caused some reduction in ossification in vertebrae and ribs, and wavy ribs. No teratogenic or foetotoxic effects were noted in exposed rabbits (10).

FIRST AID

Eyes Wash the eye with flowing water for 10 minutes.

Lungs Remove casualty from area of exposure. If unconscious, do not give anything to drink. Give artificial ventilation and chest compression or place in the recovery position as necessary. If conscious make the casualty lie or sit down quietly, give oxygen if available. Lung congestion may occur – a conscious casualty with breathing difficulties should be placed in a sitting position.

Mouth Do not make the casualty vomit. Treat unconscious casualties as for lungs, but if conscious give 1 pint of water to drink. Frost-bite of the mouth and throat may occur – give lukewarm drinks.

Skin Remove contaminated clothing immediately, drench the affected area with running water for at least 10 minutes. Frost-bite may occur – warm the affected areas carefully.

In all cases of exposure, the patient should be transferred to hospital as soon as possible.

HANDLING AND STORAGE

Propylene oxide should be handled wearing an approved respirator, chemical-resistant gloves, safety goggles, boots and other protective clothing. It should only be used in a chemical fume hood (16). It should be kept in a tightly sealed container, away from heat or sources of ignition, and protected against physical damage. It should be kept away from acids, bases, peroxides of iron or aluminium, and chlorides of iron, aluminium or tin, as contact may cause violent polymerisation. It should be kept in a flammable liquids store, but detached outside storage is preferable. The vapour may travel a considerable distance to a source of ignition and flash back. Containers should be refrigerated before opening, and should be opened carefully.

Equipment should be carbon steel or stainless steel, and tanks should be insulated, electrically grounded, diked and protected by sprinklers. New equipment must be dry and free of air and contaminants such as rust, ammonia, acetylene, or hydrogen sulphide. It should be kept under pure and dry nitrogen or methane, with sufficient inert gas pressure to keep the vapour phase out of the flammable range. Tanks and equipment should be checked often for leaks. Explosion-proof electric motors should be used, and other equipment should be vapour-tight and spark proof.

DISPOSAL

Eliminate all sources of ignition, ventilate the area and wear a laboratory coat or overalls, gloves, safety glasses and approved self-contained breathing apparatus. Absorb small spills onto paper towels, remove to open air in a bucket with lid. Evaporate on a metal tray in a safe place. NB. Explosive peroxides may be formed in the presence of air. Absorb large spills on sand or vermiculite, and transport in buckets to a safe open area for atmospheric evaporation.

FIRE PRECAUTIONS

Fires involving propylene oxide should be extinguished using carbon dioxide, dry chemical powder or alcohol foam. Water is ineffective, but should be used to cool exposed containers. Water spray may also be used to disperse vapours, protect personnel or dilute spills. It may decompose to form explosive mixtures with air under fire conditions.

FURTHER READING

Bretherick, L. *Hazards in the Chemical Laboratory* (4th edition)

Bretherick, L. *Handbook of Reactive Chemical Hazards* (4th edition)

Keith, L. H.; Walters, D. B. (editors) *Compendium of Safety Data Sheets for Research and Industrial Chemicals* (VCH, Deerfield Park, 1987)

Sax, N. Irving *Dangerous Properties of Industrial Materials* (7th edition)

Encyclopedia of Occupational Safety & Health

Patty's *Industrial Hygiene and Toxicology*

Kirk-Othmer *Encyclopedia of Chemical Technology*

National Fire Protection Association *Manual of Hazardous Reactions*

95. Methyloxirane

ACGIH Documentation of TLVs and BEIs (6th edition, 1986)

REFERENCES

1. Sheaffer, J. *CHAS Notes* 1981, **1**(4), 5
2. *MCA Case History No. 31*
3. Pogany, G. A. *Chem. & Ind.* 1979, 16-21
4. Meylan, W.; et al. *Toxicol. Ind. Health* 1986, **2**(3), 219-260
5. Gosselin, R. E.; et al. *Clinical Toxicology of Commercial Products* (5th edition) (Williams & Wilkins, Baltimore, 1984)
6. Rowe, V. K.; et al. *AMA Arch. Ind. Health* 1956, **13**, 228
7. McLaughlin, R. L. *Am. J. Ophthalmol.* 1946, **29**, 1355
8. Carpenter, C. P., Smyth, H. P. *Am. J. Ophthalmol.* 1946, **29**, 1363
9. Jensen *Contact Dermatitis* 1981, **7**(2), 148-150
10. *Environmental Health Criteria 56* Propylene oxide (WHO, Geneva, 1985)
11. IARC *Monographs on the evaluation of the carcinogenic risk of chemicals to humans* 1985, **36**, 227-243
12. Rossman, L. B.; et al. *Mutat. Res.* 1987, **189**(3), 189-204
13. Hardin, B. D.; et al. *Mutat. Res.* 1983, **117**, 337-344
14. Gulati, D. K.; et al. *Environ. Mol. Mutagen.* 1989, **13**(2), 133-193
15. Dean, B. J.; et al. *Mutat. Res.* 1985, **153**(1/2), 57-77
16. Lenga, R. E. *The Sigma-Aldrich Library of Chemical Safety Data* (2nd edition) (Sigma-Aldrich, Milwaukee, 1988)

96. Neopentyl Glycol Diacrylate

NEOPENTYL GLYCOL DIACRYLATE
$(H_2C=CHCO_2CH_2)_2C(CH_3)_2$

RISKS
Toxic if swallowed – Irritating to eyes and skin – May cause sensitisation by skin contact (R25, R36/38, R43)

SAFETY PRECAUTIONS
After contact with skin, wash immediately with plenty of water – Wear eye/face protection – If you feel unwell, seek medical advice (show label where possible) (S28, S39, S44)

IDENTIFIERS

SYNONYMS dimenthylopropane diacrylate; 2-propenoic acid, 2,2-dimethyl-1,3-propanediylester; acrylic acid, 2,2- dimethyltrimethylene ester; dimethylolpropane diacrylate; 2,2- dimethyltrimethylene acrylate; neopentanediol diacrylate; NK Ester A-NPG; SR 247

CHEMICAL ABSTRACTS No.	2223-82-7
NIOSH No.	not available
HAZCHEM CODE	not available
UN No.	not available

THRESHOLD LIMIT VALUES

US TLV (TWA)	not available
US TLV (STEL)	not available
UK EXPOSURE LIMITS (OES)	
Long-term (8 hr TWA value)	not available
Short-term (10 min TWA value)	not available
Germany	
MAK	not available
France	
VME	not available
VLE	not available
Sweden	
Short-term limit	not available
Level limit	not available

PHYSICAL PROPERTIES

Description not available.

Boiling point	not available
Melting point	not available
Density	not available
Vapour density	not available
Vapour pressure	not available
Flash point	not available
Explosive limits	not available
Autoignition temperature	not available

Solubility not available.

PACKAGING AND TRANSPORTATION

Road transportation

hazard warning sign	not available
Hazchem code	not available

Sea transportation

IMDG page No.	not available
class	not available
label	not available
packaging group	not available

Air transportation

ICAO/IATA code	not available
class	not available
label	not available
packaging group	not available
packing instructions	
cargo	not available
passenger	not available
passenger aircraft max. quantity	not available
cargo aircraft max. quantity	not available

96. Neopentyl Glycol Diacrylate

MANUFACTURE

Neopentyl glycol diacrylate may be produced by esterification of neopentyl glycol with acrylic acid.

USES

Neopentyl glycol diacrylate is used as a crosslinking agent in photocurable coatings.

CHEMICAL HAZARDS

No information is available concerning the chemical hazards of neopentyl glycol diacrylate.

BIOLOGICAL HAZARDS

Neopentyl glycol diacrylate is a skin irritant, and sensitisation and carcinogenicity have been reported following skin application in animal experiments.

Vapour Inhalation

No information is available concerning the effects of inhalation.

Eye Contact

No data specific to neopentyl glycol diacrylate appears to be available, but most multifunctional acrylates are severe eye irritants (1).

Skin Contact

Neopentyl glycol diacrylate is a strong irritant and potent skin sensitiser in the guinea pig maximisation (GPM) test (2), and cross-reactivity may occur with hydroquinone and *p*-methoxyphenol (3). Negative results have also been reported in the GPM test (4).

Swallowing

No data specific to neopentyl glycol diacrylate appears to be available, but most multifunctional acrylates have a low acute oral toxicity (1).

CARCINOGENICITY

Neopentyl glycol diacrylate had significant tumourigenic activity when applied to the skin of male mice in lifetime studies; tumour yield was 8 out of 40 mice, 3 being malignant (5).

MUTAGENICITY

Neopentyl glycol diacrylate is not mutagenic in the *Salmonella*/microsome assay with TA1535, TA1537, TA1538, TA100 and TA98 either with or without metabolic activation (6). It was not mutagenic in Chinese hamster ovary cells, or in *in vitro* sister chromatid exchange assay, and did not cause unscheduled DNA synthesis (1).

REPRODUCTIVE HAZARDS

No information is available concerning the reproductive hazards of neopentyl glycol diacrylate, however, data from other multifunctional acrylates suggest they are not potent teratogens or foetotoxins (1).

FIRST AID

Eyes Wash the eye with flowing water for 10 minutes.

Lungs Remove casualty from area of exposure. If unconscious, do not give anything to drink, give artificial ventilation and chest compression or place in the recovery position as necessary. If conscious make the/ casualty lie or sit down quietly, give oxygen if available. Lung congestion may occur – a conscious casualty with breathing difficulties should be placed in a sitting position. Convulsions may occur and may cause unconsciousness.

Mouth Do not make the casualty vomit. Treat unconscious casualties as for lungs but if conscious give 1 pint of water to drink immediately.

Skin Remove contaminated clothing immediately, drench the affected area with running water for at least 10 minutes.

In all cases of exposure, the patient should be transferred to hospital as soon as possible.

96. Neopentyl Glycol Diacrylate

HANDLING AND STORAGE

Neopentyl glycol diacrylate should be handled wearing an approved respirator, chemical-resistant gloves, safety goggles and other protective clothing. It should only be used in a chemical fume hood. It should be kept in a tightly closed container, away from heat, light or moisture.

DISPOSAL

Eliminate all sources of ignition and ventilate the area. Wearing a laboratory coat or overalls, safety glasses, and gloves, absorb the spill onto paper towels and allow to evaporate in a fume-cupboard. For large spills, absorb onto sand or vermiculite, and remove in buckets for atmospheric evaporation in a safe open area. Ideally, waste should be burned in an incinerator with afterburner.

FIRE PRECAUTIONS

Fires involving neopentyl glycol diacrylate should be extinguished using water spray, carbon dioxide, dry chemical powder, alcohol foam or polymer foam.

FURTHER READING

Sax, N. Irving *Dangerous Properties of Industrial Materials* (7th edition)

Encyclopaedia of Occupational Safety & Health

Patty's *Industrial Hygiene and Toxicology*

Kirk-Othmer *Encyclopedia of Chemical Technology*

National Fire Protection Association *Manual of Hazardous Reactions*

ACGIH Documentation of TLVs and BEIs (6th edition, 1986)

REFERENCES

1. Andrews, L. S.; Clary, J. J. *J. Toxicol. Environ. Health* 1986, **19**(2), 149-164
2. Bjoerkner, B. *Contact Dermatitis* 1984, **11**(4), 236-246
3. Van der Walle, H. B.; et al. *Contact Dermatitis* 1982, **8**(3), 147-154
4. Van der Walle, H. B.; et al. *Contact Dermatitis* 1983, **9**, 10-20
5. DePass, L. R.; et al. *J. Toxicol. Environ. Health* 1985, **16**(1), 55-60
6. Waegemakers, T. H. J. M.; Bensink, M. P. H. *Mutat. Res.* 1984, **137**(2-3), 95-102

NICKEL TETRACARBONYL
Ni(CO)$_4$

RISKS
Highly flammable – Very toxic by inhalation – Possible risk of irreversible effects (R11, R26, R40)

SAFETY PRECAUTIONS
Keep container in a well ventilated place – Do not breathe fumes – In case of accident or if you feel unwell, seek medical advice immediately (show label where possible) (S9, S23, S45)

IDENTIFIERS

SYNONYMS nickel carbonyl; tetracarbonylnickel
CHEMICAL ABSTRACTS No. 13463-39-3
NIOSH No. .... QR 6300000
HAZCHEM CODE .... not available
UN No. 1259

THRESHOLD LIMIT VALUES

US TLV (TWA)
... 0.05 ppm (0.12 mg/m^3) (as Ni) (proposed change)
US TLV (STEL) .... not available
UK EXPOSURE LIMITS (OES)
Long-term (8 hr TWA value) .... not available
Short-term (10 min TWA value) .. 0.1 ppm (0.24 mg/m^3)
Germany
MAK .... not available
France
VME .... not available
VLE .... not available
Sweden
Short-term limit .... not available
Level limit .... 0.001 ppm (0.007 mg/m^3)

97. Nickel Tetracarbonyl

PHYSICAL PROPERTIES

Description Colourless, volatile liquid or needles. The solid is white.
Boiling point .... 43°C
Melting point .... -19.3°C
Density .... 1.3185 at 17°C (liquid)
Vapour density 5.9
Vapour pressure .... 400 mm Hg at 25.8°C
Flash point .... <-20°C (closed cup)
Explosive limits .... 3 wt % – 34 wt %
Autoignition temperature .... not available
Solubility Soluble in alcohol, benzene, chloroform, acetone or carbon tetrachloride.

PACKAGING AND TRANSPORTATION

Road transportation
hazard warning sign .... 1259 toxic substance
Hazchem code .... not available

Sea transportation
IMDG page No. 6202
class 6.1
label .... poison; flammable liquid; marine pollutant
packaging group I

Air transportation
ICAO/IATA code (UN No.) 1259
class 6.1;3
label .... not available
packaging group .... not available
packing instructions
cargo .... forbidden
passenger .... forbidden
passenger aircraft max. quantity .... forbidden
cargo aircraft max. quantity .... forbidden

97. Nickel Tetracarbonyl

MANUFACTURE

Nickel tetracarbonyl is produced by passing carbon monoxide over finely divided nickel.

USES

Nickel tetracarbonyl is used in the manufacture of catalysts, in nickel vapour plating, as an intermediate in nickel refining (the Mond process), and in the manufacture of high purity nickel powder.

CHEMICAL HAZARDS

Nickel tetracarbonyl reacts explosively with bromine in the liquid state (1). Dry nickel tetracarbonyl and oxygen explode on shaking with mercury. It may also react explosively with a butane-oxygen mixture (2). It produces a deposit on exposure to atmospheric oxygen which may ignite, and mixtures with air or oxygen at low partial and total pressures explode after an induction period (3). It reacts with tetrachloropropadiene to form a shock-sensitive explosive (4). It reacts violently with dinitrogen tetraoxide. The vapour forms an explosive mixture with air; the explosion parameters of nickel tetracarbonyl-air mixtures have been reported (5). It is highly flammable and liable to explode if heated to 60°C or above.

BIOLOGICAL HAZARDS

The toxicity of nickel tetracarbonyl has be reviewed (6).

Vapour Inhalation

The vapours cause pulmonary oedema, congestion, interstitial pneumonitis, liver, kidney, spleen and pancreatic damage; signs and symptoms include headache and dizziness, followed 12-36 hours after exposure by dyspnoea, rapid pulse, cough, cyanosis, vomiting, diarrhoea, weakness, confusion and convulsions. 30 minutes exposure to 30 ppm may be lethal to humans (10). Many cases of acute nickel tetracarbonyl poisoning have been reported (11). Encephalopathy and myocardiopathy have been reported in workers with a history of acute poisoning by heavy and medium doses of nickel tetracarbonyl (12). Neurotoxicity has been reported in workers occupationally exposed to nickel tetracarbonyl. Asthma and Loeffler's syndrome have been reported following chronic exposure.

Eye Contact

Nickel tetracarbonyl may be irritating to the eyes (13). It caused a reddening of the mucosa when instilled into the eye of a rabbit (14).

Skin Contact

Nickel tetracarbonyl may cause allergic dermatitis (15). It can be absorbed through the skin.

Swallowing

Ingestion may cause irritation of the mouth, throat and stomach.

CARCINOGENICITY

The IARC evaluation is that there is limited evidence for the carcinogenicity of nickel tetracarbonyl in animal experiments (16,17). Increased incidence of malignant tumours in varied organs and tissues were reported in rats following intravenous administration, although the IARC have questioned this (18). Significant numbers of bronchial carcinomas were reported in a 2-year study using an intrabronchial pellet implantation system in rats (19). Although a high incidence of nasal and lung cancer have been reported in workers in a nickel refinery using the Mond process (20), they have also been reported in nickel refineries not using nickel tetracarbonyl (21). It is probable that nickel in some form is carcinogenic (16).

MUTAGENICITY

Increased incidence of sister chromatid exchanges, but not chromosome aberrations, has been reported in workers occupationally exposed to nickel tetracarbonyl (22).

REPRODUCTIVE HAZARDS

Nickel tetracarbonyl was teratogenic and embryotoxic in rats (23) and Syrian hamsters (24,25). Women have been excluded from work where nickel tetracarbonyl exposure may occur because of the risk of foetotoxicity of the antidote dithiocarb (26).

97. Nickel Tetracarbonyl

FIRST AID

Eyes Wash the eye with flowing water for 10 minutes.

Lungs Remove casualty from area of exposure, wearing protective clothing and positive pressure air breathing apparatus. If unconscious, do not give anything to drink. If breathing has stopped DO NOT use mouth to mouth ventilation but use a resuscitation bag and mask instead. If breathing, give 100% oxygen by mask. Convulsions may occur and may cause unconsciousness. Lung congestion may occur – a conscious casualty with breathing difficulties should be placed in a sitting position.

Mouth Do not make the casualty vomit. Treat unconscious casualties as for lungs but if conscious give 1 pint of water to drink. Lung congestion may occur – a conscious casualty with breathing difficulties should be placed in a sitting position.

Skin Wearing protective clothing and gloves, remove contaminated clothing immediately and place it in a thick plastic sack and seal. Wash affected areas with soap and copious amounts of water. Absorption through the skin may result in symptoms similar to those of inhalation.

In all cases of exposure, the patient should be transferred to hospital as soon as possible.

HANDLING AND STORAGE

Nickel tetracarbonyl should be handled wearing an approved respirator, chemical-resistant gloves, safety goggles and rubber overclothing (26). It should be kept in a tightly closed container, in a fireproof area separate from other storage, preferably in outside or detached storage. It should be kept cool and under inert gas, away from strong oxidants (13), and protected against physical damage. Laboratory or small plant storage should be in steel containers with 20% ullage space. The vapour may travel a considerable distance to a source of ignition and flash back. Explosion proof and nonsparking electrical equipment should be used. The liquid and vapour hardens natural rubber such as bungs. Protective measures have been described (27).

DISPOSAL

Wear chemical resistant overalls and gloves and compressed air breathing apparatus (essential). Eliminate all sources of ignition; ventilate the area but isolate the material. For very small spills (1-2g), treat spillage with water to dilute the chemical then very cautiously transfer to suitable containers and add an excess (50%) of calcium hypochlorite or sodium perchlorate solution. Stir cautiously and leave overnight. Carry on adding calcium hypochlorite until the pH of the mixture reaches around 7. Filter off the solids, mix them with equal parts of sand and place in a sealed metal container to be disposed of by licensed contractor.

For moderate or large spills treat as an emergency. Should only be dealt with by trained factory teams, otherwise call the fire-brigade and police.

FIRE PRECAUTIONS

Fires involving nickel tetracarbonyl should be extinguished using carbon dioxide, dry chemical powder, alcohol foam or polymer foam. Use water to cool fire exposed containers (13). Containers may explode when heated.

FURTHER READING

Bretherick, L. *Hazards in the Chemical Laboratory* (4th edition)

Bretherick, L. *Handbook of Reactive Chemical Hazards* (4th edition)

Sax, N. Irving *Dangerous Properties of Industrial Materials* (7th edition)

Encyclopedia of Occupational Safety & Health

Patty's *Industrial Hygiene and Toxicology*

Kirk-Othmer *Encyclopedia of Chemical Technology*

National Fire Protection Association *Manual of Hazardous Reactions*

ACGIH *Documentation of TLVs and BEIs* (6th edition, 1986)

Environmental Health Criteria 108. Nickel (WHO, Geneva, 1991)

Laboratory Hazards Data Sheet No. 60: Nickel and Nickel Compounds (RSC, London, 1987)

REFERENCES

1. Blanchard, A.A.; et al. *J. Am. Chem. Soc.* 1926, **48**, 872

97. Nickel Tetracarbonyl

2. Badin, E. J.; et al. *J. Am. Chem. Soc.* 1948, **70**, 2055
3. Egerton, A.; et al. *Proc. R. Soc.* 1954, **A225**, 427
4. Posey, R. G.; et al. *J. Am. Chem. Soc.* 1977, **99**, 4865 (footnote 13)
5. Dement'ev., A. A.; et al. *Tr. Proektn. Nau chrno-Issted. Inst. Gipronikel* 1974, **60**, 108-113
6. Bencko, V. *Z. Gesamte Hyg. Ihre Grenzgeb* 1984, **30**(5), 250-263
7. Mastromatteo, E. *Am. Ind. Hyg. Assoc. J.* 1986, **47**(10). 589-601
8. *IRPTC Bull.* 1987, **8**(1), 25,27
9. Arthur, J. L. *DHEW (NIOSH) Publ. (U.S.)* 1977, 77-184
10. *Hygienic Guide Series: Nickel Carbonyl* (Am. Ind. Hyg. Assoc., 1968
11. Zhicheng, S. *Br. J. Ind. Med.* 1986, **43**(6), 422-424
12. Novokhatskii, N. K.; et al. *Gig. Tr. Prof. Zabol.* 1987, **5**, 45-47
13. *International Chemical Safety Cards: Nickel Carbonyl* (CEC, Luxembourg, 1990)
14. Sanina, Yu. P. *Toksikol. Nov. Prom. Khim. Veshchestv.* 1968, (10), 144-149
15. Raithel, H. J. *Arbeitsmed. Sozialmed. Praeventivmed.* 1987, **22**(11), 268-274
16. IARC *Monographs on the evaluation of the carcinogenic risk of chemicals to humans* 1973, **11**, 126-149
17. IARC *Monographs on the evaluation of the carcinogenic risk of chemicals to humans* 1976, **22**, 75
18. Lau, T. J.; et al. *Cancer Res.* 1972, **32**, 2253-2258
19. Levy. L. S.; et al. *Br. J. Ind. Med.* 1986, **43**, 243-256
20. Doll, R. *Br. J. Ind. Med.* 1958, **15**, 217
21. Pederson, E.; et al. *Int. J. Cancer* 1973, **12**, 32-41
22. Cai, D. C.; et al. *Mutat. Res.* 1987, **188**(2), 149-152
23. Sunderman, F. W., Jr.; et al. *Science* 1979, **203**
24. Sunderman, F. W., Jr.; et al. *Teratol. Carcinogen Mutagen.* 1980, **1**, 223-233
25. Sunderman, F. W., Jr.; et al. *Nickel Toxicol. (Proc. Int. Conf.) 2nd* 1980, 113-116
26. *Dangerous Prop. Ind. Mater. Rep.* 1985, **5**(4), 76-82
27. *IRPTC Bull.* 1989, **9**(2), 14-35

NICOTINE

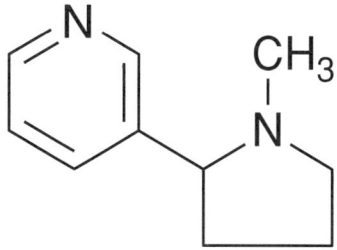

RISKS
Very toxic by inhalation, in contact with skin and if swallowed (R26/27/28)

SAFETY PRECAUTIONS
Keep locked up – Keep away from food, drink and animal feeding stuffs – After contact with skin, wash immediately with plenty of water – In case of accident or if you feel unwell, seek medical advice immediately (show label where possible) (S1, S13, S28, S45)

IDENTIFIERS

SYNONYMS pyridine,3-(1-methyl-2-pyrrolidinyl)-,*s*-; Flux Maag; 3-(*N*-methylpyrollidino)pyridine; nicotin; (-)-nicotine; l- nicotine; L-nicotine; *S*-nicotine; XL all insecticide

CHEMICAL ABSTRACTS No.	54-11-5
NIOSH No.	QS 5250000
HAZCHEM CODE	2X
UN No.	1654

THRESHOLD LIMIT VALUES

US TLV (TWA)	0.5 mg/m^3
US TLV (STEL)	not available

UK EXPOSURE LIMITS (OES)
Long-term (8 hr TWA value)	0.5 mg/m^3
Short-term (10 min TWA value)	1.5 mg/m^3

Germany
MAK	0.07 ppm (0.5 mg/m^3)

France
VME	0.5 mg/m^3
VLE	not available

Sweden
Short-term limit	not available
Level limit	not available

PHYSICAL PROPERTIES

Description Thick colourless or pale yellow oil with a slight fishy smell when warm.

Boiling point	247.3°C
Melting point	<80°C
Density	1.0097 at 20°C
Vapour density	5.61
Vapour pressure	1 mm Hg at 61.8°C
Flash point	95°C (closed cup)
Explosive limits	0.75%-4%
Autoignition temperature	243.9°C

Solubility Soluble in water and organic solvents.

PACKAGING AND TRANSPORTATION

Road transportation
hazard warning sign	1654 toxic substance
Hazchem code	2X

Sea transportation
IMDG page No.	6203
class	6.1
label	poison
packaging group	II

Air transportation
ICAO/IATA code (UN No.)	1654
class	6.1
label	poison
packaging group	II
packing instructions	
cargo	611
passenger	609
passenger aircraft max. quantity	5 litres
cargo aircraft max. quantity	60 litres

98. Nicotine

MANUFACTURE

Nicotine is obtained as a by-product of the tobacco industry.

USES

Nicotine was used in insecticides, but rarely so now. It is used as an ectoparasiticide, and has been used in animal tranquillising darts. The vitamin nicotinic acid is synthesised by oxidation of nicotine.

CHEMICAL HAZARDS

Nicotine can react with oxidising materials, and the vapour is moderately explosive on exposure to heat or flame.

BIOLOGICAL HAZARDS

Nicotine is absorbed from the gut, lungs and skin. It is one of the most toxic of poisons, and acts very rapidly. In fatal cases death often occurs within an hour (1), usually due to respiratory paralysis. The major effects are stimulation followed by depression or paralysis of the central nervous system, and excitation of smooth muscle. Symptoms include nausea, vomiting, bowel and bladder evacuation, confusion, twitching, and convulsions. Its behavioural effects have been reviewed (2,3). Its metabolism has been reviewed (4). Tolerance or habituation is well recognised.

Vapour Inhalation

Nicotine is absorbed via the lungs, causing the signs and symptoms given above.

Eye Contact

Nicotine may cause eye irritation.

Skin Contact

Nicotine can absorbed through intact skin. Workers handling uncured tobacco may suffer from green-tobacco sickness, which is characterised by vomiting, pallor, prostration, headache and giddiness (5). Dermatosis has been reported in tobacco processing and nicotine production workers (6).

Swallowing

The oral LD_{50} in rats and mice is 53 mg/kg and 3.3 mg/kg respectively (7). The mean lethal dose in adult humans is estimated to be 30-60 mg (0.5-1.0 mg/kg) (8), although ingestion of an insecticide solution with as much as 2 g nicotine has reportedly been survived. Congestion and hyperaemia of the brain and other organs, lung and intestinal haemorrhage have been reported on autopsy (9).

CARCINOGENICITY

Nicotine did not have a promoting effect on bladder carcinogenesis in rats (10). Nicotine as a precursor of N-nitrosonornicotine, a suspected lung carcinogen, has been reviewed (11).

MUTAGENICITY

Nicotine did not affect DNA formation or repair, or sister chromatid exchange in HeLa cells or human fibroblasts (12). It was not mutagenic in the Ames test, but induced repairable DNA damage in an assay with *Eschericia coli polA+/polA-* (13). Negative results have also been reported in *E. coli* (14). High concentrations of nicotine moderately increased the baseline frequency of sister chromatid exchanges in Chinese hamster ovary cells (15).

REPRODUCTIVE HAZARDS

Nicotine is a suspected behavioural teratogen (16,17). It increased perinatal mortality when administered subcutaneously to pregnant mice (18). Deformities were reported in foetuses following maternal administration to rabbits (19). Prolonged exposure to male mice resulted in a cumulative mutagenic action on spermatids and spermatozoa (20). Its effects on the female reproductive system have been reviewed (21). It gave negative results in the Chernoff/Kavlock developmental toxicity screen (22).

FIRST AID

Eyes Wash the eye with flowing water for 10 minutes.

98. Nicotine

Lungs Remove casualty from area of exposure. If unconscious, do not give anything to drink. Give artificial ventilation and chest compression or place in the recovery position as necessary. If conscious make the casualty lie or sit down quietly, give oxygen if available.

Mouth Do not make the casualty vomit. Treat unconscious casualties as for lungs but if conscious give 1 pint of water to drink. Convulsions may occur and may cause unconsciousness. Shock may result – if so do not give any drinks, and if conscious lie casualty flat with legs raised. Further artificial ventilation may be needed.

Skin Remove all contaminated clothing, wash affected areas with soap and copious amount of water. Absorption through the skin may result in symptoms similar to those of ingestion.

In all cases of exposure, the patient should be transferred to hospital as soon as possible.

HANDLING AND STORAGE

Nicotine should be handled wearing an approved respirator, rubber gloves, safety goggles and other protective clothing. It should only be used in a chemical fume hood. It should be kept in a tightly closed container, in a cool, dry, well-ventilated place (23).

DISPOSAL

Wear a laboratory coat, safety spectacles, butyl rubber gloves and suitable safety shoes, and have an approved self-contained breathing apparatus or canister respirator available. Absorb small liquid spills onto paper towels, evaporate in an iron pan in a fume cupboard, add crumpled paper and burn carefully. Brush small solid spills onto paper, put in an iron pan, cover with crumpled paper and burn carefully in a safe place outside. For large spills,
EITHER cover with sand and mix carefully. Shovel into containers, disperse in an excess solution of dilute hydrochloric acid, mix well and leave to stand for 24 hours stirring occasionally. Carefully decant acid extract into the drains, diluting with a large volume of cold tap water. Wash the sand thoroughly with cold water.
OR cover with a sand-soda ash mix (90-10), mix well, shovel into cardboard boxes, pack with crumpled paper and incinerate.

FIRE PRECAUTIONS

Fires involving nicotine should be extinguished using water, carbon dioxide or dry chemical powder.

FURTHER READING

Bretherick, L. *Hazards in the Chemical Laboratory* (4th edition)

Bretherick, L. *Handbook of Reactive Chemical Hazards* (4th edition)

Sax, N. Irving *Dangerous Properties of Industrial Materials* (7th edition)

Encyclopaedia of Occupational Safety & Health

Patty's *Industrial Hygiene and Toxicology*

Kirk-Othmer *Encyclopedia of Chemical Technology*

National Fire Protection Association *Manual of Hazardous Reactions*

ACGIH *Documentation of TLVs and BEIs* (6th edition, 1986)

Dangerous Prop. Ind. Mater. Rep. 1985, **5**(4), 82-85

REFERENCES

1. Grusz-Harday, E. *Arch. Toxikol.* 1967, **23**(1), 35-41
2. Henningfield, J. E. *Neurol. Neurobiol.* 1985, **13**(Behav. Pharmacol.: Curr. Status), 433-449
3. Emley, G. S.; Hutchinson, R. R. *Adv. Behav. Pharmacol.* 1984, **4**, 105-129
4. Leete, E. *Recent Adv. Chem. Comp. Tob. Tob. Smoke, Symp.* 1977, 365-388
5. Ghosh, S. K.; et al. *J. Soc. Occup. Med.* 1980, **30**(3), 113-117
6. Lopukhova, K. A.; et al. *Vop. Gig. Tr. Profzabol., Mater. Nauch. Konf.* 1971 (publ. 1972), 229-231
7. Lazutka, F. A.; et al. *Hyg. Sanit.* 1969, **34**(5), 187
8. Lehman, A. J. *Q. Bull. Assoc. Food Drug Off. U.S.* 1938, **13**, 65
9. Gosselin, R. E.; et al. *Clinical Toxicology of Commercial Products* (5th edition) (Williams & Wilkins, Baltimore, 1984)

98. Nicotine

10. Ito, N.; et al. *IARC Sci. Publ.* 1984, **56**(Models, Mech. Etiol. Tumour Promot.), 399-407
11. Hoffmann, D.; et al. *Cancer Lett. (Shannon)* 1985, **26**(1), 67-75
12. Altmann, H.; et al. *Oesterr. Forschungszent Seibersdorf. (Ber.)) OEFZS* 1982, OEFZS BER. No. 4184
13. Riebe, M. et al. *Mutat. Res* 1982, **101**(1), 39-43
14. De Flora, S.; et al. *Mutat. Res.* 1984, **133**(3), 161-198
15. Riebe, M.; Westphal, K. *Mutat. Res.* 1983, **124**(3-4), 281-286
16. Nelson, B. K. *JAT, J. Appl. Toxicol.* 1991, **11**(1), 33-37
17. Johns, J. M.; et al. *Neurobehav. Toxicol. Teratol.* 1982, **4**(3), 365-369
18. Nasrat, H. A.; et al. *Biol. Neonate* 1986, **49**(1), 8-14
19. Crowe, M. W. *Tob. Health Workshop Conf., Proc., 3rd* 1972, 256-266
20. Hemsworth, B. N. *IRCS Med. Sci.: Libr. Compend.* 1981, **9**(8), 728-729
21. Weathersbee, P. S. *J. Reprod. Med.* 1980, **25**(5), 243-250
22. Seidenberg, J. M.; Bekcer, R. A. *Terat. Carcinogen. Mutagen.* 1987, **7**, 17-28
23. Lenga, R. E. *The Sigma-Aldrich Library of Chemical Safety Data* (2nd edition) (Sigma-Aldrich, Milwaukee, 1988)

5-NITRO-ACENAPHTHENE

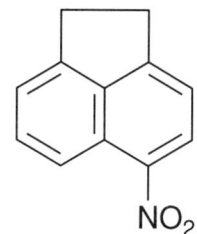

RISKS
May cause cancer (R45)

SAFETY PRECAUTIONS
Avoid exposure-obtain special instructions before use – If you feel unwell, seek medical advice (show label where possible) (S53, S44)

IDENTIFIERS

SYNONYMS acenaphthylene, 1-2-dihyro-5-nitro-; acenaphthene, 5-nitro-; 1,2-dihydro-5-nitroacenaphthylene; 5-NAN; 5- nitronaphthalene ethylene; NCI-C01967

CHEMICAL ABSTRACTS No.	602-87-9
NIOSH No.	AB 1060000
HAZCHEM CODE	not available
UN No.	not available

THRESHOLD LIMIT VALUES

US TLV (TWA)	not available
US TLV (STEL)	not available
UK EXPOSURE LIMITS (OES)	
Long-term (8 hr TWA value)	not available
Short-term (10 min TWA value)	not available
Germany	
MAK	not available
France	
VME	not available
VLE	not available
Sweden	
Short-term limit	not available
Level limit	not available

PHYSICAL PROPERTIES

Description Yellow powder.

Boiling point	not available
Melting point	103-104°C
Density	not available
Vapour density	not available
Vapour pressure	not available
Flash point	not available
Explosive limits	not available
Autoignition temperature	not available

Solubility Insoluble in water. Soluble in acetone, dimethyl sulphoxide, or ether.

PACKAGING AND TRANSPORTATION

Road transportation

hazard warning sign	not available
Hazchem code	not available

Sea transportation

IMDG page No.	not available
class	not available
label	not available
packaging group	not available

Air transportation

ICAO/IATA code	not available
class	not available
label	not available
packaging group	not available
packing instructions	
cargo	not available
passenger	not available
passenger aircraft max. quantity	not available
cargo aircraft max. quantity	not available

99. 5-Nitroacenaphthene

MANUFACTURE

5-Nitroacenaphthene is prepared by nitration of acenaphthene in acetic anhydride solution with zinc or copper nitrate at 30°C (1). It may also be prepared by nitration of acenaphthene with nitric acid in sulphuric acid (2).

USES

5-Nitroacenaphthene is used as a chemical intermediate in naphthalimide dye production.

CHEMICAL HAZARDS

5-Nitroacenaphthene emits toxic fumes of nitrogen oxides when heated to decomposition.

BIOLOGICAL HAZARDS

5-Nitroacenaphthene is carcinogenic in laboratory animals following oral or intraperitoneal administration. It is mutagenic in a bacterial assay. Little other toxicity data is available.

Vapour Inhalation

No information is available concerning the effects of inhalation.

Eye Contact

No information is available on the effects on the eyes.

Skin Contact

No information is available on the effects on the skin.

Swallowing

5-Nitroacenaphthene is carcinogenic in laboratory animals following oral administration.

CARCINOGENICITY

The IARC evaluation is that there is sufficient evidence for the carcinogenicity of 5-nitroacenaphthene in animal experiments (2). It produced adenocarcinomata in the small intestine of female rats following oral administration (3). Malignant tumours of the breast and ear duct have also been reported in female rats (4-6). Increased incidence of alveolar/bronchiolar carcinoma were observed in male and female rats, and hepatocellular carcinomas and benign ovarian tumours occurred in female mice (5). It is also carcinogenic in female hamsters following oral administration. Myeloid leukaemias, reticulum cell sarcomas and a mammary carcinoma were reported in mice following intraperitoneal administration.

MUTAGENICITY

5-Nitroacenaphthene is mutagenic in *Salmonella typhimurium* TA100 and TA98 (7-11).

REPRODUCTIVE HAZARDS

No information is available concerning the reproductive hazards of 5- nitroacenaphthene.

FIRST AID

Eyes Wash the eye with flowing water for 10 minutes.

Lungs Remove casualty from area of exposure. If unconscious, do not give anything to drink, give artificial ventilation and chest compression or place in the recovery position as necessary. If conscious make the casualty lie or sit down quietly, give oxygen if available.

Mouth Do not make the casualty vomit. Treat unconscious casualties as for lungs but if conscious give 1 pint of water to drink immediately.

Skin Remove contaminated clothing immediately, drench the affected area with running water for at least 10 minutes.

In all cases of exposure, the patient should be transferred to hospital as soon as possible.

99. 5-Nitroacenaphthene

HANDLING AND STORAGE

Exposure to 5-nitroacenaphthene should be avoided whenever possible; it should be used in enclosed systems.

DISPOSAL

No published information on the disposal of 5-nitroacenaphthene appears to be available. The sixth individual Directive within the meaning of Article 8 of Directive 80/1107/EEC relates to the prevention of exposure, information for competent authorities, protective clothing and equipment for occupational exposure to carcinogens including 5-nitroacenaphthene. It may be disposed of by incineration by a licenced contractor.

FIRE PRECAUTIONS

Fires involving 5-nitroacenaphthene should be extinguished using water, carbon dioxide, dry chemical powder, alcohol foam or polymer foam. Keep fire exposed containers cool with water.

FURTHER READING

Sax, N. Irving *Dangerous Properties of Industrial Materials* (7th edition)

Encyclopaedia of Occupational Safety & Health

Patty's *Industrial Hygiene and Toxicology*

Kirk-Othmer *Encyclopedia of Chemical Technology*

National Fire Protection Association *Manual of Hazardous Reactions*

ACGIH Documentation of TLVs and BEIs (6th edition, 1986)

REFERENCES

1. Mitoguchi, H. *Yuki Gosei Kagaku Kyokai Shi* 1969, **27**(7), 642-647
2. IARC *Monographs on the evaluation of the carcinogenic risk of chemicals to humans* 1978, **16**, 319-324
3. Takemura, N.; et al. *Br. J. Cancer* 1974, **30**(5), 481-483
4. Terasawa, M. *Tokyo Jikeikai Ika Daigaku Zasshi* 1974, **89**(4), 475-481
5. *Report* (NCI, Bethesda, 1978) DHEW/PUB/NIH-78-1373, NCI-CG-TR-118 Order No. PB-287347
6. Gold, L. S.; et al. *Fundam. Appl. Toxicol.* 1986, **6**, 677-690
7. Tokiwa, H.; et al. *Mutat. Res.* 1981, **91**(4-5), 321-325
8. Yahagi, T.; et al. *Gann* 1975, **66**(5), 581-582
9. McCann, J.; et al. *Proc. Nat. Acad. Sci. (Washington)* 1975, **72**, 5135-5139
10. McCoy, E. C.; et al. *Mutat. Res.* 1983, **111**(1), 61-68
11. Moller, M.; et al. *Mutat. Res.* 1985, **157**(2-3), 149-156

100. *m*-Nitroaniline

m-NITROANILINE

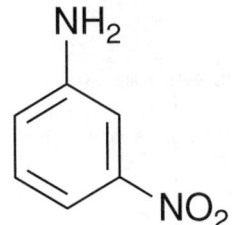

RISKS
Toxic by inhalation, in contact with skin and if swallowed – Danger of cumulative effects (R23/24/25, R33)

SAFETY PRECAUTIONS
After contact with skin, wash immediately with plenty of water – Wear protective clothing and gloves – If you feel unwell, seek medical advice (show label where possible) (S28, S36/37, S44)

IDENTIFIERS

SYNONYMS aniline, *m*-nitro-; Amarthol Fast Orange R Base; *m*-aminonitro benzene; Azobase MNA; C.I. 37030; C.I. Azoic Diazo Component 7; benzenamine, 3-nitro-; Daito Orange Base R; Devol Orange R; Diazo Fast Orange R; Fast Orange Base R; Fast Orange M Base; Fast Orange MM Base; Fast Orange R Base; Fast Orange R Salt; Hiltonil Fast Orange R Base; MNA; Naphtoelan Orange R Base; Nitranilin; *m*-nitroaminobenzene; 3- nitroaniline; 3-nitrobenzenamine; *m*-nitrophenylamine; Orange Base Irga I

CHEMICAL ABSTRACTS No.	99-09-2
NIOSH No.	BY 6825000
HAZCHEM CODE	2X
UN No.	1661

THRESHOLD LIMIT VALUES

US TLV (TWA)	not available
US TLV (STEL)	not available

UK EXPOSURE LIMITS (OES)
Long-term (8 hr TWA value)	not available
Short-term (10 min TWA value)	not available

Germany
MAK	not available

France
VME	not available
VLE	not available

Sweden
Short-term limit	not available
Level limit	not available

PHYSICAL PROPERTIES

Description	Yellow, rhombic crystals.
Boiling point	decomposes at 306.4°C
Melting point	114°C
Density	0.9011 at 25°C
Vapour density	not available
Vapour pressure	1 mm Hg at 119°C
Flash point	199°C
Explosive limits	not available
Autoignition temperature	not available

Solubility Soluble in water, alcohol, ether, methanol, acetone, dimethyl sulphoxide or chloroform. Slightly soluble in benzene.

PACKAGING AND TRANSPORTATION

Road transportation
hazard warning sign	1661 toxic substance
Hazchem code	2X

Sea transportation
IMDG page No.	6207
class	6.1
label	poison
packaging group	II

Air transportation
ICAO/IATA code (UN No.)	1661
class	6.1
label	poison
packaging group	II
packing instructions	
cargo	615
passenger	613
passenger aircraft max. quantity	25 kilograms
cargo aircraft max. quantity	100 kilograms

100. *m*-Nitroaniline

MANUFACTURE

m-Nitroaniline is prepared by the nitration of aniline, by the reduction of *m*-nitrobenzene, or from *m*-nitrobenzoic acid.

USES

m-Nitroaniline is used in the manufacture of azo dyes.

CHEMICAL HAZARDS

m-Nitroaniline is a combustible material and decomposes exothermically at 247°C. It may react explosively with ethylene oxide at 130°C. It is incompatible with acids, acid chlorides, acid anhydrides, chloroformates and strong oxidisers. On thermal decomposition it releases toxic fumes of carbon monoxide and nitrogen oxides.

BIOLOGICAL HAZARDS

m-Nitroaniline is highly toxic by ingestion, inhalation and absorption through the skin. It causes the formation of methaemoglobin, which in sufficient concentration leads to cyanosis and possible death. Symptoms of poisoning include nausea and vomiting, headache, dizziness, stupor and loss of consciousness. Liver damage may result from chronic exposure. Onset of acute symptoms may be delayed by 2-4 hours or more.

Vapour Inhalation

m-Nitroaniline is irritating to the respiratory tract. Inhalation produces the effects described above. Breathing of nitroaniline dust is hazardous.

Eye Contact

m-Nitroaniline is an eye irritant.

Skin Contact

m-Nitroaniline can be absorbed into the body through the skin, producing the toxic effects described above. It is a skin irritant.

Swallowing

The LD_{50} for *m*-nitroaniline in rats is 535 mg/kg by oral administration.

CARCINOGENICITY

No evidence has been reported that *m*-nitroaniline is carcinogenic.

MUTAGENICITY

m-Nitroaniline has been found to be mutagenic in some strains of *Salmonella typhimurium* (1,2).

REPRODUCTIVE HAZARDS

There is no information available on any reproductive hazards associated with *m*-nitroaniline.

FIRST AID

Eyes Wash the eye with flowing water for 10 minutes.

Lungs Remove casualty from area of exposure. If unconscious, do not give anything to drink. Give artificial ventilation and chest compression or place in the recovery position as necessary. If conscious make the casualty lie or sit down quietly, give oxygen if available. Convulsions may occur and may cause unconsciousness. Shock may result – if so do not give any drinks, and if conscious, lie casualty flat with legs raised.

Mouth Do not make the casualty vomit. Treat unconscious casualties as for lungs, but if conscious give 1 pint of water to drink.

Skin Remove contaminated clothing immediately, wash the affected area with soap and copious amounts of water. Absorption through the skin may cause symptoms similar to those of inhalation.

In all cases of exposure, the patient should be transferred to hospital as soon as possible.

100. *m*-Nitroaniline

HANDLING AND STORAGE

m-Nitroaniline should be handled wearing an approved respirator, chemical-resistant gloves, safety goggles and other protective clothing. Avoid getting on clothing. Wash thoroughly after handling. It should only be used in a chemical fume hood. It should be kept in a tightly closed container, in a cool, dry place or refrigerator.

DISPOSAL

Eliminate all sources of ignition and ventilate the area. Wear a laboratory coat, safety spectacles, butyl rubber gloves, and approved self-contained breathing apparatus. While stirring constantly, add contaminated material to approximately 30 times its weight of a solution prepared by dissolving 1 part sodium sulphide in 6 parts water.

FIRE PRECAUTIONS

Fires involving *m*-nitroaniline should be extinguished using water spray, carbon dioxide, dry chemical powder, alcohol foam or polymer foam. Keep fire-exposed containers cool.

FURTHER READING

Sax, N. Irving *Dangerous Properties of Industrial Materials* (7th edition)

Encyclopedia of Occupational Safety & Health

Patty's *Industrial Hygiene and Toxicology*

Kirk-Othmer *Encyclopedia of Chemical Technology*

National Fire Protection Association *Guide on Hazardous Materials* (9th edition) NFPS, USA, 1986)

ACGIH Documentation of TLVs and BEIs (6th edition, 1986)

Lenga, R. E. *The Sigma Aldrich Library of Chemical Safety Data* (2nd edition) (Sigma-Aldrich, Milwaukee, 1988)

The Merck Index (11th edition) (Merck & Co, New Jersey, 1989)

Keith, L. H.; Walters, D. B. (editors) *Compendium of Safety Data Sheets for Research and Industrial Chemicals* (VCH, Deerfield Park, 1987)

REFERENCES

1. Wai Chiu, C.; et al. *Mutat. Res.* 1978, **58**, 11-22
2. Shahin, M. M. *J. Cosmet. Sci.* 1985, **7** (6), 277-289

o-NITROANILINE

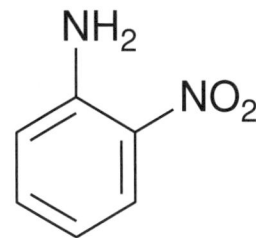

RISKS
Toxic by inhalation, in contact with skin and if swallowed – Danger of cumulative effects (R23/24/25, R33)

SAFETY PRECAUTIONS
After contact with skin, wash immediately with plenty of water – Wear protective clothing and gloves – If you feel unwell, seek medical advice (show label where possible) (S28, S36/37, S44)

IDENTIFIERS

SYNONYMS *o*-aminonitrobenzene; 2-aminonitrobenzene; aniline, *o*-nitro-; Azoene Fast Orange GR Base; Azoene Fast Orange GR Salt; Azofix Orange GR; benzenamine, 2-nitro; Brentamine Fast Orange GR Base; Brentamine Fast Orange GR Salt; C.I. 37025; C.I. Azoic Diazo Component 6; Devol Orange B; Devol Orange Salt B; Diazo Fast Orange GR; Fast Orange Base JR; Fast Orange O Base; Fast Orange O Salt; Fast Orange Salt JR; Hiltonil Fast Orange GR Base; Hiltosal Fast Orange GR Salt; Hindasol Orange GR Salt; Natasol Fast Orange GR; 2-nitroaniline; 2-nitrobenzenamine; ONA; Orange Base Ciba II; Orange Base Irga; Orange Base Salt; Orange Ciba II; Orange Salt Irga II

CHEMICAL ABSTRACTS No.	88-74-4
NIOSH No.	BY 6650000
HAZCHEM CODE	2X
UN No.	1661

THRESHOLD LIMIT VALUES

US TLV (TWA)	not available
US TLV (STEL)	not available

UK EXPOSURE LIMITS (OES)
Long-term (8 hr TWA value) .... not available
Short-term (10 min TWA value) .... not available

Germany
MAK .... not available

France
VME .... not available
VLE .... not available

Sweden
Short-term limit .... not available
Level limit .... not available

PHYSICAL PROPERTIES

Description Orange-yellow crystals.

Boiling point	284.5°C
Melting point	69-71°C
Density	0.9015 at 25°C
Vapour density	4.77
Vapour pressure	1 mm Hg at 104°C
Flash point	168°C
Explosive limits	not available
Autoignition temperature	521°C

Solubility Slightly soluble in cold water. Soluble in hot water, ether, chloroform, acetone or benzene.

PACKAGING AND TRANSPORTATION

Road transportation

hazard warning sign	1661 toxic substance
Hazchem code	2X

Sea transportation

IMDG page No.	6207
class	6.1
label	poison
packaging group	II

Air transportation

ICAO/IATA code (UN No.)	1661
class	6.1
label	poison
packaging group	II
packing instructions	
cargo	615
passenger	613
passenger aircraft max. quantity	25 kilograms
cargo aircraft max. quantity	100 kilograms

101. o-Nitroaniline

MANUFACTURE

o-Nitroaniline can be prepared from o-dinitrobenzene, o-nitroaniline-p-sulphonic acid or o-dinitrosobenzene.

USES

o-Nitroaniline is used in the manufacture of azo dyes and in the synthesis of photographic antifogging agent.

CHEMICAL HAZARDS

o-Nitroaniline is a combustible material capable of exothermic decomposition. It forms highly explosive addition compounds with hexanitroethane, and reacts vigorously with sulphuric acid above 200°C. When mixed with magnesium it forms a mixture that is hypergolic in the presence of nitric acid. It is incompatible with acids, acid chlorides, acid anhydrides, chloroformates and strong oxidising agents. When heated to decomposition it produces toxic fumes of carbon monoxide and nitrogen oxides.

BIOLOGICAL HAZARDS

o-Nitroaniline is highly toxic by ingestion, inhalation and absorption through the skin. It causes the formation of methaemoglobin, which in sufficient concentration leads to cyanosis and possible death. Symptoms of poisoning include nausea and vomiting, headache, dizziness, stupor and loss of consciousness. Liver damage may result from chronic exposure. Onset of acute symptoms may be delayed by 2-4 hours or more.

Vapour Inhalation

o-Nitroaniline is irritating to the respiratory tract. Inhalation produces the effects described above. Breathing of nitroaniline dust is hazardous.

Eye Contact

o-Nitroaniline is an eye irritant.

Skin Contact

o-Nitroaniline can be absorbed into the body through the skin, producing the toxic effects described above. It is a skin irritant.

Swallowing

The LD_{50} for o-nitroaniline in rats has been reported as 1600 mg/kg by oral administration.

CARCINOGENICITY

No evidence that o-nitroaniline is carcinogenic has been reported.

MUTAGENICITY

o-Nitroaniline produced negative results when tested for mutagenicity in *Salmonella typhimurium* (1,2).

REPRODUCTIVE HAZARDS

There is no information available on any reproductive hazards associated with o-nitroaniline.

FIRST AID

Eyes Wash the eye with flowing water for 10 minutes.

Lungs Remove casualty from area of exposure. If unconscious, do not give anything to drink. Give artificial ventilation and chest compression or place in the recovery position as necessary. If conscious make the casualty lie or sit down quietly, give oxygen if available. Convulsions may occur and may cause unconsciousness. Shock may result – if so do not give any drinks, and if conscious, lie casualty flat with legs raised.

Mouth Do not make the casualty vomit. Treat unconscious casualties as for lungs, but if conscious give 1 pint of water to drink.

Skin Remove contaminated clothing immediately, wash the affected area with soap and copious amounts of water. Absorption through the skin may cause symptoms similar to those of inhalation.

In all cases of exposure, the patient should be transferred to hospital as soon as possible.

101. o-Nitroaniline

HANDLING AND STORAGE

o-Nitroaniline should be handled wearing an approved respirator, chemical-resistant gloves, safety goggles and other protective clothing. Avoid getting on clothing. Wash thoroughly after handling. It should only be used in a chemical fume hood. It should be kept in a tightly closed container, in a cool, dry place or refrigerator. Protect from light during prolonged storage.

DISPOSAL

Eliminate all sources of ignition and ventilate the area. Wear a laboratory coat, safety spectacles, butyl rubber gloves, and approved self-contained breathing apparatus. While stirring constantly, add contaminated material to approximately 30 times its weight of a solution prepared by dissolving 1 part sodium sulphide in 6 parts water.

FIRE PRECAUTIONS

Fires involving o-Nitroaniline should be extinguished using water spray, carbon dioxide, dry chemical powder, alcohol foam or polymer foam. Keep fire-exposed containers cool.

FURTHER READING

Bretherick, L. *Handbook of Reactive Chemical Hazards* (4th edition)

Sax, N. Irving *Dangerous Properties of Industrial Materials* (7th edition)

Encyclopedia of Occupational Safety & Health

Patty's *Industrial Hygiene and Toxicology*

Kirk-Othmer *Encyclopedia of Chemical Technology*

National Fire Protection Association *Guide on Hazardous Materials* (9th edition) (NFPA, USA, 1986)

ACGIH Documentation of TLVs and BEIs (6th edition, 1986)

Lenga, R. E. *The Sigma Aldrich Library of Chemical Safety Data* (2nd edition) (Sigma-Aldrich, Milwaukee, 1988)

The Merk Index (11th edition) (Merck & Co, New Jersey, 1989)

Keith, L. H.; Walters, D. B. (editors) *Compendium of Safety Data Sheets for Research and Industrial Chemicals* (VCH, Deerfield Park, 1987)

REFERENCES

1. Wai Chiu, C.; et al. *Mutat. Res.* 1978, **58**, 11-22
2. Shahin, M. M. *J. Cosmet. Sci.* 1985, **7** (6), 277-289

102. p-Nitroaniline

p-NITROANILINE

RISKS
Toxic by inhalation, in contact with skin and if swallowed – Danger of cumulative effects (R23/24/25, R33)

SAFETY PRECAUTIONS
After contact with skin, wash immediately with plenty of water – Wear protective clothing and gloves – If you feel unwell, seek medical advice (show label where possible) (S28, S36/37, S44)

IDENTIFIERS

SYNONYMS benzenamine, 4-nitro-; aniline, p-nitro; p-aminonitrobenzene; 4-aminonitrobenzene; Azoamine Red Zh; C.I. 37035; C.I. Azoic Diazo Component 87; C.I. Developer 17; Developer P; Devol Red GG; Fast Red Base GG; Fast Red Base 2J; Fast Red 2G Base; Fast Red GG Base; Fast Red MP Base; Fast Red P Base; Napthtoelan Red GG Base; p-nitraniline; Nitrozol CF extra; 4-nitroaniline; 4- nitrobenzenamine; 4-nitrophenylamine; PNA; Red 2G Base; Shinnippan Fast Red GG Base

CHEMICAL ABSTRACTS No.	100-01-6
NIOSH No.	BY 7000000
HAZCHEM CODE	2X
UN No.	1661

THRESHOLD LIMIT VALUES

US TLV (TWA)	3 mg/m^3
US TLV (STEL)	not available

UK EXPOSURE LIMITS (OES)
Long-term (8 hr TWA value) ..... 6 mg/m^3
Short-term (10 min TWA value) ..... not available

Germany
MAK ..... 1 ppm (6 mg/m^3)

France
VME ..... 3 mg/m^3
VLE ..... not available

Sweden
Short-term limit ..... not available
Level limit ..... not available

PHYSICAL PROPERTIES

Description	Bright yellow powder.
Boiling point	332°C
Melting point	148.5°C
Density	1.424°C
Vapour density	not available
Vapour pressure	1 mm Hg at 142.4°C
Flash point	199°C (closed cup)
Explosive limits	not available
Autoignition temperature	not available

Solubility Soluble in water, alcohol, ether, benzene, methanol or dimethyl sulphoxide.

PACKAGING AND TRANSPORTATION

Road transportation
hazard warning sign	1661 toxic substance
Hazchem code	2X

Sea transportation
IMDG page No.	6207
class	6.1
label	poison
packaging group	II

Air transportation
ICAO/IATA code (UN No.)	1661
class	6.1
label	poison
packaging group	II
packing instructions	
cargo	615
passenger	613
passenger aircraft max. quantity	25 kilograms
cargo aircraft max. quantity	100 kilograms

102. *p*-Nitroaniline

MANUFACTURE

p-Nitroaniline is prepared from acetanilide.

USES

p-Nitroaniline is used as an intermediate in the manufacture of dyes, antioxidants, petrol gum inhibitor, medicinals for poultry and corrosion inhibitors.

CHEMICAL HAZARDS

p-Nitroaniline is a combustible material capable of exothermic decomposition. It is incompatible with strong acids, strong oxidisers and strong reducing agents. It reacts vigorously with sulphuric acid at or above 200°C. When reacted with sodium hydroxide at 130°C under pressure it may form sodium 4-nitrophenoxide, which is explosive. Thermal decomposition of p-nitroaniline produces toxic fumes of carbon monoxide and nitrogen oxides. In the presence of moisture it causes organic materials to undergo nitration, which may produce spontaneous ignition.

BIOLOGICAL HAZARDS

p-Nitroaniline is highly toxic by ingestion, inhalation and skin contact. It is a powerful cause of methaemoglobin formation, which in sufficient concentration leads to cyanosis and possible death. It also has a haemolytic effect. Chronic exposure can cause liver damage. The symptoms of nitroaniline poisoning are nausea and vomiting, headache, difficulty in breathing and drowsiness and loss of consciousness. The onset of symptoms may be delayed by 2-4 hours or more.

Vapour Inhalation

p-Nitroaniline is irritating to the respiratory tract. Inhalation caused the toxic effects described above. Breathing of nitroaniline dust is hazardous.

Eye Contact

p-Nitroaniline is an eye irritant.

Skin Contact

p-Nitroaniline is irritating to skin. It is rapidly absorbed via this route (1), causing the toxic effects described above.

Swallowing

The LD_{50} for *p*-nitroaniline in rats is 750 mg/kg by oral administration.

CARCINOGENICITY

No evidence has been reported that *p*-nitroaniline is carcinogenic.

MUTAGENICITY

p-Nitroaniline produced a positive result in the Ames test with *Salmonella typhimurium* TA1358 (2), and induced chromosomal changes in chinese hamster ovary cells (3). However, other papers report negative mutagenicity (4). *p*-Nitroaniline is not referred to as mutagenic in the general literature.

REPRODUCTIVE HAZARDS

There is no information available on any reproductive hazards associated with *p*-nitroaniline.

FIRST AID

Eyes Wash the eye with flowing water for 10 minutes.

Lungs Remove casualty from area of exposure. If unconscious, do not give anything to drink. Give artificial ventilation and chest compression or place in the recovery position as necessary. If conscious make the casualty lie or sit down quietly, give oxygen if available. Convulsions may occur and may cause unconsciousness. Shock may result – if so do not give any drinks, and if conscious, lie casualty flat with legs raised.

Mouth Do not make the casualty vomit. Treat unconscious casualties as for lungs, but if conscious give 1 pint of water to drink.

Skin Remove contaminated clothing immediately, wash the affected area with soap and copious amounts of water. Absorption through the skin may cause symptoms similar to those of inhalation.

102. p-Nitroaniline

In all cases of exposure, the patient should be transferred to hospital as soon as possible.

HANDLING AND STORAGE

p-Nitroaniline should be handled wearing an approved respirator, chemical-resistant gloves, safety goggles and other protective clothing. Avoid getting on clothing. Wash thoroughly after use. Avoid repeated prolonged exposures. It should only be used in a chemical fume hood. It should be kept in a tightly closed container, in a cool, dry place, protected from physical damage and away from oxidisers and reducing agents. Protect from prolonged exposure to light.

DISPOSAL

Eliminate all sources of ignition and ventilate the area. Wear a laboratory coat, safety spectacles, butyl rubber gloves, and approved self-contained breathing apparatus. While stirring constantly, add contaminated material to approximately 30 times its weight of a solution prepared by dissolving 1 part sodium sulphide in 6 parts water.

FIRE PRECAUTIONS

Fires involving p-nitroaniline should be extinguished using water spray, carbon dioxide, dry chemical powder, alcohol foam or polymer foam. Keep fire-exposed containers cool.

FURTHER READING

Sax, N. Irving *Dangerous Properties of Industrial Materials* (7th edition)

Encyclopedia of Occupational Safety & Health

Patty's *Industrial Hygiene and Toxicology*

Kirk-Othmer *Encyclopedia of Chemical Technology*

National Fire Protection Association *Guide on Hazardous Materials* (9th edition) (NFPA, USA, 1986)

ACGIH Documentation of TLVs and BEIs (6th edition, 1986)

Lenga, R. E. *The Sigma Aldrich Library of Chemical Safety Data* (2nd edition) (Sigma-Aldrich, Milwaaukee, 1988)

The Merck Index (11th edition) (Merck & Co, New Jersey, 1989)

Keith, L. H.; Walters, D. B. (editors) *Compendium of Safety Data Sheets for Research and Industrial Chemicals* (VCH, Deerfield Park, 1987)

REFERENCES

1. Bronaugh, R. L.; Maibrach, H. I. *J. Invest. Dermatol.* 1989, **84** (3), 180-183
2. Mor, L. M.; Stark, A. A. *Appl. Environ. Microbiol.* 1982, **44**, 807-808
3. Galloway, S. M.; et al. *Environ. Mol. Mutagen.* 1987, **10**, 1-175
4. Wai Chiu, C.; et al. *Mutat. Res.* 1978, **58**, 11-22

103. 2-Nitro-*p*-anisidine

2-NITRO-*p*-ANISIDINE

RISKS
Very toxic by inhalation, in contact with skin and if swallowed – Danger of cumulative effects (R26/27/28, R33)

SAFETY PRECAUTIONS
After contact with skin, wash immediately with plenty of water – Wear protective clothing and gloves – In case of accident or if you feel unwell, seek medical advice immediately (show label where possible) (S28, S36/37, S45)

IDENTIFIERS

SYNONYMS   benzenamine, 4-methoxy-2-nitro-; *p*-anisidine, 2-nitro-; Acco Fast Bordeaux GP Salt; Amarthol Fast Bordeaux GP Base; Amarthol Fast Bordeaux GP Salt; Atul Fast Bordeaux GP Base; Azobase NAS; Azoene Fast Bordeaux GP Base; Axoene fast Bordeaux GP Salt; Azofix Bordeaux GP; Azogene Fast Bordeaux G; Bordeaux Base Ciba IV; Bordeaux Base Irga IV; Bordeaux Base NGP; Bordeaux GP Base; Bordeaux GP Salt; Bordeaux GPS Salt; Bordeaux Salt Ciba IV; Bordeaux Salt NGP; Brentamine Fast Bordeaux GP Base; C.I. 37135; C.I. Azoic Diazo Component 1; Daito Bordeaux Base GP; Daito Bordeaux Salt GP; Devol Bordeaux B; Devol Boreaux GP Salt; Diabase Bordeaux GP; Diasalt Bordeaux GP; Diazo Fast Bordeaux GP; Durgasol Bordeaux GP Salt; Fast Bordeaux Base GP; Fast Bordeaux Base J; Fast Bordeaux GDN; Fast Bordeaux GND base; Fast Bordeaux GP; Fast Bordeaux GP Base; Fast Bordeaux GP Salt; Fast Bordeaux GP-T Base; Fast Bordeaux 3NA Base; Fast Bordeaux Salt GP; Fast GPN; Fast Bordeaux Salt J; Hansol Bordeaux GP Salt; Hiltonil Fast Bordeaux GP Base; Hiltosal Fast Bordeaux GP Salt; Hindasol Bordeaux GP Salt; Kako Bordeaux GP Base; Kako Bordeaux GP Salt; Kayaku Fast Bordeaux GP Base; Kayaku Fast Bordeaux Salt GP; Lake Maroon B base; Mitsui Bordeaux GP Base; Mitsui Bordeaux GP Salt; Naphthanil Bordeaux GP Base; Naphthanil Diazo Bordeaux GP; Naphthosol Fast Bordeaux GP Salt; Naphtoelan Fast Bordeaux GP Base; Naphtoelan Fast Bordeaux GP Salt; Natasol Bordeaux GP Salt; 4-methoxy-2-nitroaniline; 4-amino-3-methoxyazobenzene; 4-methoxy-2-nitroaniline; 4-amino-3-methoxyazobenzene; Pharmasol Bordeaux GP; Pharmazoid Bordeaux GP; Sanyo Fast Bordeaux GP Base; Sanyo Fast Bordeaux Salt GP; Shinnippon Fast Bordeaux GP Base; Sugai Fast Bordeaux GP Base; Symulon Bordeaux GP Base; Tulabase Fast Bordeaux GP

CHEMICAL ABSTRACTS No.	96-96-8
NIOSH No.	BY 4415000
HAZCHEM CODE	not available
UN No.	not available

THRESHOLD LIMIT VALUES

US TLV (TWA)	not available
US TLV (STEL)	not available
UK EXPOSURE LIMITS (OES)	
Long-term (8 hr TWA value)	not available
Short-term (10 min TWA value)	not available
Germany	
MAK	not available
France	
VME	not available
VLE	not available
Sweden	
Short-term limit	not available
Level limit	not available

PHYSICAL PROPERTIES

Description   Orange powder.

Boiling point	not available
Melting point	not available
Density	not available
Vapour density	not available
Vapour pressure	not available
Flash point	not available
Explosive limits	not available
Autoignition temperature	not available
Solubility	not available.

103. 2-Nitro-*p*-anisidine

PACKAGING AND TRANSPORTATION

Road transportation

hazard warning sign	not available
Hazchem code	not available

Sea transportation

IMDG page No.	not available
class	not available
label	not available
packaging group	not available

Air transportation

ICAO/IATA code	not available
class	not available
label	not available
packaging group	not available
packing instructions cargo	not available
passenger	not available
passenger aircraft max. quantity	not available
cargo aircraft max. quantity	not available

MANUFACTURE

2-Nitro-*p*-anisidine has reportedly been prepared by nitration of nitrogen-protected amino anisole in a water-immiscible organic solvent (1).

USES

2-Nitro-*p*-anisidine is used in the manufacture of dyes.

CHEMICAL HAZARDS

2-Nitro-*p*-anisidine is incompatible with acids, acid chlorides, acid anhydrides, chloroformates and strong oxidisers. It may emit toxic fumes of carbon monoxide, carbon dioxide, nitrogen oxides and hydrogen chloride when heated to decomposition (2).

BIOLOGICAL HAZARDS

2-Nitro-*p*-anisidine may be irritating to the eyes, skin and upper respiratory tract, and may be a methaemoglobin former (2).

Vapour Inhalation

2-Nitro-*p*-anisidine may be irritating to the upper respiratory tract (2).

Eye Contact

2-Nitro-*p*-anisidine may be irritating to the eyes (2).

Skin Contact

2-Nitro-*p*-anisidine may be irritating to the skin (2).

Swallowing

The oral LD_{50} in rats is 14100 mg/kg (2).

CARCINOGENICITY

No information is available concerning the carcinogenicity of 2-nitro-*p*-anisidine.

MUTAGENICITY

No information is available concerning the mutagenicity of 2-nitro-*p*-anisidine.

REPRODUCTIVE HAZARDS

No information is available concerning the reproductive hazards of 2-nitro-*p*-anisidine.

103. 2-Nitro-*p*-anisidine

FIRST AID

Eyes Wash the eye with flowing water for 10 minutes.

Lungs Remove casualty from area of exposure. If unconscious, do not give anything to drink. Give artificial ventilation and chest compression or place in the recovery position as necessary. If conscious make the casualty lie or sit down quietly, give oxygen if available. Convulsions may occur and may cause unconsciousness. Shock may result – if so do not give any drinks, and if conscious, lie casualty flat with legs raised.

Mouth Do not make the casualty vomit. Treat unconscious casualties as for lungs, but if conscious give 1 pint of water to drink.

Skin Remove contaminated clothing immediately, wash the affected area with soap and copious amounts of water. Absorption through the skin may cause symptoms similar to those of inhalation.

In all cases of exposure, the patient should be transferred to hospital as soon as possible.

HANDLING AND STORAGE

2-Nitro-*p*-anisidine should be handled wearing an approved respirator, chemical-resistant gloves, safety goggles and other protective clothing. Mechanical exhaust is required. It should be kept in a cool dry place (2).

DISPOSAL

Wear a laboratory coat, safety spectacles, butyl rubber gloves and suitable safety shoes, and have an approved self-contained breathing apparatus or canister respirator available. Absorb small liquid spills onto paper towels, evaporate in an iron pan in a fume cupboard, add crumpled paper and burn carefully. Brush small solid spills onto paper, put in an iron pan, cover with crumpled paper and burn carefully in a safe place outside. For large spills,
EITHER cover with sand and mix carefully. Shovel into containers, disperse in an excess solution of dilute hydrochloric acid, mix well and leave to stand for 24 hours stirring occasionally. Carefully decant acid extract into the drains, diluting with a large volume of cold tap water. Wash the sand thoroughly with cold water.
OR cover with a sand-soda ash mix (90-10), mix well, shovel into cardboard boxes, pack with crumpled paper and incinerate.

FIRE PRECAUTIONS

Fires involving 2-nitro-*p*-anisidine should be extinguished using water spray, carbon dioxide, dry chemical powder, alcohol foam or polymer foam.

FURTHER READING

Sax, N. Irving *Dangerous Properties of Industrial Materials* (7th edition)

Encyclopaedia of Occupational Safety & Health

Patty's *Industrial Hygiene and Toxicology*

Kirk-Othmer *Encyclopedia of Chemical Technology*

National Fire Protection Association *Manual of Hazardous Reactions*

ACGIH *Documentation of TLVs and BEIs* (6th edition, 1986)

REFERENCES

1. Schossler, W.; et al. Eur. Pat. 49,711 (Cl. C07C93/14) (1982, Bayer, A.-G.)
2. Lenga, R. E. *The Sigma-Aldrich Library of Chemical Safety Data* (2nd edition) (Sigma-Aldrich, Milwaukee, 1988)

104. Nitrobenzene

NITROBENZENE

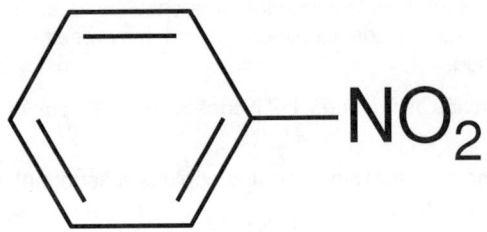

RISKS
Very toxic by inhalation, in contact with skin and if swallowed – Danger of cumulative effects (R26/27/28, R33)

SAFETY PRECAUTIONS
After contact with skin, wash immediately with plenty of water – Wear protective clothing and gloves – In case of accident or if you feel unwell, seek medical advice immediately (show label where possible) (S28, S36/37, S45)

IDENTIFIERS
SYNONYMS essence of mirbane; oil of mirbane; oil of myrbane; essence of myrbane; nitrobenzol; mirbane oil

CHEMICAL ABSTRACTS No.	98-95-3
NIOSH No.	DA 6475000
HAZCHEM CODE	2X
UN No.	1662

THRESHOLD LIMIT VALUES

US TLV (TWA)	1 ppm (5 mg/m^3)
US TLV (STEL)	not available

UK EXPOSURE LIMITS (OES)
Long-term (8 hr TWA value) .... 1 ppm (5 mg/m^3)
Short-term (10 min TWA value) 2 ppm (10 mg/m^3)

Germany
MAK .... 1 ppm (5 mg/m^3)

France
VME .... 1 ppm (5 mg/m^3)
VLE .... not available

Sweden
Short-term limit .... 1 ppm (5 mg/m^3)
Level limit .... 2 ppm (10 mg/m^3)

PHYSICAL PROPERTIES
Description Greenish yellow crystals or yellow oily liquid.

Boiling point	210°C
Melting point	6°C
Density	1.205 at 15°C
Vapour density	4.25
Vapour pressure	1 mm Hg at 44°C
Flash point	88°C (closed cup)
Explosive limits	lower limit 1.8%
Autoignition temperature	482°C

Solubility Slightly soluble in water and miscible in most organic solvents.

PACKAGING AND TRANSPORTATION

Road transportation
hazard warning sign	1662 toxic substance
Hazchem code	2X

Sea transportation
IMDG page No.	6208
class	6.1
label	poison
packaging group	II

Air transportation
ICAO/IATA code (UN No.)	1662
class	6.1
label	poison
packaging group	II
packing instructions	
cargo	611
passenger	609
passenger aircraft max. quantity	5 litres
cargo aircraft max. quantity	60 litres

104. Nitrobenzene

MANUFACTURE

Nitrobenzene is produced by the direct nitration of benzene in nitric/sulphuric acid.

USES

Nitrobenzene is used as a chemical intermediate in the manufacture of aniline dyes, benzidine, and explosives. As a solvent, it is used in shoe and floor polishes and in paints to mask unpleasant odours.

CHEMICAL HAZARDS

Nitrobenzene reacts with oxidising agents such as nitric acid, dinitrogen tetroxide and sodium chlorate to produce detonable and explosive compounds. Shock-sensitive salts are formed by reaction with silver perchlorate. It reacts exothermically with alkalis such as sodium hydroxide and potassium hydroxide to produce thermal runaway and explosions. In one incident, nitrobenzene and sulphuric acid produced a tarry mass which attacked an iron vessel, producing hydrogen which subsequently exploded. Used in the Skraup reaction to produce quinolines from aniline and glycerol, excess sulphuric acid causes runaway reactions and explosions.

BIOLOGICAL HAZARDS

Nitrobenzene affects the blood system causing cyanosis and anaemia and produces toxic effects on the central nervous system. It is absorbed into the body readily by inhalation, from skin absorption and by ingestion. The primary effect on the blood is the conversion of haemoglobin to methaemoglobin with subsequent inability of the red blood cells to transport oxygen, causing cyanosis and bluish-grey pallor to the skin. Nitrophenols and aminophenols are present in the urine of exposed individuals. Liver damage results in cases of chronic, prolonged exposures. Toxic effects are increased by the ingestion of alcohol.

Vapour Inhalation

The vapour is more toxic than the liquid with the onset of toxic symptoms from inhalation occurring rapidly. The principal toxic effect is on the blood, producing methaemoglobinaemia, the onset being insidious with cyanosis appearing only when the methaemoglobin level in the blood reaches 15%. Heinz bodies are noted in the red blood cells following prolonged exposures at low concentrations (1). Exposure to nitrobenzene causes toxic effects on the central nervous system producing headaches, vertigo, nausea, vomiting and in some cases unconsciousness. Deaths result from respiratory failure. The effects of chronic exposure include changes to the spleen, liver, and bladder; visual disturbances have been reported. Cases of poisoning by nitrobenzene and aniline in dye manufacturing have been reported (2).

Eye Contact

Liquid nitrobenzene is irritating to the eyes. The vapours cause irritation and, as a result of action on the central nervous system, visual disturbances.

Skin Contact

Nitrobenzene is rapidly absorbed through the skin (3). It may produce dermatitis by primary irritation or sensitisation.

Swallowing

Ingestion of the liquid leads to many of the effects presented by inhalation exposure, although the onset of symptoms is often delayed for several hours. It is irritant to the throat and the gastrointestinal tract. A fatal case has been described in a woman following ingestion of about 100 ml nitrobenzene (4).

CARCINOGENICITY

Nitrobenzene was selected by the National Cancer Institute as a candidate for the carcinogenesis bioassay (5). Conversion of nitrobenzene to nitrosobenzene (a known carcinogen) has been reported under physiological conditions (6). Benign and malignant bladder tumours have been reported in humans (7).

MUTAGENICITY

Nitrobenzene is not mutagenic in *Salmonella typhimurium* TA98 or TA100 (8,9). It was not mutagenic in mice after intragastric administration using the micronuclear test and metaphase analysis of chromosomal aberrations in bone marrow cells (10).

REPRODUCTIVE HAZARDS

Degeneration of the testes occurred in rats exposed to 50 ppm, but this effect was not noted in mice at the same dose (11).

104. Nitrobenzene

FIRST AID

Eyes Wash the eye with flowing water for 10 minutes.

Lungs Remove casualty from area of exposure. If unconscious, do not give anything to drink. Give artificial ventilation and chest compression or place in the recovery position as necessary. If conscious make the casualty lie or sit down quietly, give oxygen if available. Convulsions may occur and may cause unconsciousness. Shock may result – if so do not give any drinks, and if conscious, lie casualty flat with legs raised.

Mouth Do not make the casualty vomit. Treat unconscious casualties as for lungs, but if conscious give 1 pint of water to drink.

Skin Remove contaminated clothing immediately, wash the affected area with soap and copious amounts of water. Absorption through the skin may cause symptoms similar to those of inhalation.

In all cases of exposure, the patient should be transferred to hospital as soon as possible.

HANDLING AND STORAGE

Nitrobenzene should be handled wearing eye protection, butyl rubber, PVA, Teflon or Viton gloves (12), under good conditions of ventilation, away from sources of ignition. Respiratory protection is advised for conditions of inadequate ventilation or high concentration levels. Nitrobenzene is transported and stored in steel, stainless steel or aluminium drums, which should be protected from physical damage.

DISPOSAL

Eliminate all sources of ignition and ventilate the area. Wear a laboratory coat, safety spectacles, butyl rubber gloves, and approved self-contained breathing apparatus. While stirring constantly, add contaminated material to approximately 30 times its weight of a solution prepared by dissolving 1 part sodium sulphide in 6 parts water.

Nitrobenzene has been added to the list of compounds regulated as toxic in wastes under the Resource, Conservation and Recovery Act (RCRA) (13).

FIRE PRECAUTIONS

Fires involving Nitrobenzene should be fought with carbon dioxide, dry chemical powder, alcohol-resistant foam or vaporising liquids.

FURTHER READING

Bretherick, L. *Hazards in the Chemical Laboratory* (4th edition)

Bretherick, L. *Handbook of Reactive Chemical Hazards* (4th edition)

Sax, N. Irving *Dangerous Properties of Industrial Materials* (7th edition)

Encyclopedia of Occupational Safety & Health

Patty's *Industrial Hygiene and Toxicology*

Kirk-Othmer *Encyclopedia of Chemical Technology*

National Fire Protection Association *Manual of Hazardous Reactions*

ACGIH *Documentation of TLVs and BEIs* (6th edition, 1986)

Solvents in common use: health risks to workers (RSC, London, 1988)

Dangerous Prop. Ind. Mater. Rep. 1985, **5**(6), 77-81

Hazard Databank. Sheet No. 69: Nitrobenzene *Saf. Pract.* 1985, **3**(9), 14-15

REFERENCES

1. Pacseri, L.; et al. *Arch. Ind. Health* 1958, **18**(1), 1
2. Fesenko, I. T.; et al. *Vrach. Delo.* 1982, **7**, 112-113
3. Bronaugh, R. L.; Maibach, H. I. *J. Invest. Dermatol.* 1986, **84**(3), 180-183
4. Burmistrov, V. V. *Kazan. Med. Zh.* 1967, **1**, 63
5. Helmes, C. T.; et al *J. Environ. Sci. Health, part A*, 1982, **A17**(1), 75-128
6. Becker, A. R.; Stevenson, L. A. *Proc. Natl. Acad. Sci. USA* 1981, **78**(4), 2003-2007

7. Arena, J. M. *Poisoning* (4th Edition) (Charles C. Thomas, Springfield, 1979)
8. Kawai, A. *Jpn. J. Ind. Health* 1987, **29**(1), 34-54
9. Chiu, C. W.; et al. *Mutat. Res.* 1978, **58**(1), 11-22
10. Fel'dt, E. G. *Gig. Sanit.* 1985, **7**, 21-23
11. Hamm, T. E.; et al. *Toxicologist* 1984, **4**, 181
12. Forsberg, K.; Mansdorf, S. Z. *Quick selection guide to chemical protective clothing* (Van Nostrand Reinhold, New York, 1989)
13. Hanson, D. *Chem. Eng. News* 1990, **68**(11), 4

105. 2-Nitronaphthalene

2-NITRO-NAPHTHALENE

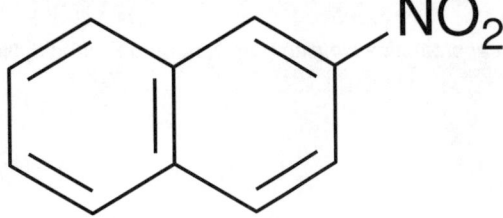

RISKS
May cause cancer (R45)

SAFETY PRECAUTIONS
Avoid exposure-obtain special instructions before use – If you feel unwell, seek medical advice (show label where possible) (S53, S44)

IDENTIFIERS

SYNONYMS	β-nitronaphthalene; naphthalene, 2-nitro-
CHEMICAL ABSTRACTS No.	581-89-5
NIOSH No.	QJ 9760000
HAZCHEM CODE	not available
UN No.	2538

THRESHOLD LIMIT VALUES

US TLV (TWA)	not available
US TLV (STEL)	not available
UK EXPOSURE LIMITS (OES)	
Long-term (8 hr TWA value)	not available
Short-term (10 min TWA value)	not available
Germany	
MAK	not available
France	
VME	not available
VLE	not available
Sweden	
Short-term limit	not available
Level limit	not available

PHYSICAL PROPERTIES

Description	Yellow crystalline solid.
Boiling point	165°C at 15 mm Hg
Melting point	79°C
Density	not available
Vapour density	not available
Vapour pressure	not available
Flash point	not available
Explosive limits	not available
Autoignition temperature	not available
Solubility	Insoluble in water. Very soluble in alcohol and ether.

PACKAGING AND TRANSPORTATION

Road transportation

hazard warning sign	2538 flammable solid
Hazchem code	not available

Sea transportation

IMDG page No.	4163
class	4.1
label	flammable solid
packaging group	III

Air transportation

ICAO/IATA code (UN No.)	2538
class	4.1
label	solid flammable
packaging group	III
packing instructions	
cargo	420
passenger	419
passenger aircraft max. quantity	25 kilograms
cargo aircraft max. quantity	100 kilograms

105. 2-Nitronaphthalene

MANUFACTURE

2-Nitronaphthalene is produced by indirect methods such as the Bucherer reaction starting with 2-naphthalenol. This method is rarely used in the U.S.A. because the 2-naphthylamine formed by this route is carcinogenic.

USES

2-Nitronaphthalene is used in the manufacture of dyes.

CHEMICAL HAZARDS

2-Nitronaphthalene is incompatible with strong oxidising agents and strong bases. It is combustible on exposure to heat or flame, and emits toxic fumes of carbon monoxide and nitrogen oxides when heated to decomposition (1).

BIOLOGICAL HAZARDS

2-Nitronaphthalene is metabolised in humans to 2-naphthylamine, a carcinogen.

Vapour Inhalation

2-Nitronaphthalene is a lung irritant.

Eye Contact

No information is available concerning the effects of 2-nitronaphthalene.

Skin Contact

2-Nitronaphthalene is a skin irritant.

Swallowing

The oral LD_{50} in rats is 4400 mg/kg (1).

CARCINOGENICITY

According to NIOSH, 2-nitronaphthalene is a suspect occupational carcinogen. It caused bladder tumours in dogs following oral administration; the human bladder carcinogen β-naphthylamine was detected in the urine.

MUTAGENICITY

2-Nitronaphthalene is mutagenic in *Salmonella typhimurium* TA102, TA96 (2), TA1535 and TA100 (3,4). It showed direct-acting base substitution activity in TA1538 and TA1535 (5). Its activity in a DNA repair test in *Escherichia coli* was decreased in the presence of S9 (6). In another study it was mutagenic in *S. typhimurium* TA1535, TA100, TA1538, TA98, TA1537, TA97 and *E. coli* (7).

REPRODUCTIVE HAZARDS

No information is available concerning the reproductive hazards of 2-nitronaphthalene.

FIRST AID

Eyes   Wash the eye with flowing water for 10 minutes.

Lungs   Remove casualty from area of exposure. If unconscious, do not give anything to drink. Give artificial ventilation and chest compression or place in the recovery position as necessary. If conscious make the casualty lie or sit down quietly, give oxygen if available. Convulsions may occur and may cause unconsciousness. Shock may result – if so do not give any drinks, and if conscious, lie casualty flat with legs raised.

Mouth   Do not make the casualty vomit. Treat unconscious casualties as for lungs, but if conscious give 1 pint of water to drink.

Skin   Remove contaminated clothing immediately, wash the affected area with soap and copious amounts of water. Absorption through the skin may cause symptoms similar to those of inhalation.

In all cases of exposure, the patient should be transferred to hospital as soon as possible.

105. 2-Nitronaphthalene

HANDLING AND STORAGE

2-Nitronaphthalene should be handled wearing an approved respirator, chemical-resistant gloves, safety goggles and other protective clothing. It should only be used in a chemical fume hood. It should be kept in a tightly closed container, in a cool dry place, away from heat, sparks and flame (1).

DISPOSAL

Eliminate all sources of ignition and ventilate the area. Wear a laboratory coat, safety spectacles, butyl rubber gloves, and approved self-contained breathing apparatus. While stirring constantly, add contaminated material to approximately 30 times its weight of a solution prepared by dissolving 1 part sodium sulphide in 6 parts water.

FIRE PRECAUTIONS

Fires involving 2-nitronaphthalene should be extinguished using water spray, carbon dioxide, dry chemical powder, alcohol foam or polymer foam.

FURTHER READING

Sax, N. Irving *Dangerous Properties of Industrial Materials* (7th edition)

Encyclopaedia of Occupational Safety & Health

Patty's *Industrial Hygiene and Toxicology*

Kirk-Othmer *Encyclopedia of Chemical Technology*

National Fire Protection Association *Manual of Hazardous Reactions*

ACGIH *Documentation of TLVs and BEIs* (6th edition, 1986)

REFERENCES

1. Lenga, R. E. *The Sigma-Aldrich Library of Chemical Safety Data* (2nd edition) (Sigma-Aldrich, Milwaukee, 1988)
2. Massaro, M.; et al. *Mutat. Res.* 1983, **122**, 243-249
3. DeFlora, S. *Carcinogenesis (London)* 1981, **2**(4), 283-298
4. Simmon, V. F. *JNCI, J. Natl. Cancer Inst.* 1979, **62**, 893-899
5. Rosenkranz, H. S.; Poirer, L. A. *JNCI, J. Natl. Cancer Inst.* 1979, **62**, 873-892
6. DeFlora, S. *Toxicol. Pathol.* 1984, **12**(4), 337-343
7. DeFlora, S. *Mutat. Res.* 1984 **133**(3), 161-198

2-NITROPROPANE
$(CH_3)_2CHNO_2$

RISKS
May cause cancer – Flammable – Harmful by inhalation and if swallowed (R45, R10, R20/22)

SAFETY PRECAUTIONS
Avoid exposure-obtain special instructions before use – Keep container in a well ventilated place – If you feel unwell, seek medical advice (show label where possible) (S53, S9, S44)

IDENTIFIERS

SYNONYMS	dimethylnitromethane isonitropropane; propane, 2-nitro-
CHEMICAL ABSTRACTS No.	79-46-9
NIOSH No.	TZ 5250000
HAZCHEM CODE	2Y
UN No.	2608

THRESHOLD LIMIT VALUES

US TLV (TWA)	10 ppm (36 mg/m^3)
US TLV (STEL)	not available

UK EXPOSURE LIMITS (OES)
Long-term (8 hr TWA value) .... 10 ppm (36 mg/m^3) under review
Short-term (10 min TWA value) .... 20 ppm (72 mg/m^3) under review

Germany
MAK .... not available

France
VME .... not available
VLE .... not available

Sweden
Ceiling limit .... 10 ppm (35mg/m^3)
Level limit .... 5 ppm (18 mg/m^3)

106. 2-Nitropropane

PHYSICAL PROPERTIES

Description	Colourless liquid.
Boiling point	120°C
Melting point	-93°C
Density	0.992 at 20°C
Vapour density	3.06
Vapour pressure	10 mm Hg at 15.8°C
Flash point	24°C (closed cup)
Explosive limits	2.6%– 11.0%
Autoignition temperature	428°C

Solubility Soluble in water, alcohol, ether, acetone, benzene or dimethyl sulphoxide.

PACKAGING AND TRANSPORTATION

Road transportation

hazard warning sign	2608 flammable liquid
Hazchem code	2Y

Sea transportation

IMDG page No.	3370
class	3.3
label	flammable liquid
packaging group	III

Air transportation

ICAO/IATA code (UN No.)	2608
class	3
label	liquid flammable
packaging group	III
packing instructions	
cargo	310
passenger	309
passenger aircraft max. quantity	60 litres
cargo aircraft max. quantity	220 litres

106. 2-Nitropropane

MANUFACTURE

2-Nitropropane is produced by reacting nitric acid with excess propane at 370-450°C and 0.81-1.2 MPa. The reaction products are cooled and the nitroparaffins condense. These are then washed to remove oxygenated impurities and then fractionated to commercial-grade nitroparaffins including 2-nitropropane.

USES

2-Nitropropane is a solvent which is used in the manufacture of vinyl and epoxy coatings, in chemical synthesis, as a rocket propellant, a petrol additive, and in paint and varnish removers.

CHEMICAL HAZARDS

2-Nitropropane is flammable and poses a serious fire hazard in the presence of heat, flame or oxidising agents. The vapour forms an explosive mixture with air. It is incompatible with strong bases, strong acids, lead and copper. It undergoes violent reactions with chlorosulphonic acid and oleum, and may react with amines and heavy metal oxides to produce explosive salts. 2-Nitropropane decomposes on heating to form toxic fumes of carbon monoxide and nitrogen oxides.

BIOLOGICAL HAZARDS

2-Nitropropane is moderately toxic by ingestion and inhalation. Symptoms of acute poisoning include anorexia, nausea, vomiting, diarrhoea, coughing and shortness of breath. At high concentrations there may be methaemoglobinaemia and cyanosis. 2-nitropropane is toxic to the kidneys and liver. It is thought to have caused the death from liver failure of a man who became ill after being exposed to high concentrations whilst applying epoxy-based paint (1). The effects of chronic exposure are uncertain. A retrospective mortality study of workers at a 2-nitropropane production plant found no increase in mortality from liver disease (2). Rats chronically exposed by inhalation developed liver abnormalities at 207 ppm, but not at lower concentrations (3, 4).

Vapour Inhalation

2-Nitropropane is toxic by inhalation (see above). It is irritating to the lungs and upper respiratory tract, and may cause coughing, wheezing and laryngitis. The lowest reported toxic concentration for man is 20 ppm (5).

Eye Contact

2-Nitropropane is an eye irritant.

Skin Contact

2-Nitropropane is a skin irritant.

Swallowing

2-Nitropropane is toxic if swallowed (see above). The LD_{50} in rats is 720 mg/kg.

CARCINOGENICITY

The IARC has concluded that there is sufficient evidence that 2-nitropropane is an animal carcinogen (6). Of 10 rats exposed to 207 ppm for 6 months, all developed neoplasms of the liver (7). There is no definite evidence of carcinogenicity in man; 2-nitropropane was given a Human Inadequate Evidence rating by the IARC. In view of the animal evidence, the ACGIH has classified it as a suspected human carcinogen.

MUTAGENICITY

2-Nitropropane is mutagenic in *Salmonella typhimurium* TA98 and TA100 with or without metabolic activation (8), and induces changes in DNA *in vitro*. However, it has proved negative in mouse micronucleus tests (9), and failed to induce chromosome aberrations or sister-chromatid exchange in chinese hamster ovary cells (10). A weak clastogenic effect has been observed in human lymphocytes *in vitro* (11).

REPRODUCTIVE HAZARDS

Foetuses of rats intraperitoneally injected with 170 mg/kg 2-nitropropane showed delayed development in the absence of maternal toxicity, but there was no evidence of gross malformations (12).

FIRST AID

Eyes Wash the eye with flowing water for 10 minutes.

Lungs Remove casualty from area of exposure. If unconscious, do not give anything to drink, give artificial ventilation and chest compression or place in the recovery position as necessary. If conscious make the casualty lie or sit down quietly, give oxygen if available. Lung congestion may occur – a conscious casualty with breathing difficulties should be placed in a sitting position. Convulsions may occur and may cause unconsciousness. Shock may result – do not give any drinks, and if conscious lie casualty flat with legs raised.

Mouth Do not make the casualty vomit. Treat unconscious casualties as for lungs but if conscious give 1 pint of water to drink immediately; give repeated drinks of water (1 cupful every 10 minutes). Convulsions may occur and may cause unconsciousness.

Skin Remove contaminated clothing immediately, drench the affected area with running water for at least 10 minutes.

In all cases of exposure, the patient should be transferred to hospital as soon as possible.

HANDLING AND STORAGE

2-Nitropropane should be handled wearing an approved respirator, butyl rubber or PVA gloves (13), safety goggles and other protective clothing. It should only be used in a chemical fume hood. Avoid getting on clothing. It should be kept in a tightly closed container, stored in a cool, dry place away from heat, flames, sparks and oxidising materials. Outside storage is preferable.

DISPOSAL

Eliminate all sources of ignition, and wear a laboratory coat, safety spectacles, neoprene gloves, approved self-contained breathing apparatus and safety shoes. EITHER cover with dry sand or soda ash and mix well, adding a small amount of water. Move the slurry to a safe place outside, neutralise with 6M hydrochloric acid and stir well. Alllow to stand for 2-3 hours , then run to waste with a large excess of water. Wash the spill site well with soap and water. OR absorb onto vermiculite or dry soft sand, mix well and shovel into a cardboard box. Burn in an incinerator fitted with an afterburner and scrubber.

FIRE PRECAUTIONS

Fires involving 2-nitropropane should be extinguished using water spray, carbon dioxide, dry chemical powder, alcohol foam or polymer foam. Keep fire-exposed containers cool by spraying with water.

FURTHER READING

Sax, N. Irving *Dangerous Properties of Industrial Materials* (7th edition)

Kirk-Othmer *Encyclopedia of Chemical Technology*

National Fire Protection Association *Fire Protection Guide on Hazardous Materials* (9th edition, NFPA, USA, 1986)

ACGIH Documentation of TLVs and BEIs (6th edition, 1986)

Dangerous Prop. Ind. Mater. Rep. 1982, (2), 58-59

Lenga, R. E. *The Sigma-Aldrich Library of Chemical Safety Data* (2nd edition) (Sigma-Aldrich, Milwaukee, 1988)

Keith, L. H.; Walters, D. B. (editors) *Compendium of Safety Data Sheets for Research and Industrial Chemicals* (VCH, Deerfield Park, 1987)

REFERENCES

1. Harrison, R.; et al. *Ann. Intern. Med.* 1987, **107**(4), 466-468
2. Bolender, F. L. Report to the International Minerals and Chemical Corporation, Northbrook. 1L, (1983)
3. Beall, J. R.; et al. National Institute of Occupational Safety and Health Report (DHHS/PUP/NIOSH-80-142, USA, 1980)
4. Griffin, T. B.; et al. *Ecotoxicol. Environ. Saf.* 1980, **4**(3), 261-281
5. *Industrial Med* 1947, **16**, 441
6. *IARC Monographs on the Evaluation of the Carcinogenic Risk of Chemicals to Humans* 1982, **29**, 331
7. Lewis, T. R.; et al. *J. Environ. Pathol. Toxicol.* 1979, **2**(5), 233-249

106. 2-Nitropropane

8. Speck, W. T.; et al *Mutat. Res.* 1982, **104**, 49-54
9. Hite, M.; Skeggs, H. *Environ. Mutagen.* 1979, **1**, 383-389
10. Galloway, S. M.; et al. *Environ. Mol. Mutagen.* 1987, **10** (Suppl. 10), 1-175
11. Gogglemann, W.; et al. *Mutagenesis* 1988, **3**(2), 137-140
12. Hardin, B. D.; et al. *Scand. J. Work Environ. Health* 1981, **7** (Suppl. 4), 66-75
13. Forsberg, K.; Mansdorf, S. Z. *Quick selection guide to chemical protective clothing* (Van Nostrand Reinhold, New York, 1989)

o-NITROTOLUENE

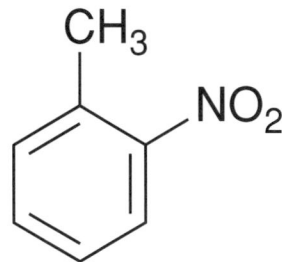

RISKS
Toxic by inhalation, in contact with skin and if swallowed – Danger of cumulative effects (R23/24/25, R33)

SAFETY PRECAUTIONS
After contact with skin, wash immediately with plenty of water – Wear suitable gloves – If you feel unwell, seek medical advice (show label where possible) (S28, S37, S44)

IDENTIFIERS
SYNONYMS benzene, 1-methyl-2-nitro-; toluene, o-nitro-; o-methylnitrobenzene; 2-methylnitrobenzene; 2-nitrotoluene

CHEMICAL ABSTRACTS No.	88-72-2
NIOSH No.	XT 3150000
HAZCHEM CODE	2X
UN No.	1664

THRESHOLD LIMIT VALUES
US TLV (TWA)	2 ppm (11 mg/m^3)
US TLV (STEL)	not available

UK EXPOSURE LIMITS (OES)
Long-term (8 hr TWA value)	5 ppm (30 mg/m^3)
Short-term (10 min TWA value)	10 ppm (60 mg/m^3)

Germany
MAK	5 ppm (30 mg/m^3)

France
VME	not available
VLE	not available

Sweden
Short-term limit	not available
Level limit	not available

PHYSICAL PROPERTIES
Description Clear yellow liquid.

Boiling point	221.7°C
Melting point	α-9.55°C; β-3.85°C
Density	1.162 at 19°C
Vapour density	4.72
Vapour pressure	1.6 mm Hg at 60°C
Flash point	106°C (closed cup)
Explosive limits	lel 2.2
Autoignition temperature	305°C

Solubility Slightly soluble in water. Soluble in benzene or ether.

PACKAGING AND TRANSPORTATION

Road transportation
hazard warning sign	1664 toxic substance
Hazchem code	2X

Sea transportation
IMDG page No.	6211
class	6.1
label	poison
packaging group	II

Air transportation
ICAO/IATA code (UN No.)	1664
class	6.1
label	poison
packaging group	II

packing instructions
cargo	
liquid	611
solid	615
passenger	
liquid	609
solid	613

passenger aircraft max. quantity
liquid	5 litres
solid	25 kilograms

cargo aircraft max. quantity
liquid	60 litres
solid	100 kilograms

107. o-Nitrotoluene

MANUFACTURE

Mononitration of toluene results in a mixture of o-, m- and p-nitrotoluene. The isomers are separated by fractional distillation and crystallisation. Pure o-nitrotoluene may be obtained by treating 2,4-dinitrotoluene with ammonium sulphide, followed by diazotisation and boiling with ethanol.

USES

o-Nitrotoluene is used in the manufacture of azo and sulphur dyes, rubber chemicals, explosives, agrochemicals, toluidine, fuchsin, and o-tolidine.

CHEMICAL HAZARDS

o-Nitrotoluene may be shock sensitive and explode on heating (1). It is a moderate fire hazard on exposure to heat or flame. Reaction with alkali in the preparation of 1,2-bis(2- nitrophenyl)ethane (2,2'-dinitrobibenzyl) is very exothermic, and cases of dangerous rises in temperature due to agitator failure have been reported (2). Crude o-nitrotoluene (containing some hydrochloric and acetic acids) may react explosively with sodium hydroxide (3). A dark brown sodium derivative of treating o- nitrotoluene in ether with sodium ignites in air (4). Solutions of o-nitrotoluene in 93% sulphuric acid decompose violently on heating to 160°C, which occurred on plant scale due to failure of the automatic temperature control (5).

BIOLOGICAL HAZARDS

Cases of poisoning appear uncommon (6). o-Nitrotoluene is a low grade methaemoglobin former. Chronic exposure may cause anaemia.

Vapour Inhalation

o-Nitrotoluene is a low grade methaemoglobin former. Chronic exposure may cause anaemia.

Eye Contact

No information is available concerning the effects of o-nitrotoluene on the eyes.

Skin Contact

No information is available concerning the effects of o-nitrotoluene on the skin.

Swallowing

The oral LD_{50} in rats is 891 mg/kg (1).

CARCINOGENICITY

No information is available concerning the carcinogenicity of o-nitrotoluene.

MUTAGENICITY

Nitrotoluenes do not induce DNA repair in rat hepatocytes *in vitro*; intestinal bacteria seem necessary for metabolic activation of o-nitrotoluene to a hepatic genotoxin. o-Nitrotoluene is active in an *in vivo-in vitro* hepatocyte DNA repair assay. Isomeric differences may be due to different metabolism; 2-nitrobenzyl alcohol or conjugates may mediate the genotoxicity of o-nitrobenzene (7). Without metabolic activation it gave a weakly positive result in an investigation of sister chromatid exchanges in Chinese hamster ovary cells; results were positive with metabolic activation. It did not induce chromosome aberrations (8). It did not induce chromosome aberrations in Chinese hamster lung cells (9). It is not mutagenic in *Salmonella typhimurium* TA98, TA100 (10), TA1535, TA1537 or TA1538 either with or without metabolic activation (11).

REPRODUCTIVE HAZARDS

No information is available concerning the reproductive hazards of o-nitrotoluene.

FIRST AID

Eyes Wash the eye with flowing water for 10 minutes.

Lungs Remove casualty from area of exposure. If unconscious, do not give anything to drink. Give artificial ventilation and chest compression or place in the recovery position as necessary. If conscious make the casualty lie or sit down quietly, give oxygen if available. Convulsions may occur and may cause unconsciousness. Shock may result – if so do not give any drinks, and if conscious, lie casualty flat with legs raised.

107. o-Nitrotoluene

Mouth Do not make the casualty vomit. Treat unconscious casualties as for lungs, but if conscious give 1 pint of water to drink.

Skin Remove contaminated clothing immediately, wash the affected area with soap and copious amounts of water. Absorption through the skin may cause symptoms similar to those of inhalation.

In all cases of exposure, the patient should be transferred to hospital as soon as possible.

HANDLING AND STORAGE

o-Nitrotoluene should be handled wearing an approved respirator, heavy butyl rubber or PVA gloves (12), and safety goggles. It should only be used in chemical fume hood. It should be kept in a tightly closed container, in a cool dry place (1).

DISPOSAL

Eliminate all sources of ignition and ventilate the area. Wear a laboratory coat, safety spectacles, butyl rubber gloves, and approved self-contained breathing apparatus. While stirring constantly, add contaminated material to approximately 30 times its weight of a solution prepared by dissolving 1 part sodium sulphide in 6 parts water.

FIRE PRECAUTIONS

Fires involving o-nitrotoluene should be extinguished using water spray, carbon dioxide, alcohol foam or polymer foam.

FURTHER READING

Bretherick, L. *Hazards in the Chemical Laboratory* (4th edition)

Bretherick, L. *Handbook of Reactive Chemical Hazards* (4th edition)

Sax, N. Irving *Dangerous Properties of Industrial Materials* (7th edition)

Encyclopedia of Occupational Safety & Health

Patty's *Industrial Hygiene and Toxicology*

Kirk-Othmer *Encyclopedia of Chemical Technology*

National Fire Protection Association *Manual of Hazardous Reactions*

ACGIH Documentation of TLVs and BEIs (6th edition, 1986)

REFERENCES

1. Lenga, R. E. *The Sigma-Aldrich Library of Chemical Safety Data* (2nd edition) (Sigma-Aldrich, Milwaukee, 1988)
2. Anon. *Loss Prev. Bull.* 1977, (13), 2
3. Anon. *ABCM Quart. Safety Summ.* 1945, **16**, 2
4. Schmidt, J. *Ber.* 1899, **32**, 2920
5. Hunt, J. K. *Chem. Eng. News* 1949, **27**, 2504
6. Von Oettingen, W. F. *Public Health Bull.* 1941, **271**, 106
7. Doolittle, D. J.; et al. *Cancer Res.* 1983, **43**, 2836-2842
8. Galloway, S. M.; et al. *Environ. Mol. Mutagen.* 1987, **10**(Suppl. 10), 1-175
9. Ishidate, M. *Chromosome aberration test in vitro* (Realize Inc., Tokyo, 1983)
10. Chiu, C. W.; et al. *Mutat. Res.* 1978, **58**, 11-22
11. Spanggord, R. J.; et al. *Environ. Mutagen.* 1982, **4**, 163-179
12. Forsberg, K.; Mansdorf, S. Z. *Quick selection guide to chemical protective clothing* (Van Nostrand Reinhold, New York, 1989)

108. *p*-Nitrotoluene

p-NITROTOLUENE

RISKS
Toxic by inhalation, in contact with skin and if swallowed – Danger of cumulative effects (R23/24/25, R33)

SAFETY PRECAUTIONS
After contact with skin, wash immediately with plenty of water – Wear suitable gloves – If you feel unwell, seek medical advice (show label where possible) (S28, S37, S44)

IDENTIFIERS

SYNONYMS benzene, 1-methyl-4-nitro; toluene, *p*-nitro-; 4-methylnitrobenzene; 4-nitrotoluene; 4-nitrotoluol; *p*-methylnitrobenzene; NCI-C60537; PNT

CHEMICAL ABSTRACTS No.	99-99-0
NIOSH No.	XT 3325000
HAZCHEM CODE	2X
UN No.	1664

THRESHOLD LIMIT VALUES

US TLV (TWA)	2 ppm (11 mg/m^3)
US TLV (STEL)	not available

UK EXPOSURE LIMITS (OES)
Long-term (8 hr TWA value)	5 ppm (30 mg/m^3)
Short-term (10 min TWA value)	10 ppm (60 mg/m^3)

Germany
MAK	5 ppm (30 mg/m^3)

France
VME	not available
VLE	not available

Sweden
Short-term limit	not available
Level limit	not available

PHYSICAL PROPERTIES

Description	Colourless or yellowish crystals.
Boiling point	238.5°C
Melting point	51.7°C
Density	1.286 at 20°C
Vapour density	4.72
Vapour pressure	1.3 mm Hg at 65°C
Flash point	106°C (closed cup)
Explosive limits	lel 1.6%
Autoignition temperature	390°C

Solubility Slightly soluble in water. Moderately soluble in methanol or ethanol. Readily soluble in acetone, benzene or diethyl ether.

PACKAGING AND TRANSPORTATION

Road transportation
hazard warning sign	1664 toxic substance
Hazchem code	2X

Sea transportation
IMDG page No.	6211
class	6.1
label	poison
packaging group	II

Air transportation
ICAO/IATA code (UN No.)	1664
class	6.1
label	poison
packaging group	II

packing instructions
cargo	
liquid	611
solid	615
passenger	
liquid	609
solid	613

passenger aircraft max. quantity
liquid	5 litres
solid	25 kilograms

cargo aircraft max. quantity
liquid	60 litres
solid	100 kilograms

108. *p*-Nitrotoluene

MANUFACTURE

Mononitration of toluene results in a mixture of 2-, 3-and 4-nitrobenzene. The isomers are separated by fractional distillation and crystallisation.

USES

p-Nitrotoluene is used in the manufacture of azo and sulphur dyes, explosives, toluidine, fuchsin, and *m*-tolidine. *p*-Nitrotoluene is the most widely used isomer.

CHEMICAL HAZARDS

It may be shock sensitive and explode on heating or standing. Dust explosions may occur. It is a moderate fire hazard on exposure to heat or flame (1). A violent explosion occurred when the residue from large scale vacuum distillation of *p*-nitrotoluene. It has been suggested that this was due to excessive proportion of dinitrotoluenes in the residue, or traces of the alkali in the crude material (2,3). Mixture with tetranitromethane is a highly sensitive high explosive which is spark detonable (4). Solutions of *o*-nitrotoluene in 93% sulphuric acid decompose violently on heating to 160°C, which occurred on plant scale due to failure of the automatic temperature control (5). This explosion temperature is 22°C lower than that reported for onset of decomposition of *p*-nitrotoluene and 93% sulphuric acid heated at 100°C per hour. Mixtures of *p*- nitrotoluene and 98% sulphuric acid or 20% oleum start to decompose at 180°C and 190°C respectively, after which decomposition accelerates to eruption (6). Runaway decomposition during sulphonation of *p*-nitrotoluene at 32°C with 24% oleum has been reported (6).

BIOLOGICAL HAZARDS

Cases of poisoning appear uncommon (7). Symptoms of poisoning include headache, dizziness, nausea, vomiting, breathing difficulties, flushed face, raised pulse and breathing rate, irritability and convulsions. It is a low grade methaemoglobin former, and chronic exposure may cause anaemia. It is a cumulative poison. It is particularly hazardous to sufferers of anaemia, heart or lung diseases (8).

Vapour Inhalation

Inhalation of *p*-nitrotoluene may result in symptoms outlined above.

Eye Contact

Chronic exposure may cause eye irritation (8).

Skin Contact

p-Nitrotoluene may be absorbed through the skin.

Swallowing

The oral LD_{50} in rats and mice is 2144 mg/kg and 1231 mg/kg respectively.

CARCINOGENICITY

No information is available concerning the carcinogenicity of *p*-nitrobenzene.

MUTAGENICITY

Nitrotoluenes do not induce DNA repair in rat hepatocytes *in vitro*. *p*-Nitrotoluene is not active in an *in vivo-in vitro* hepatocyte DNA repair assay (9). It induced sister chromatid exchanges (with and without metabolic activation), and chromosome aberrations in Chinese hamster ovary cells (10). It did not induce chromosome aberrations in Chinese hamster lung cells (11). It did not induce unscheduled DNA synthesis in rodent hepatocytes, and is not mutagenic in *Salmonella typhimurium* (12,13). Positive results have been reported in *S. typhimurium* TA100 with or without metabolic activation (14).

REPRODUCTIVE HAZARDS

No information is available concerning the reproductive hazards of *p*-nitrobenzene.

FIRST AID

Eyes Wash the eye with flowing water for 10 minutes.

Lungs Remove casualty from area of exposure. If unconscious, do not give anything to drink. Give artificial ventilation

108. p-Nitrotoluene

and chest compression or place in the recovery position as necessary. If conscious make the casualty lie or sit down quietly, give oxygen if available. Convulsions may occur and may cause unconsciousness. Shock may result – if so do not give any drinks, and if conscious, lie casualty flat with legs raised.

Mouth Do not make the casualty vomit. Treat unconscious casualties as for lungs, but if conscious give 1 pint of water to drink.

Skin Remove contaminated clothing immediately, wash the affected area with soap and copious amounts of water. Absorption through the skin may cause symptoms similar to those of inhalation.

In all cases of exposure, the patient should be transferred to hospital as soon as possible.

HANDLING AND STORAGE

p-Nitrotoluene should be handled wearing an approved respirator, butyl rubber gloves, safety goggles and other protective clothing (8). It should only be used in chemical fume hood. It should be kept in a tightly closed container, in a cool dry place (1).

DISPOSAL

Eliminate all sources of ignition and ventilate the area. Wear a laboratory coat, safety spectacles, butyl rubber gloves, and approved self-contained breathing apparatus. While stirring constantly, add contaminated material to approximately 30 times its weight of a solution prepared by dissolving 1 part sodium sulphide in 6 parts water.

FIRE PRECAUTIONS

Fires involving p-nitrotoluene should be extinguished using carbon dioxide, dry chemical powder, alcohol foam or polymer foam.

FURTHER READING

Bretherick, L. *Hazards in the Chemical Laboratory* (4th edition)

Bretherick, L. *Handbook of Reactive Chemical Hazards* (4th edition)

Sax, N. Irving *Dangerous Properties of Industrial Materials* (7th edition)

Encyclopedia of Occupational Safety & Health

Patty's *Industrial Hygiene and Toxicology*

Kirk-Othmer *Encyclopedia of Chemical Technology*

National Fire Protection Association *Manual of Hazardous Reactions*

ACGIH *Documentation of TLVs and BEIs* (6th edition, 1986)

REFERENCES

1. Lenga, R. E. *The Sigma-Aldrich Library of Chemical Safety Data* (2nd edition) (Sigma-Aldrich, Milwaukee, 1988)
2. Anon. *ABCM Quart. Safety Summ.* 1938, **9**, 65
3. Anon. *ABCM Quart. Safety Summ.* 1939, **10**, 2
4. Zotov, E. V.; et al. *Chem. Abstr.* 1982, **98**, 5965
5. Hunt, J. K. *Chem. Eng. News* 1949, **27**, 2504
6. McKeand, G. *Chem. & Ind. (London)*, 1974, 425
7. Von Oettingen, W. F. *Public Health Bull.* 1941, **271**, 106
8. *Dangerous Prop. Ind. Mater. Rep.* 1983, **3**, 85-88
9. Doolittle, D. J.; et al. *Cancer Res.* 1983, **43**, 2836-2842
10. Galloway, S. M.; et al. *Environ. Mol. Mutagen.* 1987, **10**(Suppl. 10), 1-175
11. Ishidate, M. *Chromosome aberration test in vitro* (Realize Inc., Tokyo, 1983)
12. Mirsalis, J. C.; et al. *Environ. Mol. Mutagen.* 1989, **14**(3), 155-164
13. Chiu, C. W.; et al. *Mutat. Res.* 1978, **58**, 11-22.
14. Spanggord, R. J.; et al. *Environ. Mutagen.* 1982, **4**, 163-179

5-NITRO-o-TOLUIDINE

RISKS
Toxic by inhalation, in contact with skin and if swallowed – Danger of cumulative effects (R23/24/25, R33)

SAFETY PRECAUTIONS
After contact with skin, wash immediately with plenty of water – Wear protective clothing and gloves – If you feel unwell, seek medical advice (show label where possible) (S28, S36/37, S44)

IDENTIFIERS

SYNONYMS   benzenamine, 2-methyl-5-nitro-; o-toluidine, 5-nitro-; Azoene Fast Scarlet GC Base; Azoene Fast Scarlet GC Salt; Amarthol Fast Scarlet G Base; Amarthol Fast Scarlet G Salt; Azoene Fast Scarlet GC Base; Azoene Fast Scarlet GC Salt; Azofix Scarlet G Salt; Azogene Fast Scarlet G; C.I. 37105; C.I.Azoic Diazo Component 12; Dainichi Fast Scarlet G Base; Daito Scarlet Base G; Devol Scarlet B; Devol Scarlet G Salt; Diabase Scarlet G; Diazo Fast Scarlet G; Fast Red SG Base; Fast Scarlet Base J; Fast Scarlet G; Fast Scarlet G Base; Fast Scarlet GC Base; Fast Scarlet G Salt; Fast Scarlet J Salt; Fast Scarlet M 4NT; Fast Scarlet T Base; Hiltonil Fast Scarlet G Base; Hiltonil Fast Scarlet GC Base; Hiltonil Fast Scarlet G Salt; Kayaku Scarlet G Base; Lake Scarlet G Base; Lithosol Orange R Base; Mitsui Scarlet G Base; Naphthanil Scarlet G Base; Naphtoelan Fast Scarlet G Base; Naphtoelan Fast Scarlet G Salt; PNOT; Scarlet Base Ciba II; Scarlet Base Irga II; Scarlet Base NSP; Scarlet G Base; Sugai Fast Scarlet G Base; Symulon Scarlet G Base

CHEMICAL ABSTRACTS No.	99-55-8
NIOSH No.	XU 8225000
HAZCHEM CODE	2Z
UN No.	2660

THRESHOLD LIMIT VALUES

US TLV (TWA)	not available
US TLV (STEL)	not available

UK EXPOSURE LIMITS (OES)
Long-term (8 hr TWA value)	not available
Short-term (10 min TWA value)	not available

Germany
MAK	not available

France
VME	not available
VLE	not available

Sweden
Short-term limit	not available
Level limit	not available

PHYSICAL PROPERTIES

Description	Bright yellow powder.
Boiling point	not available
Melting point	104-107°C
Density	not available
Vapour density	not available
Vapour pressure	not available
Flash point	not available
Explosive limits	not available
Autoignition temperature	not available

Solubility   Practically insoluble in water. Slightly soluble in ether and benzene. Soluble in dimethyl sulphoxide, ethanol, acetone, toluene or chloroform.

PACKAGING AND TRANSPORTATION

Road transportation
hazard warning sign	2660 harmful substance
Hazchem code	2Z

Sea transportation
IMDG page No.	6211
class	6.1
label	harmful stow away from foodstuffs
packaging group	III

Air transportation
ICAO/IATA code (UN No.)	2660
class	6.1
label	keep away from food
packaging group	III
packing instructions	
cargo	619
passenger	619
passenger aircraft max. quantity	100 kilograms
cargo aircraft max. quantity	200 kilograms

109. 5-Nitro-*o*-Toluidine

MANUFACTURE

5-Nitro-*o*-toluidine is produced by nitration of *o*-benzenesulphontoluidide followed by hydrolysis.

USES

5-Nitro-*o*-toluidine is used as a coupling component in synthetic organic pigments such as Pigment Red 18 and Pigment Yellow 1.

CHEMICAL HAZARDS

5-Nitro-*o*-toluidine decomposes exothermically when heated to 150°C, and emits toxic fumes of nitrogen oxides on decomposition.

BIOLOGICAL HAZARDS

5-Nitro-*o*-toluidine is carcinogenic in feeding studies on mice, and is mutagenic in *Salmonella typhimurium*. Little information on other toxic effects is available.

Vapour Inhalation

No information is available concerning the effects of inhalation.

Eye Contact

No information is available on the effects on the eyes.

Skin Contact

No information is available on the effects on the skin.

Swallowing

The oral LD_{50} in rats is 574 mg/kg.

CARCINOGENICITY

When fed in the diet, 5-nitro-*o*-toluidine was carcinogenic in male and female mice, causing hepatocellular carcinomas, but was not carcinogenic in rats (1,2). It gave positive results in an *in vitro* carcinogenesis assay in RLV-infected rat embryo cells (3).

MUTAGENICITY

5-Nitro-*o*-toluidine is mutagenic in *Salmonella typhimurium* TA1537, TA1538, TA98 and TA100 with or without metabolic activation, and in TA1535 only without metabolic activation (4).

REPRODUCTIVE HAZARDS

No information is available concerning the reproductive hazards of 5-nitro-*o*-toluidine.

FIRST AID

Eyes Wash the eye with flowing water for 10 minutes.

Lungs Remove casualty from area of exposure. If unconscious, do not give anything to drink. Give artificial ventilation and chest compression or place in the recovery position as necessary. If conscious make the casualty lie or sit down quietly, give oxygen if available. Convulsions may occur and may cause unconsciousness. Shock may result – if so do not give any drinks, and if conscious, lie casualty flat with legs raised.

Mouth Do not make the casualty vomit. Treat unconscious casualties as for lungs, but if conscious give 1 pint of water to drink.

Skin Remove contaminated clothing immediately, wash the affected area with soap and copious amounts of water. Absorption through the skin may cause symptoms similar to those of inhalation.

In all cases of exposure, the patient should be transferred to hospital as soon as possible.

109. 5-Nitro-o-Toluidine

HANDLING AND STORAGE

5-Nitro-o-toluidine should be handled wearing an approved respirator, chemical-resistant gloves, safety goggles and other protective clothing. It should only be used in a chemical fume hood. It should be kept in a tightly closed container in a cool, dry place.

DISPOSAL

Eliminate all sources of ignition and ventilate the area. Wear a laboratory coat, safety spectacles, butyl rubber gloves, and approved self-contained breathing apparatus. While stirring constantly, add contaminated material to approximately 30 times its weight of a solution prepared by dissolving 1 part sodium sulphide in 6 parts water.

FIRE PRECAUTIONS

Fires involving 5-nitro-o-toluidine should be extinguished using water, carbon dioxide, dry chemical powder, alcohol foam or polymer foam. Keep fire exposed containers cool with water.

FURTHER READING

Sax, N. Irving *Dangerous Properties of Industrial Materials* (7th edition)

Encyclopaedia of Occupational Safety & Health

Patty's *Industrial Hygiene and Toxicology*

Kirk-Othmer *Encyclopedia of Chemical Technology*

National Fire Protection Association *Manual of Hazardous Reactions*

ACGIH *Documentation of TLVs and BEIs* (6th edition, 1986)

REFERENCES

1. *NCI Report* 1973, DHEW/PUB/NIH-78-1357, NCI-CG-TR-107, Order No. PB-285872
2. *Carcinog. Tech. Rep. Ser. – U.S. Natl. Cancer Inst.* 1978, NCI-CG-TR-107
3. Traul, K. A.; et al. *J. Appl. Toxicol.* 1981, **1**(3), 190-195
4. Spanggord, R. J.; et al. *Environ. Mutagen.* 1982, **4**, 163-179

110. Osmium Tetroxide

OSMIUM TETROXIDE
OsO$_4$

RISKS
Very toxic by inhalation, in contact with skin and if swallowed – Causes burns (R26/27/28, R34)

SAFETY PRECAUTIONS
Keep container tightly closed and in a well ventilated place – In case of contact with eyes, rinse immediately with plenty of water and seek medical advice – In case of accident or if you feel unwell, seek medical advice immediately (show label where possible) (S7/9, S26, S45)

IDENTIFIERS

SYNONYMS osmium oxide; osmic acid; RCRA Waste No. P087

CHEMICAL ABSTRACTS No.	20816-12-0
NIOSH No.	RN 1140000
HAZCHEM CODE	not available
UN No.	2471

THRESHOLD LIMIT VALUES

US TLV (TWA)	0.0002 ppm (0.0016 mg/m^3)
US TLV (STEL)	0.0006 ppm (0.0047 mg/m^3)

UK EXPOSURE LIMITS (OES)
Long-term (8 hr TWA value) .... 0.0002 ppm (0.002 mg/m^3)
Short-term (10 min TWA value) .... 0.0006 ppm (0.006 mg/m^3)

Germany
MAK .... 0.0002 ppm (0.002 mg/m^3)

France
VME .... 0.0002 ppm (0.002 mg/m^3)
VLE .... not available

Sweden
Short-term limit .... not available
Level limit .... not available

PHYSICAL PROPERTIES

Description Monoclinic colourless crystals (A) or yellow mass (B) with pungent chlorine-like smell.

Boiling point	sublimes at 130°C
Melting point	A 39.5°C
	B 41°C
Density	4.906 at 22°C
Vapour density	not available
Vapour pressure	A 10 mm Hg at 26°C
	B 10 mm at 31.3°C
Flash point	not available
Explosive limits	not available
Autoignition temperature	not available

Solubility Soluble in water, benzene, alcohol and ether.

PACKAGING AND TRANSPORTATION

Road transportation

hazard warning sign	2471 toxic substance
Hazchem code	not available

Sea transportation

IMDG page No.	6215
class	6.1
label	poison; marine pollutant
packaging group	I

Air transportation

ICAO/IATA code (UN No.)	2471
class	6.1
label	poison
packaging group	I
packing instructions	
cargo	608
passenger	608
passenger aircraft max. quantity	5 kilograms
cargo aircraft max. quantity	50 kilograms

110. Osmium Tetroxide

MANUFACTURE

Osmium tetroxide is prepared by heating the finely divided metal at 300-400°C in a stream of air or oxygen.

USES

Osmium tetroxide is used as a fat stain in pathology laboratories. It is also used as an oxidising agent and catalyst in organic synthesis (1).

CHEMICAL HAZARDS

Osmium tetroxide reacts explosively with 1-methylimidazole (2). It catalyses decomposition of hydrogen peroxide.

BIOLOGICAL HAZARDS

Osmium tetroxide is an eye and respiratory tract irritant; onset of symptoms may be delayed. Its health hazards and safe handling have been reviewed (3).

Vapour Inhalation

Its odour has been described as having "the kick of a mule" (4). It is irritating to the nose and throat, and this may last for as long as 12 hours after exposure has stopped. Headaches and a cough may also occur (5). A human fatality has been reported following inhalation of osmium tetroxide (6).

Eye Contact

Osmium tetroxide causes lacrimation, visual disturbances such as ground glass and halo effects, and conjunctivitis. Such effects have been reported in a precious metal refining plant where exposures ranged from 0.1 to 0.6 mg/m^3 (5).

Skin Contact

No information is available concerning the effects of osmium tetroxide on the skin.

Swallowing

The oral LD_{50} in mice is 162 mg/kg.

CARCINOGENICITY

No information is available concerning the carcinogenicity of osmium tetroxide.

MUTAGENICITY

A small increase in DNA repair synthesis was seen with osmium tetroxide in Syrian hamster embryo cells (7). Osmium tetroxide gave positive results in rec assays with *Bacillus subtilis* (8).

REPRODUCTIVE HAZARDS

Antitesticular effects are reported following intratesticular injection in rats and mice (9).

FIRST AID

Eyes Wash the eye with flowing water for 10 minutes.

Lungs Remove casualty from area of exposure. If unconscious, do not give anything to drink, give artificial ventilation and chest compression or place in the recovery position as necessary. If conscious make the casualty lie or sit down quietly, give oxygen if available. Lung congestion may occur – a conscious casualty with breathing difficulties should be placed in a sitting position.

Mouth Do not make the casualty vomit. Treat unconscious casualties as for lungs but if conscious give 1 pint of water to drink. Give repeated drinks of water (1 cupful every 10 minutes).

Skin Remove contaminated clothing immediately, drench the affected area with running water for a least 10 minutes.

In all cases of exposure, the patient should be transferred to hospital as soon as possible.

110. Osmium Tetroxide

HANDLING AND STORAGE

Osmium tetroxide should be handled wearing an approved respirator, chemical-resistant gloves, safety goggles and other protective clothing. It should only be used in a chemical fume hood (10). It should be kept refrigerated in a tightly sealed container, away from combustible materials, heat, sparks and open flame.

DISPOSAL

Ventilate the area and wear a laboratory coat or acid-proof overalls, gloves, approved self-contained breathing apparatus and safety boots. Cover spill with solid sodium bicarbonate or a 50-50 mixture of soda ash-calcium hydroxide, and slowly and carefully mix to a slurry. Carefully scoop up and wash down the drain with plenty of running water. NB. Allow 2-3 minutes between each stage as heat is produced by the neutralisation.

FIRE PRECAUTIONS

Osmium tetroxide is not combustible; an extinguisher appropriate to surrounding fire conditions should be used (10).

FURTHER READING

Bretherick, L. *Hazards in the Chemical Laboratory* (4th edition)

Bretherick, L. *Handbook of Reactive Chemical Hazards* (4th edition)

Sax, N. Irving *Dangerous Properties of Industrial Materials* (7th edition)

Encyclopedia of Occupational Safety & Health

Patty's *Industrial Hygiene and Toxicology*

Kirk-Othmer *Encyclopedia of Chemical Technology*

National Fire Protection Association *Manual of Hazardous Reactions*

ACGIH Documentation of TLVs and BEIs (6th edition, 1986)

REFERENCES

1. Rylander, P. N. *Organic Syntheses with Noble Metal Catalysts* (Academic Press, New York,1975)
2. Nielsen, A. J. *J. Chem. Soc., Dalton Trans.* 1979, 1084-1088
3. Kamycek, Z. *Prac. Lek.* 1986, **38**(5), 206-209
4. Hamilton, A.; Hardy, H. *Industrial Toxicology* (3rd edition) (Publishing Sciences Group, Acton, 1974)
5. McLaughlin, A.; et al. *Br. J. Ind. Med.* 1946, **3**, 183
6. Fairhall, L. T. *Industrial Toxicology* (Williams & Wilkins, Baltimore, 1949)
7. Robison, S. H.; et al. *Mutat. Res.* 1984, **131**, 173-181
8. Kanematsu, N.; et al. *Mutat. Res.* 1980, **77**(2), 109-116
9. Kamboj, V. P.; Kar, A. B. *J. Reprod. Fertil.* 1964, **7**, 21-28
10. Lenga, R. E. *The Sigma-Aldrich Library of Chemical Safety Data* (2nd edition) (Sigma-Aldrich, Milwaukee, 1988)

PENTA-CHLOROETHANE

Cl_3CCHCl_2

111. Pentachloroethane

RISKS
Very toxic by inhalation and in contact with skin (R26/27)

SAFETY PRECAUTIONS
Keep locked up – In case of insufficient ventilation, wear suitable respiratory equipment – In case of accident or if you feel unwell, seek medical advice immediately (show label where possible) (S1, S38, S45)

IDENTIFIERS

SYNONYMS	pentalin; RCRA Waste No. U184; NCI-C53894; ethane pentachloride; ethane, pentachloro
CHEMICAL ABSTRACTS No.	76-01-7
NIOSH No.	KI 6300000
HAZCHEM CODE	2Z
UN No.	1669

THRESHOLD LIMIT VALUES

US TLV (TWA)	not available
US TLV (STEL)	not available
UK EXPOSURE LIMITS (OES)	
Long-term (8 hr TWA value)	not available
Short-term (10 min TWA value)	not available
Germany	
MAK	5 ppm (40 mg/m^3)
France	
VME	not available
VLE	not available
Sweden	
Short-term limit	not available
Level limit	not available

PHYSICAL PROPERTIES

Description Colourless liquid with a chloroform-like smell.

Boiling point	161-162°C
Melting point	-29°C
Density	1.6728 at 25°C
Vapour density	7.2
Vapour pressure	3.4 mm Hg at 25°C
Flash point	75°C
Explosive limits	not available
Autoignition temperature	not available

Solubility Insoluble in water. Miscible with alcohol or ether.

PACKAGING AND TRANSPORTATION

Road transportation

hazard warning sign	1669 toxic substance
Hazchem code	2Z

Sea transportation

IMDG page No.	6217
class	6.1
label	poison; marine pollutant
packaging group	II

Air transportation

ICAO/IATA code (UN No.)	1669
class	6.1
label	poison
packaging group	II
packing instructions	
cargo	611
passenger	609
passenger aircraft max. quantity	5 litres
cargo aircraft max. quantity	60 litres

111. Pentachloroethane

MANUFACTURE

Pentachloroethane can prepared by chlorinating 1,1,2,2-tetrachloroethane under ultra violet light, or trichloroethylene at 70°C in the presence of ferric chloride. It can also be made by liquid phase chlorination of 1,1,2-trichloroethane in a steel reactor at low temperature using steam heat (1). A method of producing pentachloroethane suitable for educational purposes has been described (2).

USES

Other than in research, pentachloroethane is not widely used commercially. It has been used as a solvent for oil and grease in metal cleaning and as a solvent for cellulose acetate, some cellulose ethers, natural gums and resins. It is used as a chemical intermediate in some tetrachloroethylene processes. It is also used for removing impurities from coal by density difference (3).

CHEMICAL HAZARDS

Mixtures of pentachloroethane with potassium or its alloys are shock-sensitive, and can explode violently on light impact (4). Such mixtures may explode after a short delay during which a voluminous solid separates out (5). Pentachloroethane is flammable when exposed to heat or flame. Its chemical hazards have been reviewed (6).

BIOLOGICAL HAZARDS

Pentachloroethane is a local eye and respiratory tract irritant. It has a strong narcotic effect and exposure may cause liver, kidney and lung damage. Its toxicity has been reviewed (7).

Vapour Inhalation

Pentachloroethane is a local irritant. Cats inhaling 1 mg/L (120 ppm) for 8-9 hours/day for 23 days showed no signs of toxicity, but liver, lung and kidney damage was found on autopsy (8). Long-term exposure of rabbits to 10 mg/m^3 was toxic (9).

Eye Contact

Pentachloroethane is a local irritant.

Skin Contact

Pentachloroethane may cause skin irritation.

Swallowing

The lethal oral dose in dogs is 1.75 g/kg (10).

CARCINOGENICITY

The IARC evaluation states that there is "limited evidence for the carcinogenicity of technical-grade pentachloroethane (containing hexachloroethane) to experimental animals" (11). Clear evidence of carcinogenicity was reported in a National Toxicology Programme bioassay in mice but not in rats when pentachloroethane was administered by gavage. The tumours produced were of the liver and kidney, but the role of the impurity hexachloroethane could not be evaluated (12,13).

MUTAGENICITY

Pentachloroethane gave positive results in *Saccharomyces cerevisiae* D7 with metabolic activation, and induced an increase in genetic effects in cells from the stationary growth phase with S9 (14). It was not mutagenic to *Salmonella typhimurium* TA1535, TA100, TA98 or TA1537 (15). Positive results in *S. typhimurium* have been reported in an abstract, but no experimental details were available (16). It induced sister chromatid exchanges, only without S9, but not chromosome aberrations in Chinese hamster ovary cells (17).

REPRODUCTIVE HAZARDS

No information is available concerning the reproductive effects of pentachloroethane.

FIRST AID

Eyes Wash the eye with flowing water for 10 minutes.

Lungs Remove casualty from area of exposure. If unconscious, do not give anything to drink. Give artificial ventilation

111. Pentachloroethane

and chest compression or place in the recovery position as necessary. If conscious make the casualty lie or sit down quietly, give oxygen if available. Lung congestion may occur – a conscious casualty with breathing difficulties should be placed in a sitting position. Shock may result – if so do not give any drinks, and if conscious, lie casualty flat with legs raised. Convulsions may occur and may cause unconsciousness.

Mouth Do not make the casualty vomit. Treat unconscious casualties as for lungs, but if conscious give 1 pint of water to drink. Convulsions may occur and may cause unconsciousness. Shock may result – do not give any drinks, and if conscious lie casualty flat with legs raised.

Skin Remove contaminated clothing immediately. Wash the affected area with soap and copious amounts of water. Absorption through the skin may cause symptoms similar to lung congestion.

In all cases of exposure, the patient should be transferred to hospital as soon as possible.

HANDLING AND STORAGE

Pentachloroethane should be handled wearing an approved respirator, chemical-resistant gloves, safety goggles and other protective clothing. It should only be used in a chemical fume hood (18). It should be kept in a tightly closed container, in a refrigerator and away from alkalies or reactive metals (1).

DISPOSAL

Eliminate all sources of ignition, ventilate the area and wear a laboratory coat or overalls, rubber gloves, approved compressed air breathing apparatus and safety boots. Absorb small spills onto paper towels, remove to a safe open air site and allow to evaporate in a metal tray. Wash spillage site with detergent. For large spills, absorb onto vermiculite-sodium carbonate mixture (90-10) or sand-soda ash (90-10) and mix carefully. EITHER transport in dry buckets to a safe open air area for atmospheric evaporation OR shovel into paper boxes and incinerate.

FIRE PRECAUTIONS

Fires involving pentachloroethane should be extinguished using water spray, carbon dioxide, or dry chemical powder.

FURTHER READING

Bretherick, L. *Hazards in the Chemical Laboratory* (4th edition)

Bretherick, L. *Handbook of Reactive Chemical Hazards* (4th edition)

Sax, N. Irving *Dangerous Properties of Industrial Materials* (7th edition)

Encyclopedia of Occupational Safety & Health

Patty's *Industrial Hygiene and Toxicology*

Kirk-Othmer *Encyclopedia of Chemical Technology*

National Fire Protection Association *Manual of Hazardous Reactions*

ACGIH *Documentation of TLVs and BEIs* (6th edition, 1986)

REFERENCES

1. Kim, J. J.; Lee, W. Y. *Kongdae Yon'gu Pogo (Soul Taehakkyo)* 1977, **9**(2), 27-31
2. Lires, O. A.; Molinari, M. A. *J. Chem. Educ.* 1973, **50**(7), 492
3. Keith, L. H.; Walters, D. B. (editors) *Compendium of Safety Data Sheets for Research and Industrial Chemicals* (VCH, Deerfield Park, 1987)
4. Leleu, M. J. *Cah. Notes Doc.* 1975, **80**, 389-392
5. Lenze, F.; et al. *Chem. Ztg.* 1932, **56**, 921
6. Leleu, M. J. *Cah. Notes Doc.* 1979, **97**, 603-634
7. Parker, J. C.; et al. *Am. Ind. Hyg. Assoc. J.* 1979, **40**(3), A-46,A-48,A-50,A-52-A-60
8. Lehmann, H. B.; Flurry, F. *Toxicology and Hygiene of Industrial Solvents* (Williams & Wilkins, Baltimore, 1943)
9. Navrotskii, V. K.; et al. *Tr. S'ezda Gig. Ukr. SSR, 8th* 1970 (pub. 1971), 224-226
10. Barsoum, G. S.; Saad, K. *J. Pharm. Pharmacol.* 1934, **7**, 205
11. IARC *Monographs on the evaluation of the carcinogenic risk of chemicals to humans* 1986, **41**, 99

111. Pentachloroethane

12. Mennear, J. H.; et al. *Fundam. Appl. Toxicol.* 1982, **2**, 82-87
13. *Carcinogenesis Bioassay of Pentachloroethane (CAS No. 76-01-7) in F344/N Rats and B6C3F$_1$ Mice (Gavage Study)* (Technical Report Series No. 232) (NTP, 1983)
14. Bronzetti, G.; et al. *Mutat. Res.* 1990, **234**(6), 429-430
15. Haworth, S.; et al. *Environ. Mutagen.* 1983, **5**(Suppl. 3), 3-142
16. Douglas, G. R.; et al. *Environ. Mutagen.* 1985, **7**(Suppl. 3), 31
17. Galloway, S. M.; et al. *Environ. Mol. Mutagen.* 1987, **10**(Suppl. 10), 1-175
18. Lenga, R. E. *The Sigma-Aldrich Library of Chemical Safety Data* (2nd edition) (Sigma-Aldrich, Milwaukee, 1988)

PENTA-CHLOROPHENOL

RISKS
Toxic by inhalation, in contact with skin and if swallowed (R23/24/25)

SAFETY PRECAUTIONS
After contact with skin, wash immediately with plenty of water – Wear suitable protective clothing and eye/face protection – If you feel unwell, seek medical advice (show label where possible) (S28, S36/39, S44)

IDENTIFIERS

SYNONYMS Chem-tol; Chlorophen; Cryptogil OL; Dowcide 7; Dowcide EC-7; Dowcide G; Dow Pentachlorophenol DP-2 Antimicrobial; Durotox; Glazel Renta; Lauxtol A; NCI- C54933; NCI-C55378; NCI-C56655; Penta; pentachlorophenate; 2,3,4,5,6 – pentachlorophenol; pentachlorophenol, Dowcide EC-7; pentachlorophenol, DP-2; pentachlorophenol, Technical; Pentacon; Penta-Kil-Pentasol; Penwar; Peratox; Permacide; Permagard; Permatox DP-2; Permatox Penta; Permite; Priltox; RCRA Waste No. U242; Santobrite; Sinituho; Term-I-Trol; Thompson's Wood Fix; Weedone

CHEMICAL ABSTRACTS No.	87-86-5
NIOSH No.	SM 6300000
HAZCHEM CODE	2X
UN No.	2020

THRESHOLD LIMIT VALUES

US TLV (TWA)	0.5 mg/m^3
US TLV (STEL)	not available

UK EXPOSURE LIMITS (OES)
Long-term (8 hr TWA value)	0.5 mg/m^3
Short-term (10 min TWA value)	1.5 mg/m^3

Germany
MAK	not available

France
VME	0.5 mg/m^3
VLE	not available

Sweden
Short-term limit	1.5 mg/m^3
Level limit	0.5 mg/m^3

PHYSICAL PROPERTIES

Description Colourless to yellow crystalline solid with a phenolic (carbolic) odour. Technical grade is dark grey to brown.

Boiling point	decomposes at 310°C
Melting point	191°C
Density	1.978 at 20°C
Vapour density	not available
Vapour pressure	40 mm Hg at 211.2°C
Flash point	not available
Explosive limits	not available
Autoignition temperature	not available

Solubility Soluble in ether or benzene. Very soluble in alcohol. Insoluble in water. Slightly soluble in cold petroleum ether.

PACKAGING AND TRANSPORTATION

Road transportation
hazard warning sign	2020 harmful substance
Hazchem code	2X

Sea transportation
IMDG page No.	6107
class	6.1
label	harmful stow away from foodstuffs; marine pollutant
packaging group	III

Air transportation
ICAO/IATA code (UN No.)	2020
class	6.1
label	keep away from food
packaging group	III
packing instructions	
cargo	619
passenger	619
passenger aircraft max. quantity	100 kilograms
cargo aircraft max. quantity	200 kilograms

112. Pentachlorophenol

MANUFACTURE

Pentachlorophenol is manufactured by chlorination of phenol at 100-180°C in the presence of a catalyst. Aluminium trichloride is the most commonly used catalyst, although activated carbon and quinoline, tellurium and some of its salts have also been used. It is also manufactured by alkaline hydrolysis of hexachlorobenzene (the Dow process).

USES

Pentachlorophenol is used as a herbicide, fungicide, molluscicide and wood preservative. The technical grade contains 4-12% tetrachlorophenols (pesticides) and is used in cooling towers, construction materials, photographic solutions, textiles and in the oil industry.

CHEMICAL HAZARDS

Pentachlorophenol is incompatible with strong oxidising agents, strong bases, acid chlorides and acid anhydrides. It emits toxic fumes of hydrogen chloride when heated to decomposition (1).

BIOLOGICAL HAZARDS

Pentachlorophenol acts on mitochondrial electron transport, uncoupling oxidative phosphorylation thus increasing metabolic rate and heat generation. Acute poisoning causes changes in breathing, blood pressure and urinary output, a high temperature, weakness, headache, nausea, vomiting and abdominal pain. In severe cases there is a rapidly progressive coma. Death following occupational exposure, with cerebral oedema and fatty degeneration of the viscera, has been reported (2). Many other cases of fatal poisoning have been reported (3). Immunological changes (4) and aplastic anaemia (5) have also been reported in workers exposed to pentachlorophenol. Survivors of poisonings may suffer from impaired sight, circulation and autonomic function (6). The occupational health hazards of pentachlorophenol exposure have been reviewed (7,8). Impurities in commercial pentachlorophenol include chlorinated dibenzodioxin and dibenzofuran, some of which are highly toxic. In animal studies, injury found at postmortem is commonly extensive vascular system damage, with high temperature and heart failure the cause of death. People with liver or kidney diseases should not be exposed to pentachlorophenol.

Vapour Inhalation

The dust causes painful irritation at concentrations above 1 mg/m^3, although levels as low as 0.3 mg/m^3 are irritating to the nose and throat. Hardened workers can tolerate up to 2.4 mg/m^3.

Eye Contact

Levels of the dust as low as 0.3 mg/m^3 are irritating to the eyes. It may cause conjunctivitis (9) and permanent corneal injury (10).

Skin Contact

Prolonged or repeated contact with the dust or solutions can cause dermatitis, burns and systemic toxicity. It is well absorbed through the skin, this being a major route of exposure in manufacturing and wood treatment. Skin absorption may be fatal. Workers with documented episodes of direct skin contact with pentachlorophenol had significantly higher risk of chloracne (11). Chloracne is a major effect of exposure to TCDD, an impurity in commercial chlorophenol preparations. In the rabbit ear bioassay, purified pentachlorophenol did not cause chloracne, and signs of liver damage were reduced when the purified material was used (9). The dermal LD_{50} in rats is 96 mg/kg.

Swallowing

The LD_{Lo} in humans has been reported as 29 mg/kg. Chronic exposure to technical grade pentachlorophenol causes liver damage in animal feeding studies.

CARCINOGENICITY

The IARC evaluation is that there is "limited evidence for the carcinogenicity of occupational exposure to chlorophenols to humans" (12). There was inadequate evidence to evaluate the carcinogenicity or genetic activity (in short-term tests) of pentachlorophenol (13). There was no evidence of carcinogenicity following chronic oral administration to mice, but subcutaneous administration to mice caused increased incidence of hepatoma. It should be noted that 2 cases of the relatively rare non-Hodgkin's lymphoma were reported in pentachlorophenol manufacture workers. They had also suffered from chloracne, although no causal relationship with pentachlorophenol has been proved (9). Following a literature review, the EPA Carcinogen Assessment Group concluded that the data on oncogenic effects of pentachlorophenol were negative, although major criticisms were levelled at the studies' experimental design (14). More recently the EPA concluded technical grade pentachlorophenol may pose a carcinogenic risk (15).

112. Pentachlorophenol

MUTAGENICITY

Pentachlorophenol induced a very slight increase in sister chromatid exchanges without S9, and chromosome aberrations with S9 in Chinese hamster ovary cells (16). No effects on sister chromatid exchange or chromosomal aberration were found *in vivo* or *in vitro* in studies of pentachlorophenol manufacturing workers occupationally exposed to 1.2-180 microg/m^3 (17). Another study found increased frequency of chromosome aberrations but not sister chromatid exchanges in lymphocytes of workers in a pentachlorophenol production plant (18). Reviews have concluded that data from animal tests suggest it is not mutagenic; results in *Drosophila* and rodent tests were negative, with some contradictory data in Ames tests and a positive result in yeast (19).

REPRODUCTIVE HAZARDS

Pentachlorophenol is foetotoxic but not overtly teratogenic in rats (at doses above 5 mg/kg) (9). Occupational exposure hazards to the male reproductive system, including pentachlorophenol, have been reviewed (20). The EPA concluded that pentachlorophenol (and possibly its HCDD contaminant) are foetotoxic and teratogenic in experimental animals, causing increased resorbtions, skeletal anomalies and subcutaneous oedema (14).

FIRST AID

Eyes Wash the eye with flowing water for 10 minutes.

Lungs Remove casualty from area of exposure. If unconscious, do not give anything to drink. Give artificial ventilation and chest compression or place in the recovery position as necessary. If conscious make the casualty lie or sit down quietly, give oxygen if available. Lung congestion may occur – a conscious casualty with breathing difficulties should be given oxygen. Convulsions may occur and may cause unconsciousness. Shock may result – do not give any drinks, if conscious lie casualty flat with legs raised. Further artificial ventilation may be needed.

Mouth Do not make the casualty vomit. Treat unconscious casualties as for lungs but if conscious give 1 pint of water to drink.

Skin Remove all contaminated clothing immediately, wash affected areas with soap and copious amounts of water. Absorption through the skin may cause symptoms similar to those of inhalation. Use ice-cold towels to cool the body.

In all cases of exposure, the patient should be transferred to hospital as soon as possible.

HANDLING AND STORAGE

Pentachlorophenol should be handled wearing an approved respirator, neoprene gloves (21), safety goggles, rubber boots, apron and other protective clothing. It should only be used in a chemical fume hood (1). It should be kept in a refrigerator, or in a cool, dry, well-ventilated place (20), away from strong oxidants (22) and protected against physical damage. Outside or detached storage is preferable. A high standard of hygiene is necessary.

DISPOSAL

Eliminate all sources of ignition, ventilate the area and wear a laboratory coat or overalls, rubber gloves, approved compressed air breathing apparatus and safety boots. Absorb small spills onto paper towels, remove to a safe open air site and allow to evaporate in a metal tray. Wash spillage site with detergent. For large spills, absorb onto vermiculite-sodium carbonate mixture (90-10) or sand-soda ash (90-10) and mix carefully. EITHER transport in dry buckets to a safe open air area for atmospheric evaporation OR shovel into paper boxes and incinerate.

FIRE PRECAUTIONS

Fires involving pentachlorophenol should be extinguished using water spray, carbon dioxide, dry chemical powder, alcohol foam or polymer foam. Use water to cool fire-exposed containers.

FURTHER READING

Bretherick, L. *Hazards in the Chemical Laboratory* (4th edition)

Bretherick, L. *Handbook of Reactive Chemical Hazards* (4th edition)

Sax, N. Irving *Dangerous Properties of Industrial Materials* (7th edition)

Encyclopedia of Occupational Safety & Health

Patty's *Industrial Hygiene and Toxicology*

Kirk-Othmer *Encyclopedia of Chemical Technology*

112. Pentachlorophenol

National Fire Protection Association *Manual of Hazardous Reactions*

ACGIH Documentation of TLVs and BEIs (6th edition, 1986)

Dangerous Prop. Ind. Mater. Rep. 1983, **3**(4)

Dangerous Prop. Ind. Mater. Rep. 1988, **8**(1), 2-7

Environmental Health Criteria 71. Pentachlorophenol (WHO, Geneva, 1987)

REFERENCES

1. Lenga, R. E. *The Sigma-Aldrich Library of Chemical Safety Data* (2nd edition) (Sigma-Aldrich, Milwaukee, 1988)
2. Gray, R. E.; et al. *Arch. Environ. Health* 1985, **40**(3), 161-164
3. Anon. *Calif. Health* 1970, **27**(12), 13
4. Zober, A.; et al. *Int. Arch. Occup. Environ. Health* 1981, **48**, 347-356
5. Roberts, H. J. *New Engl. J. Med.* 1981, **305**, 1650-1651
6. Imaizum, K. *Atsumi. v. Gankai* 1971, **3**(7), 717
7. Santodonato, J. *Monograph on human exposure to chemicals in the workplace, pentachlorophenol* (Syracuse Res. Corp., Springfield, 1987)
8. Sterling, T. D.; et al. *Int. J. Health Serv.* 1982, **12**, 559-571
9. *Toxicity Review 5. Pentachlorophenol* (HMSO, London, 1982)
10. Gosselin, R. E.; et al. *Clinical Toxicology of Commercial Products* (5th edition) (Williams & Wilkins, Baltimore, 1984)
11. O'Malley, M. A.; et al. *Am. J. Ind. Med.* 1990, **17**(4), 411-421
12. IARC *Monographs on the evaluation of the carcinogenic risk of chemicals to humans* 1986, **41**, 319
13. IARC *Monographs on the evaluation of the carcinogenic risk of chemicals to humans* 1982, **suppl. 4**, 205-206, 249-250
14. Cirelli, D. *Pentachlorophenol, position document 1. Fed. Reg.* 1978, **40**, 48446-48477
15. *Chlorinated phenols: criteria for environmental quality* (National Res. Council Canada, 1982) Publ. No. 18578 pp. 191
16. Galloway, S. M.; et al. *Environ. Mol. Mutagen.* 1987, **10**(suppl. 10), 1-175
17. Ziemsen, B.; et al. *Int. Arch. Occup. Environ. Health* 1987, **59**(4), 413-417
18. Schmid, E.; et al. *Mutat. Res.* 1983, **113**, 304-305
19. Teaf, C.; James, R. *Pentachlorophenol: A Toxicant Profile* (Cen. Biochem. Toxicol. Res., Florida State Univ., 1984)
20. Donn, S. S.; Dixon, R. L. *Ann. Rev. Pharmacol. Toxicol.* 1985, **25**, 567-592
21. Keith, L. H.; Walters, D. B. (editors) *Compendium of Safety Data Sheets for Research and Industrial Chemicals* (VCH, Deerfield Park, 1987)
22. *International Chemical Safety Cards: Pentachlorophenol* (CEC, Luxembourg, 1990)

p-PHENETIDINE

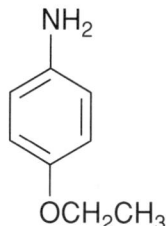

RISKS
Toxic by inhalation, in contact with skin and if swallowed – Danger of cumulative effects (R23/24/25, R33)

SAFETY PRECAUTIONS
After contact with skin, wash immediately with plenty of water – Wear protective clothing and gloves – In case of accident or if you feel unwell, seek medical advice immediately (show label where possible) (S28, S36/37, S45)

IDENTIFIERS

SYNONYMS benzenamine, 4-ethoxy-; p-aminophenetole; 4-aminophenetole; p-ethoxyaniline; 4-ethoxyaniline; 4-ethoxybenzenamine; p-phenetidin; 4-phenetidine

CHEMICAL ABSTRACTS No.	156-43-4
NIOSH No.	SI 6465500
HAZCHEM CODE	3X
UN No.	2311

THRESHOLD LIMIT VALUES

US TLV (TWA)	not available
US TLV (STEL)	not available
UK EXPOSURE LIMITS (OES)	
Long-term (8 hr TWA value)	not available
Short-term (10 min TWA value)	not available
Germany	
MAK	not available
France	
VME	not available
VLE	not available
Sweden	
Short-term limit	not available
Level limit	not available

PHYSICAL PROPERTIES

Description Colourless liquid, becomes red to brown on exposure to air and light.

Boiling point	253-255°C
Melting point	2-4°C
Density	1.0652 at 16°C
Vapour density	4.73
Vapour pressure	3.48 mm Hg at 25°C
Flash point	115°C
Explosive limits	not available
Autoignition temperature	not available

Solubility Practically insoluble in water, soluble in alcohol, acetone, dimethyl sulphoxide and ether.

PACKAGING AND TRANSPORTATION

Road transportation

hazard warning sign	2311 harmful substance
Hazchem code	3X

Sea transportation

IMDG page No.	6223
class	6.1
label	harmful stow away from foodstuffs
packaging group	III

Air transportation

ICAO/IATA code (UN No.)	2311
class	6.1
label	keep away from food
packaging group	III
packing instructions	
cargo	618
passenger	611
passenger aircraft max. quantity	60 litres
cargo aircraft max. quantity	220 litres

113. p-Phenetidine

MANUFACTURE

p-Phenetidine is prepared from nitrobenzene, ethanol, magnesium and sulphuric acid (1), or by reduction of p-nitrophenetole (2).

USES

p-Phenetidine is used in the manufacture of dyes, acetophenetidine, phencoll, synthetic sweetener (dulcin) and rubber antioxidant.

CHEMICAL HAZARDS

p-Phenetidine is air-and light-sensitive. It is incompatible with acids, acid chlorides, acid anhydrides, chloroformates and strong oxidising agents, and emits toxic fumes of carbon monoxide and nitrogen oxides when heated to decomposition (3).

BIOLOGICAL HAZARDS

p-Phenetidine is irritating to the eyes and skin, and may be absorbed through the skin. It is poisonous by inhalation, and moderately toxic intraperitoneally or if swallowed.

Vapour Inhalation

p-Phenetidine may cause upper respiratory tract irritation (3).

Eye Contact

p-Phenetidine may be irritating to the eyes (4).

Skin Contact

p-Phenetidine is a slight skin irritant in tests on rabbits (5). It may be absorbed through the skin, leading to the formation of methaemoglobin and cyanosis.

Swallowing

The oral LD_{50} in rats and mice is 580 mg/kg and 530 mg/kg respectively (6).

CARCINOGENICITY

No information is available concerning the carcinogenicity of p-phenetidine.

MUTAGENICITY

In *Salmonella typhimurium* TA100 p-phenetidine gave negative results without S9 but was mutagenic in the presence of S9 (7,8). Increased incidence of dominant lethal mutations has been reported in *Drosophila* (9).

REPRODUCTIVE HAZARDS

No information is available concerning the reproductive hazards of p-phenetidine.

FIRST AID

Eyes Wash the eye with flowing water for 10 minutes.

Lungs Remove casualty from area of exposure. If unconscious, do not give anything to drink. Give artificial ventilation and chest compression or place in the recovery position as necessary. If conscious make the casualty lie or sit down quietly, give oxygen if available. Convulsions may occur and may cause unconsciousness. Shock may result – if so do not give any drinks, and if conscious, lie casualty flat with legs raised.

Mouth Do not make the casualty vomit. Treat unconscious casualties as for lungs, but if conscious give 1 pint of water to drink.

Skin Remove contaminated clothing immediately, wash the affected area with soap and copious amounts of water. Absorption through the skin may cause symptoms similar to those of inhalation.

In all cases of exposure, the patient should be transferred to hospital as soon as possible.

113. *p*-Phenetidine

HANDLING AND STORAGE

p-Phenetidine should be handled wearing an approved respirator, chemical-resistant gloves, safety goggles and other protective clothing. Mechanical exhaust is required. It should be kept in a closed container, in a cool dry place (3).

DISPOSAL

Eliminate all sources of ignition and ventilate the area. Wearing a laboratory coat or overalls, safety glasses and gloves, absorb the spill onto paper towels and allow to evaporate in a fume-cupboard. For large spills, absorb onto sand or vermiculite, and remove in buckets for atmospheric evaporation in a safe open area. Ideally, waste should be burned in an incinerator with afterburner.

FIRE PRECAUTIONS

Fires involving *p*-phenetidine should be extinguished using water spray, carbon dioxide, dry chemical powder, alcohol foam or polymer foam.

FURTHER READING

Sax, N. Irving *Dangerous Properties of Industrial Materials* (7th edition)

Encyclopedia of Occupational Safety & Health

Patty's *Industrial Hygiene and Toxicology*

Kirk-Othmer *Encyclopedia of Chemical Technology*

National Fire Protection Association *Manual of Hazardous Reactions*

ACGIH *Documentation of TLVs and BEIs* (6th edition, 1986)

REFERENCES

1. Yukawa, Y. *J. Chem. Soc. Japan* 1950, **71**, 547
2. West, R. W. *J. Chem. Soc.* 1925, **127**, 494
3. Lenga, R. E. *The Sigma-Aldrich Library of Chemical Safety Data* (2nd edition) (Sigma-Aldrich, Milwaukee, 1988)
4. Guillot, J. P.; et al. *Food Chem. Toxicol.* 1982, **20**(5), 573-582
5. Guillot, J. P.; et al. *Food Chem. Toxicol.* 1982, **20**(5), 563-572
6. Vasilenko, N. M.; Zvezdai, V. I. *Gig. Tr. Prof. Zabol.* 1981, (8), 50-52
7. Nohmi, T.; et al. *Chem. Pharm. Bull.* 1985, **33**, 2877-2885
8. Thompson, C. Z.; et al. *Environ. Mutagen.* 1983, **5**(6), 803-811
9. Ilichkina, A. G.; et al. *Mol. Mekh. Genet. Protessov* 1976, 291-295

PHENOL

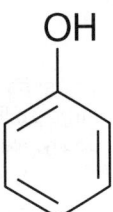

RISKS
Toxic in contact with skin and if swallowed – Causes burns (R24/25, R34)

SAFETY PRECAUTIONS
Keep out of reach of children – After contact with skin, wash immediately with plenty of water – If you feel unwell, seek medical advice (show label where possible) (S2, S28, S44)

IDENTIFIERS

SYNONYMS   Baker's P & S liquid ointment; carbolic acid; hydroxybenzene; monohydroxybenzene; NCI-C50124; oxybenzene; phenic acid; phenyl hydrate; phenyl hydroxide; phenylic acid; phenylic alcohol

CHEMICAL ABSTRACTS No.	108-95-2
NIOSH No.	SJ 3325000
HAZCHEM CODE	2X
UN No.	
solid	1671
molten	2312
aqueous solution	2821

THRESHOLD LIMIT VALUES

US TLV (TWA)	5 ppm (19 mg/m^3)
US TLV (STEL)	not available
UK EXPOSURE LIMITS (OES)	
Long-term (8 hr TWA value)	5 ppm (19 mg/m^3)
Short-term (10 min TWA value)	10 ppm (38 mg/m^3)
Germany	
MAK	5 ppm (19 mg/m^3)
France	
VME	5 ppm (19 mg/m^3)
VLE	not available
Sweden	
Short-term limit	1 ppm (4 mg/m^3)
Level limit	1 ppm (4 mg/m^3)

PHYSICAL PROPERTIES

Description   White cystalline mass or hygoscopic, translucent, needle-shaped crystals which turn pink or red if not pure, with a distinctive sweet, tarry odour, and burning taste.

Boiling point	181.9°C
Melting point	40.6°C
Density	1.072 at 20°C
Vapour density	3.24
Vapour pressure	0.35 mm Hg at 25°C
Flash point	80°C (closed cup)
Explosive limits	1.7%-8.6%
Autoignition temperature	715°C

Solubility   Soluble in water, alcohol, ether, chloroform, ethyl acetate, toluene, glycerol, and olive oil.

PACKAGING AND TRANSPORTATION

Road transportation

hazard warning sign	
solid	1671
molten	2312
aqueous solution	2821 toxic substance
Hazchem code	2X

Sea transportation

IMDG page No.	
solid	6225
molten	6224
solutions	6225
class	6.1
label	poison
packaging group	II

Air transportation

ICAO/IATA code (UN No.)	
solid	1671
molten	2312
aqueous solution	2821
class	6.1

114. Phenol

packaging group
solid II
solutions .... II, III

passenger aircraft max. quantity
solid .... 25 kilograms
molten .... forbidden
solutions .... II, 5 litres; III, 60 litres

cargo aircraft max. quantity
solid .... 100 kilograms
molten .... forbidden
solutions .... II, 60 litres; III, 220 litres

MANUFACTURE

The major methods for production of phenol are the cumene- hydroperoxide process, and the toluene-benzoic acid process. In the former, benzene is alkylated to cumene, which is then oxidised to cumene hydroperoxide. This is then cleaved to phenol and acetone. In the toluene-benzoic acid process, toluene is first oxidised to benzoic acid, which is then purified and oxydecarboxylated to phenol. Other methods for the manufacture of phenol include chlorination or sulphonation of benzene, and oxidation/dehydrogenation of cyclohexane.

USES

The most important uses of phenol are in the production of phenolic resins, bisphenol A and caprolactam. Phenol is also used in the production or manufacture of asbestos goods, bakelite, coke, drugs, explosives, fertilizers, illuminating gas, lampblack, paints, paint removers, perfumes, pharmaceutical preparations, rubber textiles, wood preservatives; and in the petroleum, leather, paper, soap, toy, tanning, dye and agricultural industries.

CHEMICAL HAZARDS

Phenol is a volatile combustible solid which gives off flammable vapours and carbon monoxide when heated. Mixtures of air and 3-10% phenol are explosive. Violent or explosive reactions occur with acetaldehyde (1); aluminium chloride plus nitrobenzene (2); aluminium chloride-nitromethane (3); butadiene (4); calcium hypochlorite (5); peroxomonosulphuric acid (6); peroxodisulphuric acid (7,8); sodium nitrite (9); and sodium nitrate-trifluoroacetic acid (10). There have been a number of runaway reactions with sudden pressure development and failure of bursting disks or reactors in the preparation of phenol-formaldehyde resins (11).

BIOLOGICAL HAZARDS

The biological hazards of phenol have been reviewed (12). Poisoning may result from inhalation, ingestion or skin absorption. Absorption is rapid and symptoms of toxicity may appear within minutes of exposure. Acute poisoning results in muscle weakness, convulsions and coma. Toxic doses act mainly through the higher centres of the central nervous system, resulting in sudden collapse. Chronic poisoning is characterised by digestive disturbances, including vomiting, difficulty in swallowing, salivation, diarrhoea and anorexia; by nervous disorders, with headache, fainting, vertigo and mental disturbances; and possibly by ochronosis (yellowing of the skin) and skin eruption. Damage to the liver and kidneys may occur. Phenol is readily conjugated and excreted in the urine. Phenol has been reported as a possible factor in cardiovascular disease (13).

Vapour Inhalation

Phenol is irritant to the upper respiratory tract, and is readily absorbed into the circulation from the lungs. A study of humans exposed to controlled conditions of phenol (1.5-5.2 ppm for 8 h with two 30 minute breaks) showed no adverse effects from inhalation or skin absorption, and urinary phenol returned to normal within 16 hours of exposure (14). A number of poisonings were reported in workers exposed to phenol at 2-3 ppm, through contaminated quenching water in a coke plant. It is unlikely, though that these poisonings were produced only by phenol; they were more likely produced by some other contaminant in the waste water (15). A case of phenol marasmus was reported in a laboratory technician who was exposed to the vapours and through spills on the skin for 13.5 years. On examination he was found to be emaciated, with an enlarged liver and altered liver function, and dark urine. Recovery was gradual after removal from exposure (16). Chronic exposures in rats (0.02-1 ppm for 2 months) produced changes in blood enzyme activity and time for excitation of extensor muscles and, at the higher exposures, decreases in body weight (17). In another study, various species (rats, mice, monkeys) exposed to 5 ppm, 8 h/day, 5 days/week showed excessive mortality (42%) after 28 days exposure. Rabbits exposed to the same concentrations for 88 days showed no external signs of toxicity, but pathological changes were noted in the lungs, liver and kidney. Rats appeared to show no signs of toxicity, internally or externally after 74 days exposure (19).

Eye Contact

The fumes are irritating to the eyes. Contact of phenol with the eyes causes severe damage, including conjunctival

114. Phenol

swelling, opacification and hypesthesia of the cornea and blindness (20). 5% phenol instilled into the eyes of rabbits produced severe damage, including corneal opacities. The duration of these opacities was reduced by washing the chemical out of the eye 30 seconds after installation (21).

Skin Contact

Phenol is irritating to, and readily absorbed through, the skin. It is absorbed with an efficiency equal to that of inhalation (22). It is a local anaesthetic, so that upon initial contact no pain is felt. By the time that pain is felt, serious burns and absorption through the skin may have occurred. Toxic or even fatal symptoms may occur from absorption though relatively small areas. A case of fatal skin absorption was reported in a technical assistant, who was not wearing protective clothing, and who was sprayed with liquid phenol over approximately 25% of his body while adjusting a faulty valve. A futile attempt was made to wash if off under a nearby tap, but he collapsed shortly afterwards and died 10 minutes after the exposure (23). A case of accidental death was reported in a person who died after being painted with benzyl benzoate as a scabicide with a brush that had been soaked in 80% phenol as an antiseptic, and which had not been thoroughly cleaned before use (24). A case of acute renal failure was reported in a man after accidental skin absorption of phenol (25). A case of dermatitis is reported in a sailor who spilt a small quantity of epoxy resin stripper containing 10% phenol onto his arm, causing a necrotic lesion which flared up again after 3 months and 7 months. This was thought to be due to the formation of an initial phenol-protein complex followed by gradual release of the phenol (26).

Swallowing

Phenol is readily absorbed from the gastrointestinal tract. After swallowing, intense burning of the mouth and throat is felt, with abdominal pain. The face is usually pale and sweaty, the pupils may be contracted or dilated; cyanosis is usually marked; the pulse is usually weak and slow, but occasionally it may be racing; respiration may initially be increased in rate, but later decreased in rate and magnitude; body temperature may fluctuate; occasionally isolated twitching of muscles or convulsions may be observed. Death is by respiratory failure. Doses as low as 1 g have resulted in death (27). The estimated LD_{LO} from reported oral toxicity data is 140 mg/kg (12). Ingestion is usually fatal. Chronic oral exposures were reported in a group of people exposed after an accidental spill caused ground water contamination. Estimated doses were 10-240 mg/day for approximately 1 month. The most significant symptoms were diarrhoea, mouth sores and dark urine. No long-term effects were noted after six months (28). Haematuria was reported to be probably due to exposure to phenol and cresol through skin contact in a group of Duo-Sol operators and maintenance workers (29).

CARCINOGENICITY

Phenol has not be shown to exhibit myeloclastogenicity like that of benzene (30). Papillomas and carcinomas have been observed in specially bred sensitive mice pretreated with DMBA, indicating a strong tumour promoting activity (31-32), whilst Swiss mice, pretreated with benzo(a)pyrene, demonstrated only a weak tumour promoting activity at best (33-34). In a recent chronic study by the National Cancer Institute in 1980, male and female rats and mice were administered 2500 or 5000 ppm phenol in their drinking water for 103 weeks. Statistically significant increases in pheochromocytomas, leukaemias and lymphomas were observed in the low dose male rats, but not the high dose group, nor in the females rats or in the mice of either sex. It was concluded that there was insufficient evidence to classify phenol as a carcinogen. Exposure to phenol in the wood- working industry has been associated with increased incidence of respiratory cancer (35). Atrophy of the papilla of the tongue has been observed in workers exposed to phenol and formaldehyde in the plastics industry, and it is suggested that the papilla may even undergo malignant transformation (36). No relationship could be found in an investigation into a possible link between exposure to phenol and cresol and bladder cancer (29).

MUTAGENICITY

Phenol has been reported to be mutagenic in the *E. coli*B/Sd-4 assay (0.1-0.2%phenol) (37), in *Salmonella* TA 1538 with metabolic activation (38), and non-mutagenic in *Salmonella* strains TA1535, TA1537, TA98 and TA100 (39). It requires metabolic activation to show genotoxicity in the *umu* test (40). It did not induce mutations in *Drosophila* or micronuclei in mouse bone marrow (38). It was mutagenic to the HGPRT locus of the V79 Chinese hamster fibroblast cell line (250-500µg/ml) (41). A significant increase in sister chromatid exchanges was observed in metabolically activated human lymphocytes (3 mM phenol) (42). Increases in sister chromatid exchanges were observed in human peripheral blood T-cell lymphocytes along with decreases in mitotic indices and inhibition of cell-cycle kinetics (43). Inhibition of DNA synthesis was produced in metabolically activated HeLa cells (2 mM phenol) (44). Ultrastructural changes were observed in HeLa cells administered ≥0.5% phenol for ≥10 mins (45).

REPRODUCTIVE HAZARDS

Increased incidence of preimplantation loss and early postnatal death has been reported in the offspring of rats exposed throughout pregnancy at concentrations of 1.3 ppm and 0.13 ppm (46). CD rats given phenol orally in water at doses of 0, 30, 60, or 120 mg/kg/day on days 6-15 of gestation exhibited dose related decreases in foetal body weight but no

114. Phenol

signs of structural malformations (47). Similar results were obtained in Sprague-Dawley rats given 20, 63, or 200 mg/kg/day intraperitoneally on days 9-11 or 12-14 gestation (48). Placental changes were recorded in women exposed to phenol as an air pollutant at 14 times the control limit. Children and pregnancies were otherwise normal (49).

FIRST AID

Eyes Wash the eye with flowing water for 10 minutes. Do NOT use Macrogol 300 in the eye.

Lungs Remove casualty from area of exposure. If unconscious, do not give anything to drink, give artificial ventilation and chest compression or place in the recovery position as necessary. If conscious make the casualty lie or sit down quietly, give oxygen if available. Lung congestion may occur – a conscious casualty with breathing difficulties should be placed in a sitting position. Convulsions may occur and may cause unconsciousness. Shock may result – do not give any drinks, and if conscious lie casualty flat with legs raised.

Mouth Do not make the casualty vomit. Treat unconscious casualties as for lungs but if conscious give 1 pint of water to drink immediately. Convulsions may occur and may cause unconsciousness.

Skin Wearing protective gloves, remove contaminated clothing immediately, flush excess chemical off the skin with water, then wash with polyethylene glycol molecular weight 300 (Macrogol 300) for at least 30 minutes. First aid treatment of patients exposed to phenol is discussed in an article (50).

In all cases of exposure, the patient should be transferred to hospital as soon as possible.

HANDLING AND STORAGE

Protective clothing should be appropriate to the amount of phenol handled. Phenol should be handled wearing an approved respirator; viton, butyl rubber or neoprene gloves (do not use nitrile or PVA gloves), safety goggles and other protective clothing. Safety showers which provide a flood of water should be installed where phenol is to be used. Polyethylene glycol 300 should be kept wherever phenol is used. Avoid contact with the skin. Wash thoroughly after use. It should be kept in a tightly closed container, in a cool, dry place, away from heat, flame and oxidising agents. It is light-sensitive, and should be kept in the dark.

DISPOSAL

Eliminate all sources of ignition and ventilate the area. Wearing a laboratory coat or overalls, safety glasses, gloves and self-contained breathing apparatus [if lacrimatory], absorb the spill onto paper towels and allow to evaporate in a fume-cupboard. For large spills, absorb onto sand or vermiculite, and remove in buckets for atmospheric evaporation in a safe open area. Ideally, waste should be burned in an incinerator with afterburner.

FIRE PRECAUTIONS

Fires involving phenol should be extinguished using water spray, carbon dioxide, dry chemical powder, alcohol foam or polymer foam. Water sprays may be used to cool exposed containers. Flammable/explosive vapours may be produced when heated.

FURTHER READING

Bretherick, L. *Hazards in the Chemical Laboratory* (4th edition)

Bretherick, L. *Handbook of Reactive Chemical Hazards* (4th edition)

Sax, N. Irving *Dangerous Properties of Industrial Materials* (7th edition)

Encyclopedia of Occupational Safety & Health

Patty's *Industrial Hygiene and Toxicology*

Kirk-Othmer *Encyclopedia of Chemical Technology*

National Fire Protection Association *Manual of Hazardous Reactions*

ACGIH Documentation of TLVs and BEIs (6th edition, 1986)

REFERENCES

1. *Chem. Saf. Data Sheet* SD-43, 1952
2. *Chem. Eng. News* 1953, **31**, 4915
3. Webb, H. F. *Chem. Eng. News* 1977, **55**(12),4

114. Phenol

4. *MCA Case History* 790, 1962
5. Fawcett, H. *Ind. Eng. Chem.* 1959, **51**, 89A-90A
6. Sidgwick, N. V. *The Chemical Elements and Their Compounds* (Oxford University Press, 1950) pp. 939
7. D'Ans, J.; et al. *Ber.* 1910, **43**, 1880
8. *Z. Anorg. Chem.* 1911, **73**, 1911
9. Bretherick, L. *Handbook of Reactive Chemical Hazards* (Butterworths, 3rd Edition, 1985) pp. 1281
10. Spitzer, U. A.; et al. *J. Org. Chem.* 1974, **39**, 3936
11. Taylor, H. D.; et al. *Major Loss Prevention in Process Industries* (Symp. Ser. No. 34) pp. 46 (IChemE, Rugby, 1971)
12. Bruce, R. M.; et al. *Toxicol. Ind. Health* 1987, **3**(4), 535-568
13. Wilcosky, T. C.; Tyroler, H. A. *JOM. J. Occup. Med.* 1983, **25**(12), 879-885
14. Piowtrowski, J. K. *Br. J. Ind. Med.* 1971, **28**, 172-178
15. Petrov, V. I. *USSR Literature on Air Pollution and Related Occupational Diseases* 1963, pp. 219-221 (available in translated form from NTIS, Springfield, VA, USA 63:11570)
16. Merliss, R. R. JOM, *J. Occup. Med. 1972,* **14***(1), 55*
17. Mukhitov *USSR Literature on Air Pollution and Related Occupational Diseases* 1964, pp. 185-199 (available in translated form from NTIS, Springfield, VA, USA TT-64-11574)
18. Sandage, C. *ASD Technical Report 61-519* Wright Patterson Airforce Base, OH: US Air Force Systems Command, Aeronautical Services Division, 1961 (available from NTIS, Springfield, VA, USA PB85-185841/XAB)
19. Deichmann, W. B. *Arch. Biochem.* 1944, **3**, 345-355
20. Grant, W. M. *Toxicology of the eye* (2nd edition) (Charles C. Thomas, Springfield, 1974) pp. 809-811
21. Murphy, J. C.; et al. *Toxicology* 1982, **23**(4), 281-291
22. NIOSH: Criteria for a recommended standard. Occupational Exposure to Phenol (NIOSH) 76-196. (US Govt Printing Office, Washington, DC 1976) pp. 23-69
23. Griffiths, G. J. *Med. Sci. Law* 1973, **13**, 46
24. Lewin, J. F.; Cleary, W. T. *Forensic Sci. Int.* 1982, **19**(2), 177-180
25. Foxall, P. J. D.; et al. *Human Toxicol.* 2989, **9**, 491-496
26. Schmidt, R.; Maibach, H. *Contact Dermatitis* 1981, **7**(4), 199-202
27. Encyclopedia of Occupational Health and Safety (ILO, Geneva, 1985)
28. Baker, E. L.; et al. *Arch. Environ. Health* 1978, **33**, 89-94
29. Hervin, R. L.; Froneburg, B. *Health Hazard Evaluation Report No. 80-020-1054* (Cities Service Company, Lake Charles, Louisiana)
30. Gad-El-Karim, M. M.; et al. *Am. J. Ind. Med.* 1985, **7**, 475-484
31. Boutwell, R. K.; Bosch, D. K. *Cancer Res.* 1959, **19**, 413-424
32. Salaman, M. H.; Glendenning, O. M. *Br. J. Cancer* 1957, **11**, 434-444
33. Van Duuren, B. L.; et al. *Towards a less harmful cigarette: a workshop* Sep 1967, pp. 173-180 (Wynder, E.; Hoffman, D. editors)
34. Van Duuren, B. L.; et al. *JNCI, J. Natl. Cancer Inst.* 1971, **46**, 1039-1044
35. Kauppinen, T. P.; et al. *Br. J. Ind. Med.* 1906, **43**(2), 84-90
36. Korycinska-Wronska, W.; et al. *Czas-Stomatol.* 1983, **36**(10), 729-736
37. Demerec, M.; et al. *Am. Nat.* 1951, **85**, 119-136
38. Gocke, E.; et al. *Mutat. Res.* 1981, **90**, 91-109
39. Haworth, S. T.; et al. *Environ. Mutagen.* 1983, **5**, (3-142)
40. Sakagami, Y.; et al. *Mutat. Res.* 1988, **209**(3-4), 155-160
41. Paschin, Yu. V.; Bahitova, L. M. *Mutat. Res.* 1982, **104**, 389-393
42. Morimoto, K.; et al. *Mutat. Res.* 1983, **119**, 355-360

43. Erexson. G. L.; et al. *Cancer Res.* 1985, **45**(6), 2471-2477
44. Painter, R. B.; Howard, R. *Mutat. Res.* 1982, **92**, 427-437
45. Nakagawa, F. *Shigaku* 1984, **71**(5), 721-734
46. Korshunov, S. F. *Industrial hygiene and condition of specific functions of working women in the petrochemical and chemical industry* R. A. Malysheva (Editor),(Sverdlovskii Nauchno-Issledovatel'skii Institut Okrhany Materinstva i Mladenchestva Mindrava RSFSR, Sverdlosk, USSR, 1974) pp. 149-153
47. Jones-Price, C.; et al. *Teratologic studies of phenol in CD-1 mice and in CD rats* Research Triangle Institute, Research Triangle Park, NC, USA (available from NTIS, Springfield, VA, USA, report nos. PB83-247726 and PB85-104461/XAB)
48. Minor, J. L.; Becker, B. A. *Toxicol. Appl. Pharmacol.* 1971, **19**, 373
49. Bonashevskaya, T. I. *Arkh. Anat., Gistol. Embriol.* 1985, **88**(2), 72-76
50. Mozingo, D. W.; et al. *J. Trauma*, 1988, **28**(5), 642-647

114. Phenol

m-PHENY-LENEDIAMINE

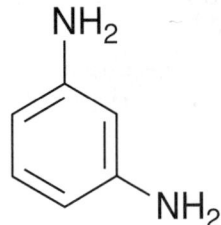

RISKS
Toxic by inhalation, in contact with skin and if swallowed – May cause sensitisation by skin contact (R23/24/25, R43)

SAFETY PRECAUTIONS
After contact with skin, wash immediately with plenty of water – If you feel unwell, seek medical advice (show label where possible) (S28, S44)

IDENTIFIERS

SYNONYMS *m*-aminoaniline; 3-aminoaniline; *m*-benzenediame; 1,3-benzenediamine; C.I. Developer 11; Developer C; Developer H; Developer M; *m*-diaminobenzene; 1, 3-diaminobenzene; Direct Brown BR; Direct Brown GG; 1,3-phenylenediaminte

CHEMICAL ABSTRACTS No.	108-45-2
NIOSH No.	SS 7700000
HAZCHEM CODE	2X
UN No.	1673

THRESHOLD LIMIT VALUES

US TLV (TWA)	0.1 mg/m^3
US TLV (STEL)	not available

UK EXPOSURE LIMITS (OES)
Long-term (8 hr TWA value) .... not available
Short-term (10 min TWA value) .... not available

Germany
MAK .... not available

France
VME .... not available
VLE .... not available

Sweden
Short-term limit .... 0.3 mg/m^3
Level limit .... 0.1 mg/m^3

PHYSICAL PROPERTIES

Description White crystals which become red on exposure to air.

Boiling point	286°C
Melting point	63°C
Density	1.139
Vapour density	3.7
Vapour pressure	1 mm Hg at 99.8°C
Flash point	187°C
Explosive limits	not available
Autoignition temperature	560°C

Solubility Soluble in water, methanol, ethanol, chloroform or acetone. Slightly soluble in ether or carbon tetrachloride. Very slightly soluble in benzene and toluene.

PACKAGING AND TRANSPORTATION

Road transportation

hazard warning sign	1673 harmful substance
Hazchem code	2X

Sea transportation

IMDG page No.	6227
class	6.1
label	harmful stow away from foodstuffs
packaging group	III

Air transportation

ICAO/IATA code (UN No.)	1673
class	6.1
label	keep away from food
packaging group	III

packing instructions
cargo 619
passenger 619

passenger aircraft max. quantity	100 kilograms
cargo aircraft max. quantity	200 kilograms

115. *m*-Phenylenediamine

MANUFACTURE

m-Phenylenediamine is produced by catalytic reduction of *m*-dinitrobenzene.

USES

m-Phenylenediamine is used in the synthesis of dyestuffs and dyestuff intermediates, and in the photographic industry. Its derivatives are used in polyurethane manufacture. It is also used in rubber chemicals, ion exchange resins, formaldehyde condensates, textile fibres, petroleum additives and corrosion inhibitors, and as a reagent for gold and bromine.

CHEMICAL HAZARDS

m-Phenylenediamine is incompatible with acids, acid chlorides, acid anhydrides, chloroformates, iron and strong oxidising agents (1,2). It is combustible on exposure to heat or flame, and emits toxic fumes of nitrogen oxides when heated to decomposition. It is air-and light-sensitive.

BIOLOGICAL HAZARDS

Occupational exposure to *m*-phenylenediamine has reportedly caused pathological changes in the kidneys and liver (3). Exposure may cause skin irritation, staining and sensitisation, asthma, headache, methaemoglobinaemia, and cyanosis (2).

Vapour Inhalation

m-Phenylenediamine may be irritating to the upper respiratory tract (2).

Eye Contact

m-Phenylenediamine is an eye irritant (2).

Skin Contact

Yellow skin staining by curing agent Z is considered to be due to *m*-phenylenediamine (4,5). It caused skin sensitisation but no local irritation in rats, rabbits and guinea pigs (6).

Swallowing

The oral LD_{50} in rats is 650 mg/kg (7).

CARCINOGENICITY

The IARC evaluation is that there is inadequate evidence to assess the carcinogenicity of *m*-phenylenediamine (8). It gave negative results in NCI carcinogenicity tests on animals (9). There was no evidence of carcinogenicity in long-term oral studies on rats and mice (10). Oral administration to mice (0.02-0.04% in drinking water for 78 weeks) induced no neoplastic changes other than deposition of brown pigment in some tissues (11). It was not carcinogenic when applied to the skin of mice (12).

MUTAGENICITY

m-Phenylenediamine is mutagenic in *Salmonella typhimurium* (13). It is mutagenic in *S. typhimurium* TA98 with metabolic activation (14,15). Significant mutagenic activity was found in urine of rats dermally treated with *m*-phenylenediamine only after metabolic activation (16). There was no evidence of a dominant lethal effect in rats following intraperitoneal administration of 20 mg/kg *m*-phenylenediamine 3 times weekly for 8 weeks (17). It induced weak dominant lethality in rats but on retesting negative results were obtained (18).

REPRODUCTIVE HAZARDS

No information is available concerning the reproductive hazards of *m*-phenylenediamine.

FIRST AID

Eyes Wash the eye with flowing water for 10 minutes.

Lungs Remove casualty from area of exposure. If unconscious, do not give anything to drink, give artificial ventilation and chest compression or place in the recovery position as necessary. If conscious make the casualty lie or sit down quietly, give oxygen if available. Lung congestion may occur – a conscious casualty with breathing difficulties should be placed in a sitting position. Convulsions may occur and may cause unconsciousness. Shock may result – do not give any drinks, and if conscious lie casualty flat with legs raised.

115. *m*-Phenylenediamine

Mouth Do not make the casualty vomit. Treat unconscious casualties as for lungs but if conscious give 1 pint of water to drink immediately. Convulsions may occur and may cause unconsciousness.

Skin Remove contaminated clothing immediately, drench the affected area with running water for at least 10 minutes.

In all cases of exposure, the patient should be transferred to hospital as soon as possible.

HANDLING AND STORAGE

m-Phenylenediamine should be handled wearing an approved respirator, neoprene gloves (19), safety goggles and other protective clothing. It should only be used in a chemical fume hood. It should be kept in a closed container, in a cool, dry place. It is light sensitive and should be stored and handled under nitrogen (1).

DISPOSAL

Eliminate all sources of ignition and ventilate the area. Wear a laboratory coat, safety spectacles, butyl rubber gloves, and approved self-contained breathing apparatus. While stirring constantly, add contaminated material to approximately 30 times its weight of a solution prepared by dissolving 1 part sodium sulphide in 6 parts water.

FIRE PRECAUTIONS

Fires involving *m*-phenylenediamine should be extinguished using water spray, carbon dioxide, dry chemical powder, alcohol foam or polymer foam.

FURTHER READING

Sax, N. Irving *Dangerous Properties of Industrial Materials* (7th edition)

Encyclopedia of Occupational Safety & Health

Patty's *Industrial Hygiene and Toxicology*

Kirk-Othmer *Encyclopedia of Chemical Technology*

National Fire Protection Association *Manual of Hazardous Reactions*

ACGIH Documentation of TLVs and BEIs (6th edition, 1986)

REFERENCES

1. Lenga, R. E. *The Sigma-Aldrich Library of Chemical Safety Data* (2nd edition) (Sigma-Aldrich, Milwaukee, 1988)
2. Keith, L. H.; Walters, D. B. (editors) *Compendium of Safety Data Sheets for Research and Industrial Chemicals* (VCH, Deerfield Park, 1987)
3. Kachalai, D. P.; et al. *Farmakol. Toksikol. (Kiev)* 1973, **8**, 180-183
4. Cohen, S. R.; Ross, P. M. *Arch. Dermatol.* 1988, **124**(1), 19-20
5. Barr, R.; et al. *Arch. Dermatol.* 1987, **123**(6), 713
6. Lopatneva, Zh. Ya.; et al. *Kauch. Rezina* 1978, (7), 25-27 Rusakov, N. V.; et al. *Organizm. Sreda, Mater. Nauch. Konf. Gig. Kafedr, 6th* 1970, **1**, 78-81
7. Burnett, C.; et al. *J. Toxicol. Environ. Health* 1977, **2**, 657
8. IARC *Monographs on the evaluation of the carcinogenic risk of chemicals to humans* 1978, **16**, 111
9. *Proc. of the House Subcommittee on Oversight and Investigation* (Washington, D.C., 19 July 1979)
10. Weisburger, E. K.; et al. *J. Environ. Pathol. Toxicol.* 1978, **2**(2), 325-356
11. Amo, H.; et al. *Food. Chem. Toxicol.* 1988, **26**(11/12), 893-897
12. Holland, J. M.; et al. *Cancer Res.* 1979, **39**(5), 1718-1725
13. Shahin, M.; et al. *Mutat. Res.* 1980, **78**, 25
14. Nomi, T.; et al. *Mutat. Res.* 1981, **91**(1), 37-39
15. Yoshikawa, K.; et al. *Eisei Shikensho Hokoku* 1976, (94), 28-32
16. Clemmensen, S.; Lam, H. R. *Mutat. Res.* 1984, **138**(2-3), 137-143
17. Burnett, C.; et al. *J. Toxicol. Environ. Health* 1977, **2**(3), 657-662

115. *m*-Phenylenediamine

18. Sheu, C. J. W.; Green, S. *Mutat. Res.* 1979, **68**, 85-98
19. *Occupational Safety and Health Data Sheet No. H5. Hand Protection* (Canada Safety Council, 1984)

116. p-Phenylenediamine

p-PHENYLENE-DIAMINE

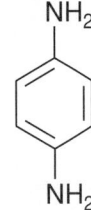

RISKS
Toxic by inhalation, in contact with skin and if swallowed – May cause sensitisation by skin contact (R23/24/25, R43)

SAFETY PRECAUTIONS
After contact with skin, wash immediately with plenty of water – If you feel unwell, seek medical advice (show label where possible) (S28, S44)

IDENTIFIERS

SYNONYMS 4-aminoaniline; *p*- aminoaniline; 1,4-benzenediamine; BASF Ursal D; *p*- benzenediamine; Benzofur D; C.I. 76060; C.I. Developer 13; C.I. Oxidation Base 10; Developer PF; *p*- diaminobenzene; 1,4-diaminobenzene; 1,4diaminobenzol; Durafur Black R; Fouramine D; Fourrine I; Fourrine L; Fur Black 41867; Fur Brown 41866; Furro D; Fur Yellow; Futramine D; Nako H; Orsin; Pelagol D; Pelagol DR; Pelagol Grey D; Peltol D; 1,4-phenylendeiamine; Renal PF; Tertrol D; Ursol D; Zoba Black D

CHEMICAL ABSTRACTS No.	106-50-3
NIOSH No.	SS 8050000
HAZCHEM CODE	2X
UN No.	1673

THRESHOLD LIMIT VALUES

US TLV (TWA)	0.1 mg/m^3
US TLV (STEL)	not available
UK EXPOSURE LIMITS (OES)	
Long-term (8 hr TWA value)	0.1 mg/m^3
Short-term (10 min TWA value)	not available
Germany	
MAK	0.1 mg/m^3
France	
VME	0.1 mg/m^3
VLE	not available
Sweden	
Short-term limit	not available
Level limit	not available

PHYSICAL PROPERTIES

Description White-slightly red crystals. Darken on exposure to air.

Boiling point	267°C
Melting point	146°C
Density	not available
Vapour density	3.72
Vapour pressure	1.08 mmHg at 100°C
Flash point	156°C
Explosive limits	not available
Autoignition temperature	not available

Solubility Soluble in alcohol, chloroform or ether.

PACKAGING AND TRANSPORTATION

Road transportation

hazard warning sign	1673 harmful substance
Hazchem code	2X

Sea transportation

IMDG page No.	6227
class	6.1
label	harmful stow away from foodstuffs
packaging group	III

Air transportation

ICAO/IATA code (UN No.)	1673
class	6.1
label	keep away from food
packaging group	III
packing instructions	
cargo	619
passenger	619
passenger aircraft max. quantity	100 kilograms
cargo aircraft max. quantity	200 kilograms

116. p-Phenylenediamine

MANUFACTURE

p-Phenylenediamine is produced by reduction of p-chloronitrobenzene.

USES

p-Phenylenediamine derivatives are widely used as antioxidants for oils and polymers, as antiozonants for rubbers, and for speciality polymers. It is also used in the dyeing of furs, the manufacture of azo dyes, in photochemical measurements, and photographic developing.

CHEMICAL HAZARDS

The finely powdered p-phenylenediamine is a significant dust explosion hazard. It is combustible on exposure to heat or flame, and can react vigorously with oxidising agents. It is incompatible with acids, acid chlorides, acid anhydrides, and chloroformates. It emits toxic acrid smoke and irritating fumes when heated to decomposition (1).

BIOLOGICAL HAZARDS

p-Phenylenediamine is more toxic and irritating than the o- or m-isomers. Systemic effects include gastritis, asthma, raised blood pressure, tremors, convulsions and coma.

Vapour Inhalation

p-Phenylenediamine has caused asthma and other respiratory symptoms in the fur dyeing industry.

Eye Contact

p-Phenylenediamine may cause eye irritation (1), protrusion of the eyeball, conjunctival inflammation and even permanent blindness.

Skin Contact

p-Phenylenediamine is mildly irritant to rabbits skin (2). Allergic contact dermatitis due to p-phenylenediamine has been reported from occupational exposure to rubber compounds (3), in the metallurgic industry (4), in beauticians (5) and in other industries (6). Its use as a hair dye has also caused vertigo, gastritis and anaemia, although the incidence of allergic sensitisation by p-phenylenediamine-containing hair dyes is infrequent and usually mild. It causes sensitisation in guinea pigs (7), and cross sensitisation to benzoquinone (8), and to o- and m-phenylenediamine has been reported in guinea pigs and rats (9).

Swallowing

2 cases of acute poisoning are reported following accidental ingestion of p-phenylenediamine; oedema, respiratory distress, kidney damage and death occurred in one, and skeletal muscle lesions were found on biopsy (10). The oral LD_{50} in rats is 98 mg/kg (2).

CARCINOGENICITY

The IARC evaluation is that there is inadequate evidence to assess the carcinogenicity of p-phenylenediamine (11). No evidence of carcinogenicity was found in rats fed up to 0.1% p-phenylenediamine in their diet for 80 weeks (12). It gave negative results in NCI carcinogenicity tests on animals (13). It gave positive results in RLV-infected rat embryo cells *in vitro* (21).

MUTAGENICITY

There was no evidence of a dominant lethal effect in rats following intraperitoneal administration of 20 mg/kg p-phenylenediamine 3 times weekly for 8 weeks (14). It is mutagenic in *Salmonella typhimurium* (15,16). It was weakly mutagenic in *Drosophila melanogaster* (17).

REPRODUCTIVE HAZARDS

Women occupationally exposed to p-phenylenediamine have been reported to have higher incidence of complications during pregnancy and childbirth (18). p-Phenylenediamine was not teratogenic in rats following administration of 5 to 30 mg/kg/day by gavage from day 6-15 of gestation (19). Repeated topical application hair dyes containing p-phenylenediamine did not have adverse reproductive effects nor increase the risk of cancer in rats (20).

FIRST AID

Eyes Wash the eye with flowing water for 10 minutes.

116. p-Phenylenediamine

Lungs Remove casualty from area of exposure. If unconscious, do not give anything to drink, give artificial ventilation and chest compression or place in the recovery position as necessary. If conscious make the casualty lie or sit down quietly, give oxygen if available. Lung congestion may occur – a conscious casualty with breathing difficulties should be placed in a sitting position. Convulsions may occur and may cause unconsciousness. Shock may result – do not give any drinks, and if conscious lie casualty flat with legs raised.

Mouth Do not make the casualty vomit. Treat unconscious casualties as for lungs but if conscious give 1 pint of water to drink immediately. Convulsions may occur and may cause unconsciousness.

Skin Remove contaminated clothing immediately, drench the affected area with running water for at least 10 minutes.

In all cases of exposure, the patient should be transferred to hospital as soon as possible.

HANDLING AND STORAGE

p-Phenylenediamine should be handled wearing an approved respirator, chemical-resistant gloves, safety goggles and other protective clothing. It should only be used in a chemical fume hood. It should be kept in a closed container, in a cool, dry place (1).

DISPOSAL

Eliminate all sources of ignition and ventilate the area. Wear a laboratory coat, safety spectacles, butyl rubber gloves, and approved self-contained breathing apparatus. While stirring constantly, add contaminated material to approximately 30 times its weight of a solution prepared by dissolving 1 part sodium sulphide in 6 parts water.

FIRE PRECAUTIONS

Fires involving p-phenylenediamine should be extinguished using water spray, carbon dioxide, or dry chemical powder.

FURTHER READING

Bretherick, L. *Hazards in the Chemical Laboratory* (4th edition)

Bretherick, L. *Handbook of Reactive Chemical Hazards* (4th edition)

Sax, N. Irving *Dangerous Properties of Industrial Materials* (7th edition)

Encyclopedia of Occupational Safety & Health

Patty's *Industrial Hygiene and Toxicology*

Kirk-Othmer *Encyclopedia of Chemical Technology*

National Fire Protection Association *Manual of Hazardous Reactions*

ACGIH *Documentation of TLVs and BEIs* (6th edition, 1986)

Gosselin, R. E.; et al. *Clinical Toxicology of Commerical Products* (5th edition) (Williams & Wilkins, Baltimore, 1984)

REFERENCES

1. Lenga, R. E. *The Sigma-Aldrich Library of Chemical Safety Data* (2nd edition) (Sigma-Aldrich, Milwaukee, 1988)
2. Lloyd, G. K.; et al. *Food Cosmet. Toxicol.* 1977, **15**(6), 607-610
3. Laubstein, H.; Moennich, H. I. *Dermatol. Monatsschr.* 1985, **17**(1), 2-13
4. Alomar, A.; et al. *Contact Dermatitis* 1985, **12**(3), 129-138
5. Edwards, E. K., Jr.; Edwards, E. K. *Cutis* 1984, **34**(1), 87-88
6. Fowler, J. F. *Contact Dermatitis* 1987, **16**(1), 38
7. Schaefer, U.; et al. *Arch. Dermatol. Res.* 1978, **261**(2), 153-161
8. Rajka, G.; Blohm, S. G. *Acta Dermatol. Venereol.* 1970, **50**(1), 51-54
9. Rusakov, N. V.; et al. *Organizm. Sreda, Mater. Nauch. Konf. Gig. Kafedr, 6th* 1970, **1**, 78-81
10. Baud, F. J.; et al. *J. Toxicol. Med.* 1984, **4**(3), 273-283
11. IARC *Monographs on the evaluation of the carcinogenic risk of chemicals to humans* 1978, **16**, 124
12. Imaida, K.; et al. *Toxicol. Lett.* 1983, **16**(3-4), 259-269
13. *Proc. of the House Subcommittee on Oversight and Investigation* (Washington, D.C., 19 July 1979)

14. Burnett, C.; et al. *J. Toxicol. Environ. Health* 1977, **2**(3), 657-662
15. Dunkel, V. C.; Simmon, V. F. *IARC Sci. Publ.* 1980, **27**(Mol. Cell. Aspects Carcinog. Screening Test), 283-302
16. Byeon, W. H.; et al. *Misaengmul Hakhoe Chi.* 1975, **13**(2), 51-58
17. Blijleven, W. G. H. *Mutat. Res.* 1977, **48**, 181-186
18. Guranda, S. V. *Ginekol., Mater. Sovmestnoi Konf.* 1972 (publ. 1973), 66-70
19. Re, T. A.; et al. *Fundam. Appl. Toxicol.* 1981, **1**(6), 421-425
20. Burnett, C. M.; Goldenthal, E. I. *Food Cosmet. Toxicol.* 1988, **26**(5), 467-474
21. Traul, K. A.; et al. *J. Appl. Toxicol* 1981, **1**(3), 190-195.

117. Phenylhydrazine

PHENYLHYDRAZINE

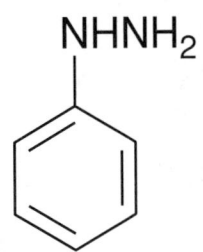

RISKS
Toxic by inhalation, in contact with skin and if swallowed – Irritating to eyes (R23/24/25, R36)

SAFETY PRECAUTIONS
After contact with skin, wash immediately with plenty of water – If you feel unwell, seek medical advice (show label where possible) (S28, S44)

IDENTIFIERS

SYNONYMS hydrazine-benzene; hydrazinobenzene; monophenylhydrazine; hydrazine, phenyl

CHEMICAL ABSTRACTS No.	100-63-0
NIOSH No.	MV 8925000
HAZCHEM CODE	3X
UN No.	2572

THRESHOLD LIMIT VALUES

US TLV (TWA)	0.1 ppm (0.44 mg/m^3)
US TLV (STEL)	not available

UK EXPOSURE LIMITS (OES)
Long-term (8 hr TWA value) .... 5 ppm (20 mg/m^3) under review
Short-term (10 min TWA value) .... 10 ppm (45 mg/m) under review

Germany
MAK .... 5 ppm (22 mg/m^3)

France
VME	not available
VLE	not available

Sweden
Short-term limit	not available
Level limit	not available

PHYSICAL PROPERTIES

Description	Monoclinic prisms or oil.
Boiling point	243.5°C
Melting point	19.6°C
Density	1.0978 at 20°C
Vapour density	3.7
Vapour pressure	1 mm Hg at 71.8°C
Flash point	88.9°C (closed cup)
Explosive limits	not available
Autoignition temperature	not available

Solubility Sparingly soluble in water. Miscible with alcohol or chloroform.

PACKAGING AND TRANSPORTATION

Road transportation

hazard warning sign	2572 toxic substance
Hazchem code	3X

Sea transportation

IMDG page No.	6227
class	6.1
label	poison
packaging group	II

Air transportation

ICAO/IATA code (UN No.)	2572
class	6.1
label	poison
packaging group	II
packing instructions	
cargo	611
passenger	609
passenger aircraft max. quantity	5 litres
cargo aircraft max. quantity	60 litres

117. Phenylhydrazine

MANUFACTURE

Phenylhydrazine is produced by diazotisation of aniline with sodium nitrite and hydrochloric acid, followed by treatment with sodium sulphite and then sodium hydroxide.

USES

Phenylhydrazine is used in the manufacture of dyes, pharmaceuticals and nitron (a stabiliser for explosives), and as a reagent for sugars, aldehydes and ketones. It is also used in the synthesis of indoles.

CHEMICAL HAZARDS

Phenylhydrazine reacts vigorously with lead(IV) oxide (1), and violently with 2-phenylamino-3-phenyloxazirane. Interaction with perchloryl fluoride, in the presence of diluent below 0°C, caused separation of explosive solids (2). It is air- and light-sensitive, and emits highly toxic fumes of nitrogen oxides when heated to decomposition.

BIOLOGICAL HAZARDS

Phenylhydrazine is a strong skin sensitiser, and its primary toxic effects are haemolytic anaemia (3,4), and liver and kidney damage. Positive results have been reported in some carcinogenicity studies.

Vapour Inhalation

The LC_{50} in mice is 2.12 mg/l (5).

Eye Contact

A 50% solution of phenylhydrazine caused septic conjunctivitis when applied to rabbits' eyes (5).

Skin Contact

Phenylhydrazine is a skin irritant and a strong skin sensitiser, causing eczematous dermatitis with redness, swelling and vesiculation on repeated exposure (6). Many cases of contact dermatitis have been reported from occupational exposure. It is readily absorbed from the skin, but fatalities have not been reported.

Swallowing

The oral LD_{50} in rats, mice, rabbits and guinea pigs is 188, 175, 80 and 80 mg/kg respectively (7).

CARCINOGENICITY

Positive (8), weakly positive (9) and negative results (10,11) have been reported in carcinogenicity studies in rodents.

MUTAGENICITY

Phenylhydrazine gave positive results in the Ames test, and induced a significant DNA fragmentation in the liver or lung of intraperitoneally treated male mice (12). Positive results are reported in the Ames test for *Salmonella typhimurium* TA1537 (13) and TA97 (14), and phenylhydrazine induced a direct DNA damage in *Escherichia coli* (14). Phenylhydrazine is highly toxic and its mutagenicity is detected at doses resulting in low survival of tester organisms (in this case TA1530) (15).

REPRODUCTIVE HAZARDS

No information is available concerning the reproductive hazards of phenylhydrazine.

FIRST AID

Eyes Wash the eye with flowing water for 10 minutes.

Lungs Remove casualty from area of exposure. If unconscious, do not give anything to drink. Give artificial ventilation and chest compression or place in the recovery position as necessary. If conscious make the casualty lie or sit down quietly, give oxygen if available. Lung congestion may occur – a conscious casualty with breathing difficulties should be placed in a sitting position. Convulsions may occur and may cause unconsciousness.

Mouth Do not make the casualty vomit. Treat unconscious casualties as for lungs but if conscious give 1 pint of water to drink.

Skin Remove contaminated clothing immediately, wash affected areas with soap and copious amounts of water. Absorption through skin may cause symptoms similar to those of inhalation.

In all cases of exposure, the patient should be transferred to hospital as soon as possible.First Aid.60

117. Phenylhydrazine

HANDLING AND STORAGE

Phenylhydrazine should be handled wearing an approved respirator, chemical-resistant gloves, safety goggles and other protective clothing. It should only be used in a chemical fume hood. It should be kept in a tightly closed container, away from heat and flame, in a cool dry place and under nitrogen (16).

DISPOSAL

Eliminate all sources of ignition and ventilate the area. Wear laboratory coat, gloves, approved self-contained breathing apparatus or canister respirator, and safety boots. Absorb small spills onto paper towels or vermiculite. Carefully scoop into a plastic bag and incinerate. Cover large spills with sand and shovel into enamel or polythene buckets with lids. Carefully add an excess of dilute hydrochloric acid or dilute 2M sulphuric acid and stir well. Allow to stand for 24 hours. Run the acid extract to waste with plenty of running water. Wash the sand with cold water and treat as normal refuse.

FIRE PRECAUTIONS

Fires involving phenylhydrazine should be extinguished using alcohol foam.

FURTHER READING

Bretherick, L. *Hazards in the Chemical Laboratory* (4th edition)

Bretherick, L. *Handbook of Reactive Chemical Hazards* (4th edition)

Keith, L. H.; Walters, D. B. (editors) *Compendium of Safety Data Sheets for Research and Industrial Chemicals* (VCH, Deerfield Park, 1987)

Sax, N. Irving *Dangerous Properties of Industrial Materials* (7th edition)

Encyclopaedia of Occupational Safety & Health

Patty's *Industrial Hygiene and Toxicology*

Kirk-Othmer *Encyclopedia of Chemical Technology*

National Fire Protection Association *Manual of Hazardous Reactions*

ACGIH *Documentation of TLVs and BEIs* (6th edition, 1986)

Toxicology card No. 109 Phenylhydrazine *Cah. Notes Doc.* 1991, **142**, 129-132

REFERENCES

1. Leleu, J. *Cah. Notes Doc.* 1976, **83**, 281-294
2. Scott, F. L.; et al. *Chem. & Ind. (London)* 1960, 258
3. Itano, H. A.; et al. *Br. J. Haematol.* 1976, **32**(1), 99-104
4. Ikuo, Y.; Yamada, S. *J. Toxicol. Sci.* 1977, **2**(3), 251-259
5. Pham, Q. C. *Gig. Tr. Prof. Zabol.* 1979, **3**, 45-47
6. von Oettingen, W. F. The aromatic amino and nitro compounds, their toxicity and potential dangers *U.S. Public Health Bull. No. 271* 1941
7. Ekshtat, B. Y. *Hyg. Sanit.* 1965, **30**, 191
8. Toth, B.; Shimizu, H. *Krebsforsch* 1976, **87**, 267
9. Clayson, D. B.; et al. *Proc. 3rd Perugia Quadrennial International Conf. on Cancer* (Univ. Perugia, Italy, 1966) p. 869
10. Roe, F. J. C.; et al. *Nature (London)* 1967, **216**, 375
11. Kelly, M. G.; et al. *JNCI, J. Natl. Cancer Inst.* 1969, **42**, 337
12. Parodi, S.; et al. *Cancer Res.* 1981, **41**, 1469-1482
13. DeFlora, S. *Carcinogenesis (London)* 1981, **2**(4), 283-298
14. DeFlora, S.; et al. *Mutat. Res.* 1984, **133**, 161-198
15. Tosk, J.; et al. *Mutat. Res.* 1979, **66**(3), 247-252
16. Lenga, R. E. *The Sigma-Aldrich Library of Chemical Safety Data* (2nd edition) (Sigma-Aldrich, Milwaukee, 1988)

PHENYL OXIRANE

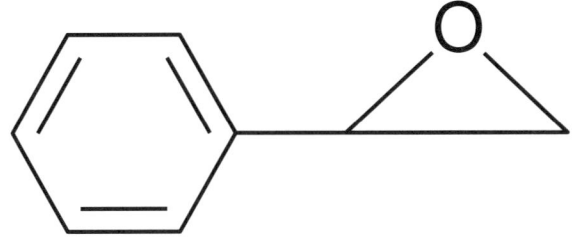

RISKS
May cause cancer – Harmful in contact with skin – Irritating to eyes (R45, R21, R36)

SAFETY PRECAUTIONS
Avoid exposure-obtain special instructions before use – If you feel unwell, seek medical advice (show label where possible) (S53, S44)

IDENTIFIERS

SYNONYMS benzene, (epoxyethyl)-; ethane, 1, 2-epoxy-1- phenyl-; (epoxysethyl)benzene; epoxystyrene; α,β- epoxystyrene; 1, 2-epoxyethylbenzene; oxirane, phenyl; phenethylene oxide; phenylethylene oxide; 2-phenyloxirane; styrene epoxide; styrene oxide; styrene 7,8-oxide; styryl oxide

CHEMICAL ABSTRACTS No.	96-09-3
NIOSH No.	CZ 9625000
HAZCHEM CODE	not available
UN No.	not available

THRESHOLD LIMIT VALUES

US TLV (TWA)	not available
US TLV (STEL)	not available

UK EXPOSURE LIMITS (OES)
Long-term (8 hr TWA value)	not available
Short-term (10 min TWA value)	not available

Germany
MAK	not available

France
VME	not available
VLE	not available

Sweden
Short-term limit	not available
Level limit	not available

PHYSICAL PROPERTIES

Description Colourless liquid.

Boiling point	194.2°C
Melting point	-36.6°C
Density	1.0469 at 25°C
Vapour density	4.14
Vapour pressure	0.3 mm Hg at 20°C
Flash point	74°C (open cup)
Explosive limits	not available
Autoignition temperature	498°C

Solubility Practically insoluble in water. Soluble in dimethyl sulphoxide, ethanol and acetone. Miscible in ether and benzene.

PACKAGING AND TRANSPORTATION

Road transportation
hazard warning sign	not available
Hazchem code	not available

Sea transportation
IMDG page No.	not available
class	not available
label	not available
packaging group	not available

Air transportation
ICAO/IATA code	not available
class	not available
label	not available
packaging group	not available
packing instructions	
cargo	not available
passenger	not available
passenger aircraft max. quantity	not available
cargo aircraft max. quantity	not available

118. Phenyl Oxirane

MANUFACTURE

Styrene oxide is produced commercially by epoxidation of styrene with peroxyacetic acid or treating styrene chlorohydrin with alkali (1,2).

USES

Styrene oxide is used as an intermediate in manufacture of styrene glycol and derivates. It is used as a reactive plasticiser or diluent in epoxy resins, and as a chemical intermediate in the production of β-phenethyl alcohol (a fragrance material), agrochemicals, and polyols (1,2). It is also used in textile and fibre treatment.

CHEMICAL HAZARDS

Styrene oxide is flammable when exposed to heat flame or oxidisers. It will polymerise exothermically or react vigorously with compounds with a labile hydrogen, such as water, in the presence of catalysts like acids, bases and certain salts. It is incompatible with acids, bases and oxidising agents (3). It emits acrid smoke when heated to decomposition.

BIOLOGICAL HAZARDS

Styrene oxide is an eye and skin irritant, and causes sensitisation. It may cause central nervous system depression and hepatic lesions. It causes stomach tumours on oral administration to experimental animals, and is mutagenic in several test systems. Reproductive effects have also been reported in animal studies. The metabolism of styrene (4) and biological monitoring of styrene oxide (5) have been discussed.

Vapour Inhalation

Inhalation is the main route of exposure, and it is well absorbed from the lungs. It is irritating to the respiratory tract, and may cause headache, fatigue, weakness, central nervous system depression, peripheral neuropathy and unsteady gait.

Eye Contact

Styrene oxide is an eye irritant. Concentrations above 1% caused corneal injury in rabbits. Serious burns or permanent blindness appear to be unlikely.

Skin Contact

Undiluted styrene oxide, or solutions as dilute as 1%, is a skin irritant and causes skin sensitisation after single or repeated exposure. Skin absorption is slow in humans. The dermal LD_{50} in rabbits is 930-1060 mg/kg (6).

Swallowing

The oral LD_{50} in rats is 4290 mg/kg (6).

CARCINOGENICITY

The IARC evaluation states that there is "sufficient evidence" for the carcinogenicity of styrene oxide in experimental animals (7). Significantly increased incidence of forestomach tumours have been reported in BDIV (8) and Sprague-Dawley rats (9) after oral administration. No increase in incidence of skin tumours has been reported in skin painting studies in mice (10).

MUTAGENICITY

Styrene oxide is mutagenic to bacteria, yeasts, insects and mammalian cells *in vitro*. It causes DNA damage in mammalian cells *in vivo* and *in vitro*, and chromosomal aberrations and sister chromatid exchanges *in vitro*. It induced chromosome aberrations and dose-dependent sister chromatid exchanges in PAH-stimulated human lymphocytes (11). It is mutagenic to *Salmonella typhimurium* TA1535 and TA100 without metabolic activation (12-14). Its genotoxicity has been reviewed (15,16).

REPRODUCTIVE HAZARDS

Reduced fecundity, foetal weight and length in rats and rabbits, and ossification defects in rats are reported following inhalation exposure to 15-100 ppm styrene oxide during gestation (17-19). Maternal mortality was high. Repeated administration to hens eggs causes reduced embryonic viability and malformations (20,21).

FIRST AID

Eyes Wash the eye with flowing water for 10 minutes.

118. Phenyl Oxirane

Lungs Remove casualty from area of exposure. If unconscious, do not give anything to drink, give artificial ventilation and chest compression or place in the recovery position as necessary. If conscious make the casualty lie or sit down quietly, give oxygen if available. Lung congestion may occur – a conscious casualty with breathing difficulties should be placed in a sitting position.

Mouth Do not make the casualty vomit. Treat unconscious casualties as for lungs but if conscious give 1 pint of water to drink immediately; give repeated drinks of water (1 cupful every 10 minutes).

Skin Remove contaminated clothing immediately, drench the affected area with running water for at least 10 minutes.

In all cases of exposure, the patient should be transferred to hospital as soon as possible.

HANDLING AND STORAGE

Styrene oxide should be handled wearing an approved respirator, chemical-resistant gloves, safety goggles and other protective clothing. It should only be used in a chemical fume hood (3). It should be kept in a tightly closed container, away from heat or flame.

DISPOSAL

Eliminate all sources of ignition and ventilate the area. Wearing a laboratory coat or overalls, safety glasses, gloves and self-contained breathing apparatus [if lacrimatory], absorb the spill onto paper towels and allow to evaporate in a fume-cupboard. For large spills, absorb onto sand or vermiculite, and remove in buckets for atmospheric evaporation in a safe open area. Ideally, waste should be burned in an incinerator with afterburner.

FIRE PRECAUTIONS

Fires involving styrene oxide should be extinguished using carbon dioxide, dry chemical powder, alcohol foam or polymer foam. Container explosion may occur under fire conditions (3).

FURTHER READING

Sax, N. Irving *Dangerous Properties of Industrial Materials* (7th edition)

Encyclopedia of Occupational Safety & Health

Patty's *Industrial Hygiene and Toxicology*

Kirk-Othmer *Encyclopedia of Chemical Technology*

National Fire Protection Association *Manual of Hazardous Reactions*

ACGIH *Documentation of TLVs and BEIs* (6th edition, 1986)

REFERENCES

1. Snyder, R. *Ethel Browning's Toxicity and Metabolism of Industrial Solvents volume 1: Hydrocarbons* (2nd edition) (Elsevier, New York, 1987)
2. IARC *Monographs on the evaluation of the carcinogenic risk of chemicals to humans* 1976, **11**, 201-208
3. Lenga, R. E. *The Sigma-Aldrich Library of Chemical Safety Data* (2nd edition) (Sigma-Aldrich, Milwaukee, 1988)
4. Loef, A. *Toxicokinetics of styrene. Biotransformation and covalent binding* (Natl. Board Occup. Saf. Health, Solna, Sweden, 1986)
5. Lof, A.; et al. *Scan. J. Work, Environ. Health* 1986, **12**(1), 70-74
6. Smyth, H. F., Jr.; et al. *Arch. Ind. Hyg. Occup. Med.* 1954, **10**, 61-68
7. IARC *Monographs on the evaluation of the carcinogenic risk of chemicals to humans* 1985, **36**, 245-263
8. Ponomarkov, V.; et al. *Cancer Lett.* 1984, **24**(1), 95-101
9. Maltoni, C.; et al. *Med. Lav.* 1979, **5**, 358
10. Van Duuren, B. L.; et al. *JNCI, J. Natl. Cancer Inst.* 1963, **31**, 41-45
11. Norppa, H.; et al. *Mutat. Res.* 1981, **91**(3), 243-250
12. Vainio, H.; et al. *Scand. J. Work, Environ. Health* 1976, **2**(3), 147-151
13. Busk, L. *Mutat. Res.* 1979, **67**(3), 201-208
14. DeFlora, S. *Carcinogenesis (London)* 1981, **2**(4), 283-298

118. Phenyl Oxirane

15. Barale, R. *Mutat. Res.* 1991, **257**(2), 107-126
16. Cooper, P. *Food Cosmet. Toxicol.* 1980, **18**(4), 434-435
17. Sikov, M. R.; et al. *DHHS (NIOSH) Publ. (U.S.)* 1981, 81-124
18. Sikov, M. R.; et al. *NTIS Report* 1980, order no. PB81-168510
19. Hardin, B. D.; et al. *Scand. J. Work, Environ. Health* 1981, **7**, 66-75
20. Vainio, H.; et al. *Teratology* 1977, **8**, 319-325
21. Kankaanpaa, J. T. J.; et al. *Acta Pharmacol. Toxicol.* 1979, **45**, 399-402

119. Phosgene

PHOSGENE

COCl$_2$

RISKS
Very toxic by inhalation (R26)

SAFETY PRECAUTIONS
Keep container tightly closed and in a well ventilated place – Avoid contact with skin and eyes – In case of accident or if you feel unwell, seek medical advice immediately (show label where possible) (S7/9, S24/25, S45)

IDENTIFIERS

SYNONYMS carbonic dichloride; carbon dichloride oxide; carbon oxychloride; carbonyl; chloride, carbonyl dichloride; CG; chloroformyl chloride; diphosgene; phosgen

CHEMICAL ABSTRACTS No.	75-44-5
NIOSH No.	SY 5600000
HAZCHEM CODE	2XE
UN No.	1076

THRESHOLD LIMIT VALUES

US TLV (TWA)	0.1 ppm (0.4 mg/m^3)
US TLV (STEL)	not available

UK EXPOSURE LIMITS (OES)
Long-term (8 hr TWA value) ..... 0.1 ppm (0.4 mg/m^3)
Short-term (10 min TWA value) .... not available

Germany
MAK .... 0.1 ppm (0.4 mg/m^3)

France
VME .... not available
VLE .... 0.1 ppm (0.4 mg/m^3)

Sweden
Ceiling limit .... 0.05 ppm (0.2 mg/m^3)
Level limit .... not available

PHYSICAL PROPERTIES

Description Colourless gas with an odour of new mown hay.

Boiling point	8.3°C
Melting point	-118°C
Density	1.37 at 20°C
Vapour density	3.4
Vapour pressure	1180 mm Hg at 20°C
Flash point	not available
Explosive limits	not available
Autoignition temperature	not available

Solubility Very slightly soluble in water. Very soluble in benzene, toluene, acetic acid and most liquid hydrocarbons. Decomposes slightly in water.

PACKAGING AND TRANSPORTATION

Road transportation
hazard warning sign	1076 toxic gas
Hazchem code	2XE

Sea transportation
IMDG page No.	2172
class	2;2.3
label	poison gas; corrosive
packaging group	not available

Air transportation
ICAO/IATA code (UN No.)	1076
class	2;6.1;8
label	not available
packaging group	not available

packing instructions
cargo .... forbidden
passenger .... forbidden

passenger aircraft max. quantity	forbidden
cargo aircraft max. quantity	forbidden

119. Phosgene

MANUFACTURE

Phosgene is prepared from carbon monoxide and chlorine or nitrosyl chloride; or from oleum and carbon tetrachloride. Most phosgene manufacture is within the plant in which it is to be used.

USES

Phosgene is used as an intermediate in the production of isocyanates, carbamates, organic carbonates and chloroformates. The polyurethane and polycarbonate industries are the largest users of phosgene. It is also used in metal-recovery operations. Following the fire at Basle, Sandoz have discontinued manufacture using phosgene (1). There has been debate as to whether the European Commission might legislate to ban phosgene (2).

CHEMICAL HAZARDS

Diphosgene may be used as a safer substitute to phosgene (3). Phosgene may react violently with hexafluororoisopropylideneaminolithium (4). Mixtures of phosgene and potassium are shock-sensitive explosives. As phosgene forms dangerously explosive carbonyl azide with sodium azide, all traces of phosgene should be removed before adding sodium azide in the preparation of *tert*-butyl azidoformate (5). Dangerous reactions may occur with aluminium, sodium, lithium, isopropyl alcohol, or 2,4-hexadiyn-1,6-diol. Phosgene decomposes to hydrochloric acid and carbon monoxide in the presence of moisture (water or steam). It also produces toxic and corrosive chloride fumes when heated to decomposition.

BIOLOGICAL HAZARDS

Occupational exposure to phosgene may result from its industrial use or from decomposition of chlorinated organic compounds in the presence of air or oxygen.

Vapour Inhalation

The odour threshold is 0.5-1 ppm, being higher after prolonged exposure. Phosgene has poor warning properties as there is little immediate respiratory tract irritation. Effects of exposure may be delayed for 2 to 24 hours, and include burning in the throat and chest, and dyspnoea. Nausea, vomiting and headache have also been reported following industrial exposure (6). Severe exposures cause coughing, tight chestedness, painful breathing and bloody spit, and pulmonary oedema may develop rapidly, death resulting in 36 hours. In less severe cases pneumonia may occur days later. It has been reported in animal studies that exposure to low concentrations produces tolerance against the acute effects of phosgene and other oedemagenic agents (7-9). Development of tolerance is, however, thought to be the trigger for chronic effects such as emphysema and fibrosis (10). Phosgene may cause chronic lung disease in humans (11). The LC_{50} in humans has been estimated as 400 ppm for a 2 minute exposure (12). The pathology, clinical course and therapy of phosgene poisoning have been reviewed (13).

Eye Contact

Phosgene is a severe eye irritant.

Skin Contact

Phosgene is a severe skin irritant.

Swallowing

Not applicable.

CARCINOGENICITY

No information is available concerning the carcinogenicity of phosgene.

MUTAGENICITY

No information is available concerning the mutagenicity of phosgene.

REPRODUCTIVE HAZARDS

No information is available concerning the reproductive hazards of phosgene.

FIRST AID

Eyes Wash the eye with flowing water for 10 minutes.

Lungs Remove casualty from area of exposure. If unconscious, do not give anything to drink. Give artificial ventilation and chest compression or place in the recovery position as necessary. If conscious make the casualty lie or sit down quietly, give oxygen if available. Lung congestion may occur – a conscious casualty with breathing difficulties should be placed in a sitting position. Shock may result – do not give any drinks, and if conscious, lie casualty flat with legs raised.

Mouth None.

Skin Remove contaminated clothing immediately and wash the affected areas with soap and copious amounts of water. Absorption through the skin may cause symptoms similar to inhalation.

In all cases of exposure, the patient should be transferred to hospital as soon as possible.

HANDLING AND STORAGE

Strict precautions are necessary in use and storage of phosgene. Phosgene is subject to the CIMAH Regulations (14). Care should be taken to prevent contamination with water which could lead to the build up of pressure. As wet phosgene is very corrosive, phosgene should not be stored with any quantity of water. It should be used or stored in outdoor installations or in a well ventilated noncombustible area. Paper soaked in alcoholic or carbon tetrachloride solution containing 10% of a mixture of equal parts of *p*-dimethylaminobenzaldehyde and diphenylamine and dried turns from yellow to deep orange in contact with levels of phosgene at around the TLV. There are several reports from industry on the safe handling of phosgene. In a dye manufacturers it is stored as a liquid in 2000 lb cylinders in a building 600 ft from the production area and 100 ft from all other buildings. Physical connections and valves are enclosed, and the enclosure and its caustic scrubber exhaust continuously monitored by an alarm system. It is piped to the reaction area in a double pipe, any leak being detected by continuously venting the annular pipe into a solution of Congo Red (15). Precautions taken by US phosgene producers have been described, and include 30-50 ft high-pressure ammonia-steam curtains to handle in-plant phosgene releases (16). A recommended procedure for working with highly toxic gases such as phosgene has been described (17). Plant design should include proper neutralisation facilities and water fog equipment.

DISPOSAL

Eliminate all sources of ignition and ventilate the area. Wearing a laboratory coat, approved breathing apparatus, heavy work gloves. Because of the resulting corrosion, water should not be used on the source of a phosgene leak. Where recycling is not possible, phosgene waste should be handled by caustic scrubbing in packed columns.

FIRE PRECAUTIONS

Phosgene containers should be cooled with water in case of fire, and water spray used to protect personnel attempting to shut-off a leak. Sodium hydroxide or anhydrous ammonia have been used to neutralise phosgene.

FURTHER READING

Bretherick, L. *Hazards in the Chemical Laboratory* (4th edition)

Bretherick, L. *Handbook of Reactive Chemical Hazards* (4th edition)

Sax, N. Irving *Dangerous Properties of Industrial Materials* (7th edition)

Encyclopedia of Occupational Safety & Health

Patty's *Industrial Hygiene and Toxicology*

Kirk-Othmer *Encyclopedia of Chemical Technology*

National Fire Protection Association *Manual of Hazardous Reactions*

ACGIH Documentation of TLVs and BEIs (6th edition, 1986)

Dangerous Prop. Ind. Mater. Rep. 1983, **3**(3)

Hazard Data Bank. Sheet No. 91: Phosgene *Saf. Pract.* 1987, **5**(7), 34-35

NT. Not. Tec. AMMA 1981, **36**(5), 6-9

Chemical Safety Data Sheet SD-95 (Manufacturing Chemists' Association)

REFERENCES

1. *Chem. Ind. (London)* 1987, **23**, 805
2. *Promosafe* 1988, **15**(2), 65
3. Anon. *Sichere Chemarb.* 1988, **40**(3), 34

119. Phosgene

4. Swindell, R. F. *Inorg. Chem.* 1972, **11**, 242
5. Yajima, H.; et al. *Chem. Pharm. Bull.* 1970, **18**, 850-851
6. Kuzelova, M.; et al. *Prac. Lek.* 1975, **27**(4), 115-117
7. Box, G. E. P.; Bullumbine, H. *Br. J. Pharm.* 1947, **2**, 38
8. Henschler, D.; Laux, W. *Arch. Exp. Pathol. u Pharmak.* 1960, **329**, 433
9. Stockinger, H. E. *Proc. 13th Int. Congr. Occup. Health* (New York, 1960)
10. Stockinger, H. E.; et al. *Arch. Ind. Health* 1957, **16**, 514
11. Caldston, M.; et al. *J. Clin. Invest.* 1947, **26**, 169-181
12. Chasis, H. *Phosgene, review of the literature on the effects of exposure to man and experimental animals* (Contract W-49-036-CWS-1, 1944)
13. Diller, W. F. *Pneumonologie* 1974, **150**(2-4), 139-148
14. *CIMAH (Amendment) Regulations 1988* (HMSO, London, 1988)
15. Alspach, J.; Bianchi, R. *J. Plant Oper. Prog.* 1984, **3**(1), 40-42
16. Watzman, A.; Bluestone, M. *Chem. Week* 1986, **138**(22), 34-36
17. Buxton, A. G. *Symp. Pap. – Inst. Chem. Eng. North West. Branch* 1980, (2), 4/1-4/3

120. Phosphorus, white (yellow)

PHOSPHORUS, WHITE (YELLOW)

P

RISKS
Spontaneously flammable in air – Very toxic by inhalation and if swallowed – Causes severe burns (R17, R26/28, R35)

SAFETY PRECAUTIONS
Keep contents under water – In case of contact with eyes, rinse immediately with plenty of water and seek medical advice – After contact with skin, wash immediately with plenty of water – In case of accident or if you feel unwell, seek medical advice immediately (show label where possible) (S5, S26, S28, S45)

IDENTIFIERS

SYNONYMS Exolite; Exolite 405; Exolit LPKN; Exolit VPK-n; phosphorus-31

CHEMICAL ABSTRACTS No.	7723-14-0
NIOSH No.	TH 3500000
HAZCHEM CODE	2WE
UN No.	1381

THRESHOLD LIMIT VALUES

US TLV (TWA)	0.02 ppm (0.1 mg/m^3)
US TLV (STEL)	not available

UK EXPOSURE LIMITS (OES)
Long-term (8 hr TWA value)	0.1 mg/m^3
Short-term (10 min TWA value)	0.3 mg/m^3

Germany
MAK	0.1 mg/m^3

France
VME	0.1 mg/m^3
VLE	0.3 mg/m^3

Sweden
Short-term limit	not available
Level limit	not available

PHYSICAL PROPERTIES

Description Pale yellow, waxy, translucent solid.

Boiling point	280°C
Melting point	44.1°C
Density	1.82
Vapour density	4.42
Vapour pressure	1 mm at 76.6°C
Flash point	spontaneously flammable in air
Explosive limits	not available
Autoignition temperature	30°C

Solubility Insoluble in water. Very soluble in carbon disulphide, ether, chloroform or benzene.

PACKAGING AND TRANSPORTATION

Road transportation
hazard warning sign	1381 flammable solid
Hazchem code	2WE

Sea transportation
IMDG page No.	4249
class	4.2
label	
. spontaneously combustible; poison; marine pollutant	
packaging group	I

Air transportation
ICAO/IATA code (UN No.)	1381
class	4.2;6.1
label	not available
packaging group	not available
packing instructions	
cargo	forbidden
passenger	forbidden
passenger aircraft max. quantity	forbidden
cargo aircraft max. quantity	forbidden

120. Phosphorus, white (yellow)

MANUFACTURE

Phosphorus does not occur free in nature but as phosphates in minerals such as chlorapatite and fluorapatite, in small amounts in granitic rocks and in fertile soils. Elemental phosphorus is produced as a byproduct or intermediate in the production of phosphate fertiliser. Phosphate rock containing the mineral apatite is heated, and elemental phosphorus is liberated as a vapour.

USES

White phosphorus is used in gas analysis, for smoke screens and in the manufacture of rat poisons and fertilisers. Phosphorus has been added to the list of substances identified in the Preliminary Assessment Information Rule requiring (as of 16 May 1988) manufacturers and importers to submit production volume, end use and exposure data to the US EPA (1).

CHEMICAL HAZARDS

Autoignition temperature is 30°C. White phosphorus reacts explosively with: chlorosulphuric acid (2); liquid bromine or chlorine; magnesium perchlorate (3); chlorates, bromates or iodates of barium, calcium, magnesium, sodium or zinc; peroxides of lead, potassium or sodium; molten chromium trioxide (at 200°C); ammonium nitrate, mercury(I) nitrate, silver nitrate; seleninyl chloride; potassium permanganate; chromyl chloride; azides of bromine, chlorine or iodine (4); dichlorine oxide; chlorine dioxide; nitrogen trichloride; nitrogen tribromide hexaammoniate; and mixtures of phosphorus and sulphur ignite and/or may explode on heating. White phosphorus ignites in air if finely divided or warmed and also ignites on contact with: fluorine, chlorine, bromine vapour or solid iodine; antimony pentachloride; molten ammonium nitrate; boron triiodide; sulphur trioxide vapour (with some delay) or liquid (immediately); boiling sulphuric acid or its vapour; chlorine dioxide; iodine trichloride; and phosphorus vapour heated in nitric acid in the presence of air may ignite. If white phosphorus is incompletely immersed while undergoing oxidation in hydrogen peroxide solutions, ignition may occur and lead to a violent reaction (5). Contact with boiling caustic alkalis or hot calcium hydroxide evolves phosphine which usually ignites in air. Violent reactions occur with: cerium, lanthanum, neodymium, praseodymium (above 400°C); powdered aluminium (6); and warm or molten phosphorus burns vigorously in nitrogen oxide, dinitrogen tetraoxide and dinitrogen pentaoxide. Phosphorus reacts incandescently with: cyanogen iodide; hexalithium disilicide; monorubium acetylide; monocaesium acetylide; metals such as osmium and, when heated, beryllium, copper, manganese, thorium and zirconium; bromine triflouride; and selenium.

BIOLOGICAL HAZARDS

White phosphorus is one of the most highly toxic inorganic substances, both the fumes and the element itself being poisonous. Absorption of toxic amounts from the lungs or gastrointestinal tract causes vomiting, weakness and has an acute effect on the liver. Necrosis of the jaw ('phossy jaw') is the most common symptom of chronic poisoning, and was observed in workers involved in manufacture of matches in 1845. Weakness, anaemia, loss of appetite, and changes in the long bones (which may become brittle and spontaneously fracture) also occur. Workers should have regular dental checks to detect adverse effects on the teeth which may occur.

Vapour Inhalation

Acute inhalation of white phosphorus vapours or smoke is highly irritating to the respiratory tract, human short term exposure (10 to 15 minutes) to over 400 mg/m^3 producing signs of irritation (7). Hepatic involvement has been reported in at least one case (8). Signs of chronic toxicity generally occur after many years of exposure and include gastrointestinal distress, a slight jaundice, lowered blood potassium levels or increased chloride concentration, leukopenia and anaemia, and 'phossy jaw' (9). Cases of the latter have been reported among workers in the chemical and fireworks industries (10-12). Hepatic and renal involvement are generally uncharacteristic (10). In animal experiments, inhalation of over 20 ppm of phosphorus vapours by rats (7 hours/day, 5 days/week) caused high mortality due to oedema of the lungs and bronchopneumonia (13). Bone changes have been reported after prolonged exposure of animals (14).

Eye Contact

White phosphorus is particularly hazardous to the eyes and can cause severe damage. External contact with the eyes can cause conjunctivitis with a yellow tint.

Skin Contact

White phosphorus can cause second and third degree burns on contact with the skin. The affected area usually turns greyish white and infection follows. Dermal application has resulted in liver and kidney damage in animal experiments (15).

120. Phosphorus, white (yellow)

Swallowing

Ingestion of white phosphorus causes gastrointestinal effects such as nausea, vomiting and diarrhoea, which may begin within 30 minutes, and vomit and faeces may have a characteristic garlic odour. Death from cardiovascular collapse may occur in about 12 hours. A period of apparent recovery may occur lasting about 2 days. The second stage is characterised by the return of gastrointestinal distress, and signs of hepatic, renal and cardiovascular problems such as jaundice, pitting oedema, high pulse rate and low blood pressure. The fatal dose in humans is about 1 mg/kg (16) and toxic symptoms may result from as little as 0.2 mg/kg (17,18). The most common pathological findings in deaths have been fatty degeneration of the liver and kidneys (19). Oral administration of 10 mg/kg to rats caused toxic hepatitis within 140 hours (20). In dogs, regular doses of 0.1 mg/kg/day resulted in chronic poisoning with liver damage (21), and hepatic degeneration leading to cirrhosis from chronic oral administration has been produced in experimental animals (22,23).

CARCINOGENICITY

No information is available concerning the carcinogenicity of white phosphorus.

MUTAGENICITY

No information is available concerning the mutagenicity of white phosphorus.

REPRODUCTIVE HAZARDS

Administration of elemental phosphorus in oil solutions at 0.0005 mg/kg to pregnant rats increased foetal mortality and altered embryonic development (24).

FIRST AID

Eyes   Wash the eye with flowing water for at least 15 minutes.

Lungs   Remove casualty from area of exposure. If unconscious, do not give anything to drink. Give artificial ventilation and chest compression or place in the recovery position as necessary. If conscious make the casualty lie or sit down quietly, give oxygen if available. Lung congestion may occur – a conscious casualty with breathing difficulties should be placed in a sitting position.

Mouth   Do not make the casualty vomit. Treat unconscious casualties as for lungs but if conscious give 1 pint of water to drink. Immediately then give repeated drinks of water (1 cupful every 10 minutes.) Convulsions may occur and may cause unconsciousness.

Skin   Wearing protective gloves remove contaminated clothing and place the burn under water immediately, then soak with 8.4% sodium bicarbonate. Wash with 2% copper sulphate and remove the black particles of phosphorus with a spatula or tweezers. Cover with a sterile dressing soaked in sodium bicarbonate.
First aid treatment of phosphorus burns have been described (25,26). Cases of necrosis following copper sulphate treatment of white phosphorus burns have been reported, and wound management and therapy described (27). The use of silver nitrate as an alternative to copper sulphate has been suggested (28).

In all cases of exposure, the patient should be transferred to hospital as soon as possible.

HANDLING AND STORAGE

White Phosphorus should be handled wearing a laboratory coat or chemical resistant overalls, safety glasses, face visor and rubber gloves; under conditions of good ventilation and away from sources of ignition. It should be kept under water or in an inert atmosphere, separate from other storage.

DISPOSAL

Eliminate all sources of ignition. Ventilate the area but isolate the material. Wear a laboratory coat or chemical resistant overalls, safety spectacles and face visor and rubber gloves. For large or small spills of yellow phosphorus, cover the spilt material with wet sand, spray with water to keep it wet and shovel into buckets. Remove to an isolated place away from buildings, dry out and burn. The phosphorus will ignite spontaneously in air. For large spills call the fire brigade.

FIRE PRECAUTIONS

Fires involving white phosphorus should be extinguished using water spray, taking care not scatter, until fire is extinguished and phosphorus has solidified, then cover with wet sand or dirt. Full flame retardant protective clothing should be worn.

120. Phosphorus, white (yellow)

FURTHER READING

Brethrick, L. *Handbook of Reactive Chemical Hazards* (4th edition)

Sax, N. Irving *Dangerous Properties of Industrial Materials* (7th edition)

Encyclopedia of Occupational Safety & Health (3rd edition, 1983)

Patty's *Industrial Hygiene and Toxicology* (3rd edition)

Kirk-Othmer *Encyclopedia of Chemical Technology* (3rd edition)

National Fire Protection Association *Manual of Hazardous Reactions*

ACGIH *Documentation of TLVs and BEIs* (6th edition, 1986)

Toxicological Index, CSST (1986)

Dangerous Prop. Ind. Mater. Rep. 1983, **3**(4)

REFERENCES

1. *TSCA Chem.-Prog. Bull.* June 1988, **9**(3), 9
2. Heumann, K.; et al. *Ber.* 1882, **15**, 417
3. *1965 Summary of Serious Accidents*, Washington, USAEC, 1966
4. Anon. *J. R. Inst. Chem.* 1957, **81**, 475
5. Denicke, K. *Angew. Chem. (Intern. Ed.)* 1967, **6**, 240
6. Matignon, C. *Comp. Prend.* 1900, **130**, 1393-1394
7. White, S. A.; Armstrong, C. C. Project A5.2-1 Chemical Warfare Service, Edgewood Arsenal, Md., 1935
8. Aisenshtadt, V. S.; et al. *Gig. Tr. Prof. Zabol.* 1971, **15**, 48
9. Hughes, J. P. W.; et al. *Br. J. Ind. Med.* 1962, **19**, 83
10. Heimann, J. *J. Ind. Hyg. Tox.* 1946, **28**, 142
11. Hughes, J. P. W.; et al. *Br. J. Ind. Med.* 1962, **19**, 83
12. Nomura, T. *J. Sci. Labour (Japan)* p. 109 (Feb 1956)
13. Tennessee Valley Authority Report: *Toxicological Studies of Phosphorus,* Part II, p. 28 (1953, unpublished)
14. Ibid.; *Phosphorus study (Excluding the Enviromental Phase)* (1950 unpublished)
15. Orcult, T. J.; Pruitt, B. A. *Major Probl. Clin. Surg.* 1976, **19**, 84
16. Smyth, H F., Jr. *Am. Ind. Hyg, Assoc. Q.* 1956, **17**, 129
17. Sollmann, T. *A Manual of Pharmacology (1939).* Referenced in Tennessee Valley Authority Report No. 13077, p. 1 (1943, unpublished)
19. Simon, F. A.; Pickering, L. K. *J. Am. Med. Assoc.* 1976, **235**, 1343
20. Strekyukhina, N. A.; Lukashev, A. A. *Gig. Tr. Prof. Zabol.* 1980, **10**, 63
21. Buchanan, D. J.; et al. *Arch. Ind. Hyg. Occup. Med* 1954, **9**, 1
22. Adams, C. O.; Sarnat, B.G. *Arch. Pathol.* 1940, **30**, 1192
23. Ashburn, L. L.; et al. *Proc. Soc. Exp. Biol. Med.* 1948, **67**, 351
24. Shortanbaeva, M. A.; Varshavskaya, S. P. *Zdravookhr. Kaz.* 1976, **5**, 87-89
25. Kaufmann, T.; et al. *J. Burn Care Rehabil.* 1988, **9**(5), 474-475
26. Polakoff, P. L. *Occup. Health Saf.* 1986, **55**(3), 24-25
27. Mozingo, D. W.; et al. *J. Trauma* 1988, **28**(5), 642-647
28. Song, Z. Y.; et al. *Scand. J. Work, Environ. Health* 1985, **11**(Suppl. 4), 33

4-PICOLINE

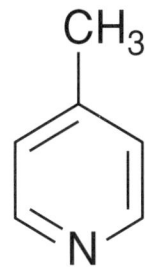

RISKS
Flammable – Harmful by inhalation and if swallowed – Toxic in contact with skin – Irritating to eyes, respiratory system and skin (R10, R20/22, R24, R36/37/38)

SAFETY PRECAUTIONS
In case of contact with eyes, rinse immediately with plenty of water and seek medical advice – Wear suitable protective clothing – If you feel unwell, seek medical advice (show label where possible) (S26, S36, S44)

IDENTIFIERS
SYNONYMS   pyridine, 4-methyl; Ba 35846; *p*-methylpyridine; γ-methylpyridine; 4-methylpyridine; *p*-picoline; γ-picoline

CHEMICAL ABSTRACTS No.	108-89-4
NIOSH No.	UT 5425000
HAZCHEM CODE	2S
UN No.	2313

THRESHOLD LIMIT VALUES

US TLV (TWA)	not available
US TLV (STEL)	not available
UK EXPOSURE LIMITS (OES)	
Long-term (8 hr TWA value)	not available
Short-term (10 min TWA value)	not available
Germany	
MAK	not available
France	
VME	not available
VLE	not available
Sweden	
Short-term limit	not available
Level limit	not available

PHYSICAL PROPERTIES
Description   Colourless liquid, disagreeable colour. Turns brown if not very pure.

Boiling point	145°C
Melting point	2.4°C
Density	0.9571 at 15 °C
Vapour density	3.21
Vapour pressure	4 mm Hg at 20°C
Flash point	57°C (open cup)
Explosive limits	not available
Autoignition temperature	500°C

Solubility   Soluble in water, alcohol or ether.

PACKAGING AND TRANSPORTATION

Road transportation
hazard warning sign	2313 flammable liquid
Hazchem code	2S

Sea transportation
IMDG page No.	3376
class	3.3
label	flammable liquid; corrosive; marine pollutant
packaging group	III

Air transportation
ICAO/IATA code (UN No.)	2313
class	3
label	liquid flammable
packaging group	II
packing instructions	
cargo	307
passenger	305
passenger aircraft max. quantity	5 litres
cargo aircraft max. quantity	60 litres

121. 4-Picoline

MANUFACTURE

4-Picoline occurs in coal tar, bone oil and horses' urine. In recent years synthetic picolines have replaced the coal tar products.

USES

4-Picoline is used as a solvent for resins, in waterproofing fabrics, and in the manufacture of isonicotinic acid and its derivatives.

CHEMICAL HAZARDS

4-Picoline is incompatible with acids, acid chlorides, oxidising agents and chloroformates (1). It is flammable on exposure to heat, flames or oxidisers, and when heated to decomposition it emits toxic fumes of nitrogen oxides.

BIOLOGICAL HAZARDS

Picolines cause symptoms similar to those of pyridine, such as nervous system depression with narcosis, weakness, ataxia, diarrhoea and unconsciousness. They are also strong eye and skin irritants. The toxicity and irritancy of pyridine derivatives tends to be greater in those substituted at the 4 position than at the 2 position (2).

Vapour Inhalation

The LC_{Lo} (4 hours) in rats is 1000 ppm (3). Inhalation causes respiratory tract irritation and symptoms of exposure as outline above.

Eye Contact

4-Picoline is a severe eye irritant (3).

Skin Contact

4-Picoline is a severe skin irritant (3), and rapidly penetrates intact guinea pig skin. The dermal LD_{50} in rabbits is 270 mg/kg (1). There is no evidence of sensitisation in guinea pigs (2).

Swallowing

The oral LD_{50} in rats is 1290 mg/kg (3).

CARCINOGENICITY

No information is available concerning the carcinogenicity of 4-picoline.

MUTAGENICITY

4-Picoline is not mutagenic in *Salmonella typhimurium* (4).

REPRODUCTIVE HAZARDS

No information is available concerning the reproductive hazards of 4-picoline.

FIRST AID

Eyes Wash the eye with flowing water for 10 minutes.

Lungs Remove casualty from area of exposure. If unconscious, do not give anything to drink. Give artificial ventilation and chest compression or place in the recovery position as necessary. If conscious make the casualty lie or sit down quietly, give oxygen if available.

Mouth Do not make the casualty vomit. Treat unconscious casualties as for lungs, but if conscious give 1 pint of water to drink.

Skin Remove contaminated clothing immediately. Wash the affected area with soap and copious amounts of water. Absorption through the skin may cause symptoms similar to those of inhalation.

In all cases of exposure, the patient should be transferred to hospital as soon as possible.

HANDLING AND STORAGE

4-Picoline should be handled wearing an approved respirator, chemical-resistant gloves, safety goggles and other

121. 4-Picoline

protective clothing. It should only be used in chemical fume hood. It should be kept in a cool dry place (1).

DISPOSAL

Wear a laboratory coat, safety spectacles, butyl rubber gloves and suitable safety shoes, and have an approved self-contained breathing apparatus or canister respirator available. Absorb small liquid spills onto paper towels, evaporate in an iron pan in a fume cupboard, add crumpled paper and burn carefully. Brush small solid spills onto paper, put in an iron pan, cover with crumpled paper and burn carefully in a safe place outside. For large spills,
EITHER cover with sand and mix carefully. Shovel into containers, disperse in an excess solution of dilute hydrochloric acid, mix well and leave to stand for 24 hours stirring occasionally. Carefully decant acid extract into the drains, diluting with a large volume of cold tap water. Wash the sand thoroughly with cold water.
OR cover with a sand-soda ash mix (90-10), mix well, shovel into cardboard boxes, pack with crumpled paper and incinerate.

FIRE PRECAUTIONS

Fires involving 4-picoline should be extinguished using alcohol foam.

FURTHER READING

Sax, N. Irving *Dangerous Properties of Industrial Materials* (7th edition)

Encyclopedia of Occupational Safety & Health

Patty's *Industrial Hygiene and Toxicology*

Kirk-Othmer *Encyclopedia of Chemical Technology*

National Fire Protection Association *Manual of Hazardous Reactions*

ACGIH Documentation of TLVs and BEIs (6th edition, 1986)

REFERENCES

1. Lenga, R. E. *The Sigma-Aldrich Library of Chemical Safety Data* (2nd edition) (Sigma-Aldrich, Milwaukee, 1988)
2. Fassett, D. W.; Roudabush, R. L. *Toxicity of Pyridine Derivatives with Relationship to Chemical Structure* (presented to Am. Ind. Hyg. Assoc., Los Angeles, 1953)
3. Smyth, H. F., Jr.; et al. *Arch. Ind. Hyg. Occup. Med.* 1954, **10**, 61
4. Ho, C. H.; et al. *Mutat. Res.* 1981, **85**(5), 335-345

122. Potassium Fluorosilicate

POTASSIUM FLUOROSILICATE
K_2SiF_6

RISKS
Toxic by inhalation, in contact with skin and if swallowed (R23/24/25)

SAFETY PRECAUTIONS
Keep locked up and out of reach of children – In case of contact with eyes, rinse immediately with plenty of water and seek medical advice – If you feel unwell, seek medical advice (show label where possible) (S1/2, S26, S44)

IDENTIFIERS

SYNONYMS silicate (2-1), hexafluoro, dipotassium; dipotassium hexafluorosilicate; dipotassium hexafluorosilicate(2-); potassium fluorosilicate (K_2SiF_6); potassium hexafluorosilicate; potassium silicafluoride; potassium silicofluoride (K_2SiF_6)

CHEMICAL ABSTRACTS No.	16871-90-2
NIOSH No.	VV 8400000
HAZCHEM CODE	1Z
UN No.	2655

THRESHOLD LIMIT VALUES

US TLV (TWA)	not available
US TLV (STEL)	not available
UK EXPOSURE LIMITS (OES)	
Long-term (8 hr TWA value)	not available
Short-term (10 min TWA value)	not available
Germany	
MAK	not available
France	
VME	not available
VLE	not available
Sweden	
Short-term limit	not available
Level limit	not available

PHYSICAL PROPERTIES

Description	White, fine powder or crystals.
Boiling point	not available
Melting point	decomposes
Density	2.27
Vapour density	not available
Vapour pressure	not available
Flash point	not available
Explosive limits	not available
Autoignition temperature	not available

Solubility Slightly soluble in cold water. Insoluble in alcohol.

PACKAGING AND TRANSPORTATION

Road transportation

hazard warning sign	2655 harmful substance
Hazchem code	1Z

Sea transportation

IMDG page No.	6250
class	6.1
label	harmful stow away from foodstuffs; marine pollutant
packaging group	III

Air transportation

ICAO/IATA code (UN No.)	2655
class	6.1
label	keep away from food
packaging group	III
packing instructions	
cargo	619
passenger	619
passenger aircraft max. quantity	100 kilograms
cargo aircraft max. quantity	200 kilograms

122. Potassium Fluorosilicate

MANUFACTURE

Potassium hexafluorosilicate may be prepared by treating sodium fluorosilicate with an aqueous potassium salt (1), or from fluorosilicic acid (a by-product of the fertiliser industry) (2,3).

USES

Potassium hexafluorosilicate is used in the manufacture of opalescent glass, insecticides, as a porcelain enamel frit, flux and sand inhibitor.

CHEMICAL HAZARDS

Potassium hexafluorosilicate is incompatible with hydrofluoric acid, and emits toxic fumes of potassium oxide and fluoride when heated to decomposition.

BIOLOGICAL HAZARDS

Potassium hexafluorosilicate is a strong irritant.

Vapour Inhalation

Potassium hexafluorosilicate is a strong irritant.

Eye Contact

Potassium hexafluorosilicate is a strong irritant.

Skin Contact

Potassium hexafluorosilicate is a strong irritant.

Swallowing

Potassium hexafluorosilicate causes vomiting and diarrhoea if swallowed. The oral LD_{50} in guinea pigs is 500 mg/kg.

CARCINOGENICITY

No information is available concerning the carcinogenicity of potassium hexafluorosilicate.

MUTAGENICITY

No information is available concerning the mutagenicity of potassium hexafluorosilicate.

REPRODUCTIVE HAZARDS

No information is available concerning the reproductive hazards of potassium hexafluorosilicate.

FIRST AID

Eyes Wash the eye with flowing water for 10 minutes, then irrigate with normal saline for at least 30 minutes. Do not use calcium gluconate gel.

Lungs Remove casualty from area of exposure. If unconscious do not give anything to drink, give artificial ventilation and chest compression or place in the recovery position as necessary. If conscious make the casualty lie or sit down quietly, give oxygen if available.

Mouth Do not make the casualty vomit. Treat unconscious casualties as for lungs but if conscious give milk or calcium gluconate by mouth.

Skin Remove contaminated clothing immediately, wearing protective gloves and clothes. If calcium gluconate gel is available, immediately rub into all affected areas and massage until all pain goes. If not, wash with soap and water for 30 minutes.

In all cases of exposure, the patient should be transferred to hospital as soon as possible.

HANDLING AND STORAGE

Potassium hexafluorosilicate should be handled wearing an approved respirator, chemical-resistant gloves, safety goggles and other protective clothing. It should be kept in a tightly closed container in a cool, dry place.

122. Potassium Fluorosilicate

DISPOSAL

Wearing a laboratory coat or overalls, safety glasses, rubber gloves and suitable safety shoes, scoop up spill into a large beaker, carefully add calcium hydroxide water and stir until dissolved. Run the solution to waste with excess water.
For larger spills add soda ash (and calcium hydroxide if a fluoride) at intervals. Decant liquid after 24 hours and neutralise with 6M hydrochloric acid. Discharge supernatant to drain with x1000 dilution of cold tap water. The sludge should be removed by a licenced contractor. For solutions cover and mix with dry soda ash, and shovel into buckets. Carefully add cold water, neutralise with 6M hydrochloric acid and wash down the drain with x1000 volume of cold water.

FIRE PRECAUTIONS

Potassium hexafluorosilicate is not combustible and an extinguisher suitable to surrounding fire conditions should be used.

FURTHER READING

Bretherick, L. *Hazards in the Chemical Laboratory* (4th edition)

Bretherick, L. *Handbook of Reactive Chemical Hazards* (4th edition)

Sax, N. Irving *Dangerous Properties of Industrial Materials* (7th edition)

Encyclopaedia of Occupational Safety & Health

Patty's *Industrial Hygiene and Toxicology*

Kirk-Othmer *Encyclopedia of Chemical Technology*

National Fire Protection Association *Manual of Hazardous Reactions*

ACGIH *Documentation of TLVs and BEIs* (6th edition, 1986)

REFERENCES

1. Abramov, O. B.; et al. *Otkrytiya, Izobret., Prom. Obraztay, Tovarnye Znaki* 1978, **55**(42), 76
2. Augustyn, W.; Switonska-Oskedra, M. Pol. Pat. 53,384 (Cl. C)1*b*, 1967
3. Augustyn, W. *Chemik* 1983, **36**(9), 227-232

1,3-PROPANE-SULTONE

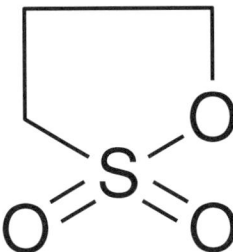

123. 1,3-Propanesultone

PHYSICAL PROPERTIES

Description Off-white solid or colourless liquid.

Boiling point	112°C
Melting point	30-33°C
Density	1.392 at 40°C
Vapour density	not available
Vapour pressure	not available
Flash point	< 110°C
Explosive limits	not available
Autoignition temperature	not available

Solubility Moderately soluble in water and most organic solvents.

RISKS

May cause cancer – Harmful in contact with skin and if swallowed (R45, R21/22)

SAFETY PRECAUTIONS

Avoid exposure-obtain special instructions before use – If you feel unwell, seek medical advice (show label where possible) (S53, S44)

IDENTIFIERS

SYNONYMS 1,2-oxathiolane, 2,2-dioxide; propane sultone; γ-propane sultone; 3-hydroxy-1-propane sulphonic acid γ-sultone; 3-hydroxy-1-propanesulphonic acid sultone; 3-hydroxy-1-propanesulphonic acid sulphone; 1-propanesulphonic acid-3-hydroxy-γ-sultone

CHEMICAL ABSTRACTS No.	1120-71-4
NIOSH No.	RP 5425000
HAZCHEM CODE	not available
UN No.	not available

THRESHOLD LIMIT VALUES

US TLV (TWA)	not available
US TLV (STEL)	not available
UK EXPOSURE LIMITS (OES)	
Long-term (8 hr TWA value)	not available
Short-term (10 min TWA value)	not available
Germany	
MAK	not available
France	
VME	not available
VLE	not available
Sweden	
Short-term limit	not available
Level limit	not available

PACKAGING AND TRANSPORTATION

Road transportation

hazard warning sign	not available
Hazchem code	not available

Sea transportation

IMDG page No.	not available
class	not available
label	not available
packaging group	not available

Air transportation

ICAO/IATA code	not available
class	not available
label	not available
packaging group	not available
packing instructions	
cargo	not available
passenger	not available
passenger aircraft max. quantity	not available
cargo aircraft max. quantity	not available

123. 1,3-Propanesultone

MANUFACTURE

1,3-Propane sultone can be produced by sulphonation of allyl alcohol, or dehydration of γ-hydroxy-propanesulphonic acid.

USES

1,3-Propane sultone is used as a chemical intermediate to introduce the sulphopropyl group into molecules, and thus give water solubility and anionic character. It is used as an intermediate in manufacture of resins, dyes, vulcanisation accelerators and pesticides. The reaction with substituted imidazolines gives acid inhibitors for mild steel.

CHEMICAL HAZARDS

1,3-Propane sultone is incompatible with strong oxidising agents, strong acids or bases. It emits toxic fumes of carbon monoxide and sulphur oxides when heated to decomposition (1).

BIOLOGICAL HAZARDS

1,3-Propane sultone is an irritant, and is carcinogenic and teratogenic in experimental animals. It is mutagenic in bacterial assays and cultured mammalian cells.

Vapour Inhalation

Vapour or mist of 1,3-propane sultone may be irritating to the upper respiratory tract (1).

Eye Contact

Vapour or mist of 1,3-propane sultone may cause eye irritation (1).

Skin Contact

1,3-Propane sultone is a skin irritant.

Swallowing

The TD_{Lo} in rats is 7840 mg/kg (2).

CARCINOGENICITY

In the studies reviewed by the IARC it was concluded that 1,3-propane sultone is carcinogenic in rats after oral, subcutaneous, intravenous or pre-natal administration, and has a local carcinogenic effect in mice (3-5). When administered to rats by gavage at doses of 28 or 56 mg/kg it induced a high incidence of mammary adenocarcinoma and malignant glioma of the cerebrum and cerebellum, and a somewhat increased incidence of leukaemia, and ear duct carcinomas. Mortality was high and the experiments were terminated early. Not enough animals survived to assess the potential of 1,3-propane sultone to induce late-appearing tumours (6).

MUTAGENICITY

1,3-Propane sultone is mutagenic in *Salmonella typhimurium* TA1535 and TA100 (7,8). It gave positive results in *in vitro* chromosome tests on Chinese hamster fibroblast cells (9).

REPRODUCTIVE HAZARDS

Intravenous administration of 1,3-propane sultone to rats on the 15th day of pregnancy caused malignant neurogenic tumours in the offspring (10).

FIRST AID

Eyes Wash the eye with flowing water for 10 minutes.

Lungs Remove casualty from area of exposure. If unconscious, do not give anything to drink. Give artificial ventilation and chest compression or place in the recovery position as necessary. If conscious make the casualty lie or sit down quietly, give oxygen if available.

Mouth Do not make the casualty vomit. Treat unconscious casualties as for lungs, but if conscious give 1 pint of water to drink.

Skin Remove contaminated clothing immediately. Wash the affected area with soap and copious amounts of water. Absorption through the skin may cause symptoms similar to those of inhalation.

123. 1,3-Propanesultone

In all cases of exposure, the patient should be transferred to hospital as soon as possible.

HANDLING AND STORAGE

Propane sultone should be handled wearing an approved respirator, surgical rubber gloves, safety goggles and other protective clothing. It should only be used in a chemical fume hood or glove box (11). It should be kept refrigerated in sealed glass ampules or screw topped bottles or vials with teflon cap liners (12), in a cool, dry place (1). It hydrolyses slowly in the presence of water to free acid which is very corrosive to metals.

DISPOSAL

Ventilate the area, and wear rubber boots, heavy rubber gloves, approved self-contained breathing apparatus and disposable overalls. Sweep the spill into a bag and hold for waste disposal. Wash the spill site thoroughly and discard overalls after use.

A validated procedure for destruction of potential carcinogens including 1,3-propane sultone has been described. Compounds were degraded using a strong base and analysis down to the limits of detection showed complete destruction (11).

FIRE PRECAUTIONS

Fires involving propane sultone should be extinguished using water spray, carbon dioxide, dry chemical powder, alcohol foam or polymer foam.

FURTHER READING

Bretherick, L. *Hazards in the Chemical Laboratory* (4th edition)

Bretherick, L. *Handbook of Reactive Chemical Hazards* (4th edition)

Sax, N. Irving *Dangerous Properties of Industrial Materials* (7th edition)

Encyclopaedia of Occupational Safety & Health

Patty's *Industrial Hygiene and Toxicology*

Kirk-Othmer *Encyclopedia of Chemical Technology*

National Fire Protection Association *Manual of Hazardous Reactions*

ACGIH Documentation of TLVs and BEIs (6th edition, 1986)

REFERENCES

1. Lenga, R. E. *The Sigma-Aldrich Library of Chemical Safety Data* (2nd edition) (Sigma-Aldrich, Milwaukee, 1988)
2. Ulland, B.; et al. *Nature (London)* 1971, **230**, 460
3. IARC *Monographs on the evaluation of the carcinogenic risk of chemicals to humans* 1974, **4**, 253-258
4. Druckrey, H.; et al. *Naturwissenschaften* 1968, **55**, 449
5. Druckrey, H.; et al. *Z. Krebsforsch* 1970, **75**, 69-84
6. Weisburger, E. K.; et al. *JNCI, J. Natl. Cancer Inst.* 1981, **67**(1), 75-88
7. Simmon, V. F. *JNCI J. Natl. Cancer Inst.* 1979, **62**(4), 893-899
8. Rosenkranz, H. S.; et al. *JNCI J. Natl. Cancer Inst.* 1979, **62**(4), 873-892
9. Ishidate, M., Jr.; Odashima, S. *Mutat. Res.* 1977, **48**(3-4), 337-353
10. *IARC Sci. Publ.* 1973, **4**, 45
11. *Dangerous Prop. Ind. Mater. Rep.* 1984, **4**(3), 82-85
12. Lunn, G.; Sansone, E. B. *J. Chem. Educ.* 1990, **67**(10), A249-A251

124. 3-Propanolide

3-PROPANOLIDE

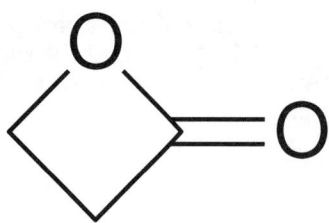

RISKS
May cause cancer – Very toxic by inhalation – Irritating to eyes and skin (R45, R26, R36/38)

SAFETY PRECAUTIONS
Avoid exposure-obtain special instructions before use – In case of accident or if you feel unwell, seek medical advice immediately (show label where possible) (S53, S45)

IDENTIFIERS

SYNONYMS 2-oxetanone; Betaprone; BPL; hydracrylic acid β-lactone; 3-hydroxypropionic acid lacton; 3-propanolide; propiolactone; β-propiolactone; 1,3-propiolactone; 3- propiolactone; β-propionolactone

CHEMICAL ABSTRACTS No.	57-57-8
NIOSH No.	RQ 7350000
HAZCHEM CODE	not available
UN No.	not available

THRESHOLD LIMIT VALUES

US TLV (TWA)	not available
US TLV (STEL)	not available

UK EXPOSURE LIMITS (OES)
Long-term (8 hr TWA value) .... not available
Short-term (10 min TWA value) .... not available

Germany
MAK .... not available

France
VME .... not available
VLE .... not available

Sweden
Short-term limit .... not available
Level limit .... not available

PHYSICAL PROPERTIES

Description Cloudless liquid.

Boiling point	decomposes at 155°C
Melting point	-33.4°C
Density	1.148 at 20°C
Vapour density	not available
Vapour pressure	3.4 mm Hg at 25°C
Flash point	75°C (closed cup)
Explosive limits	not available
Autoignition temperature	not available

Solubility Soluble in water, dimethyl sulphoxide and ethanol. Miscible in acetone and ether at 25°C.

PACKAGING AND TRANSPORTATION

Road transportation
hazard warning sign	not available
Hazchem code	not available

Sea transportation
IMDG page No.	not available
class	not available
label	not available
packaging group	not available

Air transportation
ICAO/IATA code	not available
class	not available
label	not available
packaging group	not available

packing instructions
cargo .... not available
passenger .... not available

passenger aircraft max. quantity	not available
cargo aircraft max. quantity	not available

124. 3-Propanolide

MANUFACTURE

3-Propanolide is prepared by condensation of ketene with formaldehyde.

USES

3-Propanolide is used as an intermediate in organic synthesis, and as a vapour sterilant and disinfectant.

CHEMICAL HAZARDS

3-Propanolide may react violently with 2-aminoethanol, aniline, chlorosulphonic acid, ethylene diamine, ethylene imine, hydrogen chloride, hydrogen fluoride, nitric acid, oleum, pyridine, sodium hydroxide, sulphuric acid or ammonium hydroxide (1).

BIOLOGICAL HAZARDS

3-Propanolide is carcinogenic in animal studies, and should be treated as a suspected human carcinogen.

Vapour Inhalation

3-Propanolide is irritating to the respiratory tract. Reported LC_{50} values in rats are 250 ppm and 25 ppm for 30 minute and 6 hour exposures respectively. No information is available on carcinogenicity from this route of exposure.

Eye Contact

3-Propanolide causes irritation, lacrimation, and the undiluted material causes permanent corneal opacification in rabbits' eyes.

Skin Contact

3-Propanolide is a skin irritant and can be absorbed through the skin. Application to the skin of mice caused cancer.

Swallowing

3-Propanolide causes mouth and stomach burns.

CARCINOGENICITY

From the studies reviewed by the IARC, it was concluded that 3-propanolide is carcinogenic in mice by dermal, subcutaneous or intraperitoneal routes. There was some evidence of carcinogenicity in rats, but results were equivocal in hamsters and guinea pigs (2). 3- Propanolide caused skin papillomas and carcinomas in several lifetime skin painting studies in mice (3,4). It is also carcinogenic when applied to guinea pig skin (5). It induced local carcinomas on subcutaneous injection (6) and forestomach tumours when administered by intragastric feeding to mice (6,7). Its carcinogenicity has been reviewed (8).

MUTAGENICITY

The mutagenicity of 3-propanolide has been reviewed (8). It gave positive results in the HeLa DNA synthesis inhibition test (9). It was directly active in sister chromatid exchange induction in Chinese hamster Don (lung) and ovary cells (10). 3-Propanolide is mutagenic in *Salmonella typhimurium* TA100 and TA1535 (11,12), and induces point mutation and chromosomal damage at the (TK+/-) locus of L5178Y mouse lymphoma cells (13).

REPRODUCTIVE HAZARDS

No information is available concerning the reproductive hazards of 3-propanolide.

FIRST AID

Eyes Wash the eye with flowing water for 10 minutes.

Lungs Remove casualty from area of exposure. If unconscious, do not give anything to drink, give artificial ventilation and chest compression or place in the recovery position as necessary. If conscious make the casualty lie or sit down quietly, give oxygen if available. Lung congestion may occur – a conscious casualty with breathing difficulties should be placed in a sitting position. Convulsions may occur and may cause unconsciousness. Allergic asthma (wheezy breathing) may occur and immediate treatment is needed – Ventolin may be useful.

Mouth Do not make the casualty vomit. Treat unconscious casualties as for lungs but if conscious give 1 pint of water to drink immediately. Convulsions may occur and may cause unconsciousness.

124. 3-Propanolide

Skin Remove contaminated clothing immediately, drench the affected area with running water for at least 10 minutes.

In all cases of exposure, the patient should be transferred to hospital as soon as possible.

HANDLING AND STORAGE

For protection and handling requirements, consult CFR Title 29 Part 1910 (14). It should be kept in a tightly closed container, in a refrigerator in glass containers, protected from air and light (14).

DISPOSAL

Eliminate all sources of ignition, ventilate the area and wear a laboratory coat or acid-proof overalls, gloves, approved self- contained breathing apparatus and safety boots. Carefully mix well with dry soda ash, sodium bicarbonate or sodium carbonate, adding water if necessary. Scoop slurry into plastic buckets, and wash until neutral, neutralising with 6M ammonium hydroxide or 6M hydrochloric acid. Discharge neutralised waste into drain with x1000 dilution of water. A validated method for destruction of potential carcinogens has been described (15).

FIRE PRECAUTIONS

Fires involving 3-propanolide should be extinguished using carbon dioxide, dry chemical powder, alcohol foam or polymer foam (14).

FURTHER READING

Bretherick, L. *Hazards in the Chemical Laboratory* (4th edition)

Bretherick, L. *Handbook of Reactive Chemical Hazards* (4th edition)

Sax, N. Irving *Dangerous Properties of Industrial Materials* (7th edition)

Encyclopaedia of Occupational Safety & Health

Patty's *Industrial Hygiene and Toxicology*

Kirk-Othmer *Encyclopedia of Chemical Technology*

National Fire Protection Association *Manual of Hazardous Reactions*

ACGIH Documentation of TLVs and BEIs (6th edition, 1986)

Dangerous Prop. Ind. Mater. Rep. 1983, **3**(2), 57-60

REFERENCES

1. Keith, L. H.; Walters, D. B. (editors) *Compendium of Safety Data Sheets for Research and Industrial Chemicals* (VCH, Deerfield Park, 1987)
2. IARC *Monographs on the evaluation of the carcinogenic risk of chemicals to humans* 1974, **4**, 259-269
3. Roe, F. J. C.; Glendenning, O. M. *Br. J. Cancer* 1956, **10**, 357
4. Palmes, E. D.; et al. *Am. Ind. Hyg. Assoc. J.* 1962, **23**, 257
5. Parish, D. J.; Searle, C. E. *Br. J. Cancer* 1966, **20**(1), 200-205
6. Van Duuren, B. J.; et al. *JNCI, J. Natl. Cancer Inst.* 1979, **63**(6), 1433-1439
7. Dunkelburg, H. *Br. J. Cancer* 1982, **46**, 924
8. Brusick, D. J. *Mutat. Res.* 1977, **39**(3-4), 241-255
9. Painter, R. B.; Howard, R. *Mutat. Res.* 1982, **92**, 427-437
10. Baker, R. S. U.; et al. *Mutat. Res.* 1983, **118**, 103-116
11. Simmon, V. F. *JNCI, J. Natl. Cancer Inst.* 1979, **62**(4), 893-899
12. Rosenkranz, H. S.; Poirier, L. A. *JNCI, J. Natl. Cancer Inst.* 1979, **62**(4), 873-892
13. Clive, D. *Dev. Toxicol. Environ. Sci.* 1977, **2**(Prog. Genet. Toxicol.), 241-247
14. Lenga, R. E. *The Sigma-Aldrich Library of Chemical Safety Data* (2nd edition) (Sigma-Aldrich, Milwaukee, 1988)
15. Lunn, G.; Sansone, E. B. *J. Chem. Educ.* 1990, **67**(10), A249-A251

125. Propargyl Alcohol

PROPARGYL ALCOHOL

$HC\equiv CCH_2OH$

RISKS
Flammable – Toxic by inhalation, in contact with skin and if swallowed – Causes burns (R10, R23/24/25, R34)

SAFETY PRECAUTIONS
In case of contact with eyes, rinse immediately with plenty of water and seek medical advice – After contact with skin, wash immediately with plenty of water – Wear suitable protective clothing – If you feel unwell, seek medical advice (show label where possible) (S26, S28, S36, S44)

IDENTIFIERS

SYNONYMS ethylynyl methanol; 1-propyne-3-ol; RCRA Waste No. P102; 2-propyn-1-ol; ethynylcarbinol; 2-propynol; 3-propynol; propynyl alcohol; 2-propynyl alcohol

CHEMICAL ABSTRACTS No.	107-19-7
NIOSH No.	UK 5075000
HAZCHEM CODE	not available
UN No.	not available

THRESHOLD LIMIT VALUES

US TLV (TWA)	1 ppm (2.3 mg/m^3)
US TLV (STEL)	not available

UK EXPOSURE LIMITS (OES)
Long-term (8 hr TWA value) 1 ppm (2 mg/m^3)
Short-term (10 min TWA value) 3 ppm (6 mg/m^3)

Germany
MAK 2 ppm (5 mg/m^3)

France
VME not available
VLE not available

Sweden
Short-term limit not available
Level limit not available

PHYSICAL PROPERTIES

Description Straw coloured liquid with a geranium like smell.

Boiling point	114-115°C
Melting point	-48 to -52°C
Density	0.9715 at 20°C
Vapour density	1.93
Vapour pressure	11.6 mm Hg at 20°C
Flash point	36°C (open cup)
Explosive limits	not available
Autoignition temperature	not available

Solubility Miscible with water, alcohol or ether, but not with aliphatic hydrocarbons.

PACKAGING AND TRANSPORTATION

Road transportation
hazard warning sign	not available
Hazchem code	not available

Sea transportation
IMDG page No.	not available
class	not available
label	not available
packaging group	not available

Air transportation
ICAO/IATA code	not available
class	not available
label	not available
packaging group	not available
packing instructions	
cargo	not available
passenger	not available
passenger aircraft max. quantity	not available
cargo aircraft max. quantity	not available

125. Propargyl Alcohol

MANUFACTURE

Propargyl alcohol is a by-product of butynediol manufacture. Several methods of preparing propargyl alcohol have been reported. These include preparation from formaldehyde and sodium acetylide; epichlorohydrin and sodium; acetylene and formaldehyde; and by heating β-bromoallyl alcohol with concentrated potassium hydroxide.

USES

Propargyl alcohol is used as a chemical intermediate, corrosion inhibitor, solvent stabiliser, soil fumigant, and has been used to prevent steel becoming brittle. It is a component in oil-well acidising compositions, and is also used in metal pickling and plating. Its cryoprotective action has been evaluated on bull sperm (1).

CHEMICAL HAZARDS

Propargyl alcohol may react explosively with sulphuric acid (2-4), and if dried with alkali before distillation the residue may explode. Sodium sulphate is therefore recommended as a desiccant (5). Reaction with mercury(II) sulphate and sulphuric acid to produce hydroxyacetone has been violently exothermic (6). Addition of phosphorus pentaoxide to propargyl alcohol causes ignition (7). It has a low flash point, is a fire hazard if exposed to heat or flame, and emits acrid smoke when heated to decomposition.

BIOLOGICAL HAZARDS

Propargyl alcohol is an irritant and central nervous system depressant. Chronic exposure of rodents reportedly caused central nervous system disturbances, reductions in blood pressure and haemoglobin, and changes in the lungs, liver and kidneys, although the dose and route of administration are not given (8).

Vapour Inhalation

Propargyl alcohol is irritating to the mucous membranes. The LC_{50} (1 hour) in rats is 1040-1200 ppm (9).

Eye Contact

Undiluted propargyl alcohol is severely irritating to the eyes of rabbits and caused permanent corneal injury. A 10% aqueous solution caused slight irritation but no permanent injury (10).

Skin Contact

Propargyl alcohol is irritating to the skin. There is no evidence of sensitisation. The dermal LD_{50} in rabbits is 88 mg/kg (9).

Swallowing

The oral LD_{50} in rats, mice and guinea pigs is 55, 50 and 60 mg/kg respectively.

CARCINOGENICITY

No information is available concerning the carcinogenicity of propargyl alcohol.

MUTAGENICITY

Propargyl alcohol is weakly mutagenic in *Salmonella typhimurium* hisD3052 with metabolic activation, but not in TA98 or TA100 (11).

REPRODUCTIVE HAZARDS

No information is available concerning the reproductive hazards of propargyl alcohol.

FIRST AID

Eyes Wash the eye with flowing water for 10 minutes.

Lungs Remove casualty from area of exposure. If unconscious, do not give anything to drink. Give artificial ventilation and chest compression or place in the recovery position as necessary. If conscious make the casualty lie or sit down quietly, give oxygen if available. Convulsions may occur and may cause unconsciousness.

Mouth Do not make the casualty vomit. Treat unconscious casualties as for lungs, but if conscious give 1 pint of water to drink.

Skin Remove contaminated clothing immediately. Drench the affected area with running water for at least 10 minutes.

125. Propargyl Alcohol

In all cases of exposure, the patient should be transferred to hospital as soon as possible.

HANDLING AND STORAGE

Propargyl alcohol should be handled wearing an approved respirator, chemical-resistant gloves, safety goggles and other protective clothing. It should only be used in chemical fume hood. It should be kept in a tightly closed container, refrigerated, and away from heat, sparks and flame. The vapour may travel a considerable distance to a source of ignition and flash back (12). Clean and rust-free steel may be used for short-term storage, but for longer storage stainless steel (304 and 316), glass-lined or phenolic lined vessels are suitable. Avoid using aluminium, epoxides and epoxy-phenolics. It is usually handled in standard steel pipe or braided steel hose; rubber is not suitable. Non sparking tools should be used.

DISPOSAL

Eliminate all sources of ignition and ventilate the area. Wearing a laboratory coat or overalls, safety glasses and gloves, absorb the spill onto paper towels and allow to evaporate in a fume-cupboard. For large spills, absorb onto sand or vermiculite, and remove in buckets for atmospheric evaporation in a safe open area. Ideally, waste should be burned in an incinerator with afterburner.

FIRE PRECAUTIONS

Fires involving propargyl alcohol should be extinguished using carbon dioxide, dry chemical powder, alcohol foam or polymer foam. Container explosion may occur under fire conditions.

FURTHER READING

Bretherick, L. *Hazards in the Chemical Laboratory* (4th edition)

Bretherick, L. *Handbook of Reactive Chemical Hazards* (4th edition)

Sax, N. Irving *Dangerous Properties of Industrial Materials* (7th edition)

Encyclopedia of Occupational Safety & Health

Patty's *Industrial Hygiene and Toxicology*

Kirk-Othmer *Encyclopedia of Chemical Technology*

ACGIH Documentation of TLVs and BEIs (6th edition, 1986)

REFERENCES

1. Jeyendran, R. S.; Graham, E. F. *Cryobiology* 1980, **17**(5), 458-464
2. Sauer, J. C. *Org. Synth.* 1963, **coll. vol. 4**, 813
3. Crouse, D. M. *Chem. Eng. News* 1979, **57**(41), 4
4. Hart, E. V. *Chem. Eng. News* 1980, **58**(2), 4
5. Anon. *Agnew. Chem. (Nachr.)* 1954, **2**, 209
6. Reppe, W.; et al. *Ann.* 1955, **596**, 38
7. *_National Fire Protection Association *Manual of Hazardous Reactions* p.321
8. Stasenkova, K. P.; Kochetkova, T. A. *Toksikol. Novykh. Prom. Khim. Veshchestv.* 1966, **8**, 60-70
9. Vernot, E. H.; et al. *Toxicol. Appl. Pharmacol.* 1977, **42**(2), 417-423
10. *Toxicity profile: Propargyl alcohol* (BIBRA, Carshalton, 1988)
11. Basu, A. K.; Marnett, L. J. *Cancer Res.* 1984, **44**, 2848
12. Lenga, R. E. *The Sigma-Aldrich Library of Chemical Safety Data* (2nd edition) (Sigma-Aldrich, Milwaukee, 1988)

126. Propyl Chloroformate

PROPYL CHLOROFORMATE

$ClCO_2CH_2CH_2CH_3$

RISKS
Flammable – Toxic by inhalation – Causes burns (R10, R23, R34)

SAFETY PRECAUTIONS
In case of contact with eyes, rinse immediately with plenty of water and seek medical advice – Wear suitable protective clothing – If you feel unwell, seek medical advice (show label where possible) (S26, S36, S44)

PHYSICAL PROPERTIES

Description Colourless liquid.
Boiling point .... 114-115°C at 768mm Hg
Melting point .... not available
Density .... 1.090 at 20°C
Vapour density 4.2
Vapour pressure .... 26 mm Hg at 20°C
Flash point .... -50°C (also listed as 28°C)
Explosive limits .... not available
Autoignition temperature .... not available
Solubility Insoluble and slightly decmposes in water, alcohol. Miscible with ether or benzene.

IDENTIFIERS

SYNONYMS carbonchloridic acid, propyl ester; formic acid, chloro-, propyl ester; propyl chlorocarbonate; *n*-propyl chloroformate

CHEMICAL ABSTRACTS No. 109-61-5
NIOSH No. .... LQ 6830000
HAZCHEM CODE .... not available
UN No. 2740

THRESHOLD LIMIT VALUES

US TLV (TWA) .... not available
US TLV (STEL) .... not available
UK EXPOSURE LIMITS (OES)
Long-term (8 hr TWA value) .... not available
Short-term (10 min TWA value) .... not available

Germany
MAK .... not available

France
VME .... not available
VLE .... not available

Sweden
Short-term limit .... not available
Level limit .... not available

PACKAGING AND TRANSPORTATION

Road transportation

hazard warning sign .... 2740 toxic substance, ... corrosive substance, flammable liquid
Hazchem code .... not available

Sea transportation

IMDG page No. 6244
class 6.1
label .... poison; corrosive; flammable liquid
packaging group I

Air transportation

ICAO/IATA code (UN No.) 2740
class 6.1;3.8
label .... poison; liquid flammable; corrosive
packaging group I
packing instructions
cargo 605
passenger .... forbidden
passenger aircraft max. quantity .... forbidden
cargo aircraft max. quantity .... 2.5 litres

126. Propyl Chloroformate

MANUFACTURE

Alkyl chloroformates are formed by reaction of liquid anhydrous alcohols with excess dry, chlorine-free phosgene at low temperature. Corrosion resistant equipment is necessary.

USES

Chloroformates are used as intermediates for carbonates and percarbonates. Propyl chloroformate is used as a reactive intermediate for polymerisation initiators.

CHEMICAL HAZARDS

Propyl chloroformate is flammable on exposure to heat or flame, and may react vigorously with oxidising agents. It is incompatible with acids, strong bases, alcohols and amines. Hydrolysis may occur in moist air. It emits toxic fumes of hydrogen chloride, carbon monoxide and phosgene when heated to decomposition (1).

BIOLOGICAL HAZARDS

Propyl chloroformate is irritating to the eyes, skin and mucous membranes. It is toxic if absorbed through the skin, and moderately toxic if inhaled or swallowed. Range finding toxicity data for propyl chloroformate is available (2).

Vapour Inhalation

Propyl chloroformate vapours are strongly irritating to the mucous membranes.

Eye Contact

Propyl chloroformate vapours are strongly irritating to the eyes.

Skin Contact

Propyl chloroformate is a corrosive skin irritant. The dermal LD_{50} in mice is 10 mg/kg.

Swallowing

The oral LD_{50} in mice is 650 mg/kg.

CARCINOGENICITY

No information is available concerning the carcinogenicity of propyl chloroformate.

MUTAGENICITY

No information is available concerning the mutagenicity of propyl chloroformate.

REPRODUCTIVE HAZARDS

No information is available concerning the reproductive hazards of propyl chloroformate.

FIRST AID

Eyes Wash the eye immediately with flowing water.

Lungs Remove casualty from area of exposure. If unconscious, do not give anything to drink, give artificial ventilation and chest compression or place in the recovery position as necessary. If conscious make the casualty lie or sit down quietly, give oxygen if available. Lung congestion may occur – a conscious casualty with breathing difficulties should be placed in a sitting position.

Mouth Do not make the casualty vomit. Treat unconscious casualties as for lungs but if conscious give 1 pint of water to drink immediately.

Skin Remove contaminated clothing immediately, wash the affected area with soap and copious amounts of water. Absorption through the skin may cause symptoms similar to those of inhalation.

In all cases of exposure, the patient should be transferred to hospital as soon as possible.

HANDLING AND STORAGE

Propyl chloroformate should be handled wearing an approved respirator, chemical-resistant gloves, safety goggles and

126. Propyl Chloroformate

other protective clothing. It should only be used in a chemical fume hood. It should be kept in a tightly closed container, in a cool dry place, away from heat and sources of ignition. The vapour may travel a considerable distance to a source of ignition and flash back. It should be refrigerated on arrival, and transfer from containers to storage tanks or reactors should be via a closed system using stainless steel and nickel, or glass pumps, valves and lines. Containers should be vented from time to time, and opened carefully using a safety shield (1).

DISPOSAL

Eliminate all sources of ignition, ventilate the area and wear a laboratory coat or overalls, rubber gloves, approved compressed air breathing apparatus and safety boots. Absorb small spills onto paper towels, remove to a safe open air site and allow to evaporate in a metal tray. Wash spillage site with detergent. For large spills, absorb onto vermiculite-sodium carbonate mixture (90-10) or sand-soda ash (90-10) and mix carefully. EITHER transport in dry buckets to a safe open air area for atmospheric evaporation OR shovel into paper boxes and incinerate.

FIRE PRECAUTIONS

Fires involving propyl chloroformate should be extinguished using carbon dioxide, dry chemical powder, alcohol foam or polymer foam. DO NOT use water. Container explosion may occur under fire conditions.

FURTHER READING

Bretherick, L. *Hazards in the Chemical Laboratory* (4th edition)

Bretherick, L. *Handbook of Reactive Chemical Hazards* (4th edition)

Sax, N. Irving *Dangerous Properties of Industrial Materials* (7th edition)

Encyclopedia of Occupational Safety & Health

Patty's *Industrial Hygiene and Toxicology*

Kirk-Othmer *Encyclopedia of Chemical Technology*

National Fire Protection Association *Manual of Hazardous Reactions*

ACGIH Documentation of TLVs and BEIs (6th edition, 1986)

REFERENCES

1. Lenga, R. E. *The Sigma-Aldrich Library of Chemical Safety Data* (2nd edition) (Sigma-Aldrich, Milwaukee, 1988)
2. Smyth, H. F., Jr.; et al. *Am. Ind. Hyg. Assoc. J.* 1969, **30**(5), 470-476

127. Resorcinol

RESORCINOL

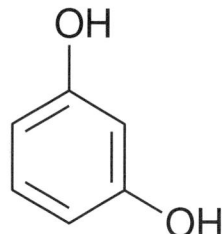

RISKS
Harmful if swallowed – Irritating to eyes and skin (R22, R36/38)

SAFETY PRECAUTIONS
In case of contact with eyes, rinse immediately with plenty of water and seek medical advice (S26)

IDENTIFIERS

SYNONYMS   1,3-benzenediol; *m*-benzenediol; C.I. 76505; C.I. Developer 4; C.I. Oxidation Base 31; Developer O; Developer R; Developer RS; *m*-dihydroxybenzene; 1,3- hydroxybenzene; Durafur Developer G; Fouramine RS; Fourrine 79; Fourrine EW; *m*-hydroqulinone; *m*-hydroxyphenol; 3- hyroxyphenol; Nako TGG; Pelagol Grey RS; Pelagol RS; resorcin

CHEMICAL ABSTRACTS No.	108-46-3
NIOSH No.	VG 9625000
HAZCHEM CODE	2X
UN No.	2876

THRESHOLD LIMIT VALUES

US TLV (TWA)	10 ppm (45 mg/m^3)
US TLV (STEL)	20 ppm (90 mg/m^3)

UK EXPOSURE LIMITS (OES)
Long-term (8 hr TWA value) ... 10 ppm (45 mg/m^3)
Short-term (10 min TWA value) ... 20 ppm (90 mg/m^3)

Germany
MAK .... not available

France
VME .... 10 ppm (45 mg/m^3)
VLE .... not available

Sweden
Short-term limit .... not available
Level limit .... not available

PHYSICAL PROPERTIES

Description Very white crystals, become pink on exposure to light when not perfectly pure, or on contact with iron, with an unpleasant sweet taste.

Boiling point	276°C
Melting point	110°C
Density	1.272 at 15°C
Vapour density	3.79
Vapour pressure	1 mm Hg at 108.4°C
Flash point	127°C (closed cup)
Explosive limits	lel:1.4 vol at 200°C
Autoignition temperature	607°C

Solubility Very soluble in alcohol, ether or glycerol. Slightly soluble in chloroform. Soluble in water.

PACKAGING AND TRANSPORTATION

Road transportation

hazard warning sign	2876 harmful substance
Hazchem code	2X

Sea transportation

IMDG page No.	6248
class	6.1
label	harmful stow away from foodstuffs
packaging group	III

Air transportation

ICAO/IATA code (UN No.)	2876
class	6.1
label	keep away from food
packaging group	III
packing instructions cargo	619
passenger	619
passenger aircraft max. quantity	100 kilograms
cargo aircraft max. quantity	200 kilograms

127. Resorcinol

MANUFACTURE

Resorcinol is produced by fusion of the sodium salt of *m*-benzenedisulphonic acid with excess sodium hydroxide, followed by acidification.

USES

Resorcinol is used in the manufacture of adhesives for wood and rubber, and in the preparation of dyes (xanthene and azo types), pharmaceuticals and ultraviolet absorbers (benzophenones). It is also used as a cross-linking agent for epoxy resins, and to stabilise and plasticise cellulose esters and ethers.

CHEMICAL HAZARDS

An explosion occurred when a tarry material, formed by the use of too low a concentration of nitric acid (82%) during preparation of dinitroresorcinol, came into contact with higher strength acid. It is incompatible with acetanilide, alkalies, ferric salts, spirit nitrous ether or urethane (1).

BIOLOGICAL HAZARDS

The toxicity of resorcinol is similar to that of phenol, although convulsions are a more common effect. It may cause siderosis of the spleen, kidney damage, changes in the liver and heart, and lung oedema and emphysema (2). Its industrial toxicology has been reviewed (3).

Vapour Inhalation

No toxic effects occurred in animals exposed to 8 ppm, 6 hours/day for 2 weeks. Workers exposed to 10 ppm did not complain of irritation (3). It has reportedly caused respiratory problems in the rubber industry (4).

Eye Contact

Resorcinol may cause eye irritation.

Skin Contact

A 3-25% solution of resorcinol may cause itching, redness, dermatitis, and oedema or corrosion (5). Sensitisation and cross-reactivity to other phenolics have been reported (6). No skin sensitisation was reported in guinea pigs (7,8). It may be absorbed through the skin in a suitable solvent causing symptoms similar to those of ingestion. Skin absorption under normal conditions is slight (9).

Swallowing

The oral LD_{50} in rats and guinea pigs is 370 mg/kg. Hypothermia, hypotension, decreased respiration, tremors, icterus and haemoglobinuria was reported in a child after ingestion of 8 g resorcinol (5). In another case of ingestion in a child, 4 g reportedly caused dizziness and somnolence.

CARCINOGENICITY

In 1977 the available data did not allow an evaluation of the carcinogenicity of resorcinol by the IARC (5). It was not tumorigenic in mice in skin painting studies (10,11).

MUTAGENICITY

Resorcinol is not mutagenic in *Salmonella typhimurium* TA1535, TA1537, TA98 or TA100 either with or without metabolic activation (5,12). Mutagenicity has been reported in *Salmonella typhimuirum* only on ZLM medium (13). No increased incidence of micronucleated erythrocytes was reported after oral dosing in rats (14). Increased frequency of chromosome aberrations have been reported in cultured human lymphocytes, but resorcinol did not induce sister chromatid exchanges. It was negative in an *in vitro* test with human diploid fibroblasts. It did not induce chromosome aberrations or sister chromatid exchanges in Chinese hamster ovary cells. It did not increase the frequency of chromosome aberrations in mice after *in vivo* administration (15,16).

REPRODUCTIVE HAZARDS

At levels which were not overtly maternally toxic, resorcinol was not teratogenic in rats following administration by gavage on days 6-15 of pregnancy (17). It was not teratogenic when applied topically to pregnant rats (18).

FIRST AID

Eyes Wash the eye with flowing water for 10 minutes.

127. Resorcinol

Lungs Remove casualty from area of exposure. If unconscious, do not give anything to drink. Give artificial ventilation and chest compression or place in the recovery position as necessary. If conscious make the casualty lie or sit down quietly, give oxygen if available.

Mouth Do not make the casualty vomit. Treat unconscious casualties as for lungs but if conscious give 1 pint of water to drink. Convulsions may occur and may cause unconsciousness.

Skin Remove contaminated clothing immediately, wash affected areas with soap and copious amounts of water. Absorption through the skin may cause symptoms similar to those of ingestion.

In all cases of exposure, the patient should be transferred to hospital as soon as possible.First Aid.07

HANDLING AND STORAGE

Resorcinol should be handled wearing an approved respirator, chemical-resistant gloves, safety goggles and other protective clothing. It should only be used in chemical fume hood. It should be kept in a tightly closed container, in a cool dry place (1).

DISPOSAL

Eliminate all sources of ignition and ventilate the area. Wearing a laboratory coat or overalls, safety glasses and gloves, absorb the spill onto paper towels and allow to evaporate in a fume-cupboard. For large spills, absorb onto sand or vermiculite, and remove in buckets for atmospheric evaporation in a safe open area. Ideally, waste should be burned in an incinerator with afterburner.

FIRE PRECAUTIONS

Fires involving resorcinol should be extinguished using carbon dioxide, dry chemical powder, alcohol foam or polymer foam. Container explosion may occur under fire conditions.

FURTHER READING

Bretherick, L. *Hazards in the Chemical Laboratory* (4th edition)

Bretherick, L. *Handbook of Reactive Chemical Hazards* (4th edition)

Sax, N. Irving *Dangerous Properties of Industrial Materials* (7th edition)

Encyclopedia of Occupational Safety & Health

Patty's *Industrial Hygiene and Toxicology*

Kirk-Othmer *Encyclopedia of Chemical Technology*

National Fire Protection Association *Manual of Hazardous Reactions*

ACGIH Documentation of TLVs and BEIs (6th edition, 1986)

Dangerous Prop. Ind. Mater. Rep. 1980, **1**(2), 58-59

Toxicology card No. 178: 1,3-benzenediol *Cah. Notes Doc.* 1989, **135**, 389-392

REFERENCES

1. Lenga, R. E. *The Sigma-Aldrich Library of Chemical Safety Data* (2nd edition) (Sigma-Aldrich, Milwaukee, 1988)
2. von Oettingen, W. F. *Natl. Inst. Health Bull.* 1949, **190**
3. Flickinger, C. W. *J. Am. Ind. Hyg. Assoc.* 1976, **37**(10), 596-606
4. Mastromatteo, E. *J. Occup. Med.* 1965, **1**, 502
5. IARC *Monographs on the evaluation of the carcinogenic risk of chemicals to humans* 1977, **15**, 155-175
6. Fisher, A. A. *Contact Dermatitis* (Lea & Febiger, Philadelphia, 1973)
7. Baer, H.; et al. *Immunochemistry* 1966, **3**(6), 479-485
8. Masamoto, Y.; Takase, Y. *Shinshu Igaku Zasshi* 1983, **31**(6), 522-528
9. Maibach, H. I. *Labo-Pharma-Probl. Tech.* 1983, **335**, 727-729
10. Stenbach, F.; Shibik, P. *Toxicol. Appl. Pharmacol.* 1974, **30**, 7-13
11. Stenbach, F. *Acta Pharmacol. Toxicol.* 1977, **41**(5), 417-431

127. Resorcinol

12. Shahin, M. M.; et al. *Mutat. Res.* 1980, **78**(3), 213-218
13. Gocke, E.; et al. *Mutat. Res.* 1981, **90**, 91-109
14. Hossack, D. J. N.; Richardson, J. C. *Experientia* 1977, **33**(3), 377-378
15. Darroudi, F.; Natarajan, A. T. *Mutat. Res.* 1983, **124**(2), 179-189
16. Dean, B. J. *Mutat. Res.* 1985, **154**(3), 153-181
17. DiNardo, J. C.; et al. *Toxicol. Appl. Pharmacol.* 1985, **78**(1), 163-166
18. Burnett, C.; et al. *J. Toxicol. Environ. Health* 1976, **1**(6), 1027-1040

128. Selenium

SELENIUM
Se

RISKS
Toxic by inhalation and if swallowed – Danger of cumulative effects (R23/25, R33)

SAFETY PRECAUTIONS
When using do not eat, drink or smoke – After contact with skin, wash immediately with plenty of water – If you feel unwell, seek medical advice (show label where possible) (S20/21, S28, S44)

PHYSICAL PROPERTIES

Description Steel grey, non-metallic element. Exists in several allotropic forms: amorphous, crystalline or red and grey or metallic. Liquid is brownish red.

Boiling point	690°C
Melting point	170°C-217°C
Density	4.81
Vapour density	not available
Vapour pressure	1 mm Hg at 356°C
Flash point	not available
Explosive limits	not available
Autoignition temperature	not available

Solubility Insoluble in water or alcohol. Very slightly soluble in ether.

IDENTIFIERS

SYNONYMS	colloidal selenium; Vandex; C.I. 77805
CHEMICAL ABSTRACTS No.	7782-49-2
NIOSH No.	VS 7700000
HAZCHEM CODE	2Z
UN No.	2658

PACKAGING AND TRANSPORTATION

Road transportation
hazard warning sign	2658 harmful substance
Hazchem code	2Z

Sea transportation
IMDG page No.	6250
class	6.1
label	harmful stow away from foodstuffs
packaging group	III

Air transportation
ICAO/IATA code (UN No.)	2658
class	6.1
label	keep away from food
packaging group	III
packing instructions	
cargo	619
passenger	619
passenger aircraft max. quantity	100 kilograms
cargo aircraft max. quantity	200 kilograms

THRESHOLD LIMIT VALUES

US TLV (TWA)	0.2 mg/m^3
US TLV (STEL)	not available
UK EXPOSURE LIMITS (OES)	
Long-term (8 hr TWA value)	0.1 mg/m^3
Short-term (10 min TWA value)	not available
Germany	
MAK	0.1 mg/m^3
France	
VME	not available
VLE	not available
Sweden	
Short-term limit	not available
Level limit	0.1 mg/m^3

128. Selenium

MANUFACTURE

Selenium is widely distributed in nature, occurring in sulphide ores of the heavy metals. It is obtained as a by-product of precious metal recovery from copper refinery slimes.

USES

Selenium is used as a decolouriser for glass and ceramics, in photoelectric cells, as a vulcanising agent in rubber manufacture, as a catalyst, in photographic toning baths, as a rectifier in radio and television sets, and in electrodes. In the U.S.A. the PAIR applies to selenium (1).

CHEMICAL HAZARDS

Selenium may react explosively with nitrogen trichloride, and forms explosive mixtures with sodium peroxide. Explosive products are also formed with alkali metal amides or alkaline earth metal amides. An explosion occurred on conversion of recovered selenium metal to the dioxide by heating in oxygen, probably due to selenium catalysed oxidation of organic impurities in the selenium; oxidation with nitric acid is, however, safe (2). Mixtures with barium peroxide ignite at 265°C (3), and selenium ignites on grinding with silver(I) oxide. Contact with bromine pentafluoride or chlorine trifluoride at ambient or slightly elevated temperatures is violent, ignition often occurring. It ignites with fluorine at ambient temperature. It also reacts violently with chromium trioxide, and with aqueous solutions of potassium bromate. It reacts incandescently with hexalithium disilicide, and metal acetylides incandesce on heating in selenium vapour. Slightly moist mixtures of selenium and chlorates (except alkali metal chlorates) incandesce. Selenium reacts incandescently with uranium, zinc, platinium, phosphorus, sodium, potassium and nickel on warming. The reaction with potassium is mildly explosive. Powdered tin and selenium react very exothermically at 350°C. Particles of cadmium and selenium must be under a critical size to avoid explosions during synthesis of cadmium selenide by heating together. Reaction of cadmium or zinc with sulphur, selenium or tellurium is similar.

BIOLOGICAL HAZARDS

Occupational exposure to selenium compounds causes bronchial irritation, irritation of the nose and throat, garlic-smelling breath (due to respiratory excretion of dimethylselenium), and gastrointestinal symptoms. Symptoms of chronic exposure include a metallic taste, pallor, irritability, giddiness, excess dental caries, depression, nervousness, fatigue, dermatitis, and hair and fingernail loss (4). Liver and kidney damage, myocarditis, anaemia and pancreatitis have also been reported in animals. There appear to be no reports of disabling chronic disease or death from industrial exposure (5). The health effects of selenium have been reviewed (6,7), although relatively little is known about the health effects of industrial exposure (8).

Vapour Inhalation

Exposure to elemental selenium causes sneezing, coughing and mucous membrane irritation. Heavy exposure to selenium fume may cause headache, sore throat, dyspnoea and tracheobronchitis (9,10). As selenium dioxide may also be present, these symptoms may be due to exposure to selenium dioxide. Interstitial pneumonitis has been reported in rats exposed to selenium dust.

Eye Contact

Exposure to elemental selenium may cause redness of the eye and conjunctivitis (9-11).

Skin Contact

Elemental selenium does not cause skin injury. Dermatitis has been reported in industry but may be due to the dioxide.

Swallowing

Elemental selenium has a low oral toxicity due to its insolubility. The oral LD_{50} in rats is 6700 mg/kg. Excess dietary selenium causes blind staggers and/or alkali disease (anorexia and collapse) in livestock.

CARCINOGENICITY

The WHO in 1987 felt that there was insufficient animal data available to make an evaluation of the carcinogenicity of selenium compounds (8). No excess mortality from cancer was found in workers exposed to selenium in rectifier production. Evidence from epidemiological studies suggest an inverse relationship between cancer and selenium ingestion in humans. Trace amounts of selenium appear to be antitumourigenic (12-14). The epidemiology of selenium and cancer have been reviewed (15,16).

MUTAGENICITY

Selenium compounds are reportedly weak mutagens in the Ames test, and have produced sister chromatid exchanges and chromosome aberrations in mammalian cells *in vitro* (8).

REPRODUCTIVE HAZARDS

It has been suggested that excess selenium has caused reproductive problems in livestock. It should be noted that selenium in the diet may have different characteristics from those of industrial exposure (8).

FIRST AID

Eyes Wash the eye with flowing water for 10 minutes.

Lungs Remove casualty from area of exposure. If unconscious, do not give anything to drink. Give artificial ventilation and chest compression or place in the recovery position as necessary. If conscious make the casualty lie or sit down quietly, give oxygen if available. Lung congestion may occur – a conscious casualty with breathing difficulties should be placed in a sitting position. Allergic asthma (wheezy breathing) may occur and immediate treatment is needed – Ventolin may be useful.

Mouth Do not make the casualty vomit. Treat unconscious casualties as for lungs but if conscious give 1 pint of water to drink. Shock may result – if so do not give any drinks, and if conscious lie casualty flat with legs raised. Convulsions may occur and may cause unconsciousness.

Skin Remove contaminated clothing immediately, drench the affected area with running water for at least 10 minutes.

In all cases of exposure, the patient should be transferred to hospital as soon as possible.

HANDLING AND STORAGE

Selenium should be handled wearing an approved respirator, long rubber or neoprene gloves or gauntlets, safety goggles and other protective clothing. It should only be used in a chemical fume hood (17). It should be kept in a tightly closed container, in a cool, dry place, away from strong acids or strong oxidising agents and in a fireproof area (18). Commercial elemental selenium is relatively inert and can be handled without special precautions.

DISPOSAL

Ventilate the area, but isolate the material. Wear a laboratory coat, safety spectacles, compressed air breathing apparatus and rubber gloves. Mix powdered spillages with large quantities of sand then treat as normal refuse.

FIRE PRECAUTIONS

Selenium is not combustible; an extinguisher suitable for surrounding fire conditions should be used. Various symptoms have been described in fire fighters at a selenium photocell plant (19).

FURTHER READING

Bretherick, L. *Hazards in the Chemical Laboratory* (4th edition)

Bretherick, L. *Handbook of Reactive Chemical Hazards* (4th edition)

Sax, N. Irving *Dangerous Properties of Industrial Materials* (7th edition)

Encyclopedia of Occupational Safety & Health

Patty's *Industrial Hygiene and Toxicology*

Kirk-Othmer *Encyclopedia of Chemical Technology*

National Fire Protection Association *Manual of Hazardous Reactions*

ACGIH Documentation of TLVs and BEIs (6th edition, 1986)

Rosenfeld, I,; Beath, O. A. *Selenium* (Academic Press, New York, 1964)

Bagnall, K. W. *The Chemistry of Selenium, Tellurium and Polonium* (Elsevier, New York, 1966)

Merian, E.; et al. (editors) *Current topics in environmental and toxicological chemistry, Vol. 11*

Carcinogenic and mutagenic metal compounds 2 (Gordon & Breach, Montreux, 1986)

Dangerous Prop. Ind. Mater. Rep. 1981, **1**(3)

128. Selenium

Lacasse, Y.; Richier, C. Toxicity of selenium and its derivativatives *Union Med. Can.* 1976, **105**(8), 1192-1199

Morel c.; et al. *Cah. Notes Doc.* 1980, **98**, 181-185

REFERENCES

1. *TSCA Chem. Prog. Bull.* 1988, **9**(3), 9
2. Astin, S.; et al. *J. Chem. Soc.* 1933, 391
3. Johnson, L. B. *Ind. Eng. Chem.* 1960, **52**, 241-244
4. Cooper, C. W.; Glover, J. R. *Zingaro and Cooper's Selenium* (Van Nostrand Reinhold Co., New York, 1974) pp. 654-673
5. Glover, J. R. *Ind. Med. Surg.* 1970, **39**, 50
6. *NTIS Report* 1984, EPA/540/1-86/058 Order No. PB86-134699/GAR
7. Olson, O. E. *J. Am. Coll. Toxicol.* 1986, **5**(1), 45-70
8. *Environmental Health Criteria 58: Selenium* (WHO, Geneva, 1987)
9. Clinton, M. *J. Ind. Hyg. Toxicol.* 1947, **29**, 225
10. Dudley, H. C. *Pub. Health Repts.* 1938, **53**, 94
11. Holness, D. L.; et al. *Arch. Environ. Health* 1989, **44**(5), 291-297
12. Newberne, P. M.; Suphakarn, V. *Nutr. Cancer* 1983, **5**(2), 107-119
13. Vernie, L. N. *Biochim. Biophys. Acta* 1984, **738**(4), 203-217
14. Gehardsson, L.; et al. *Acta Pharm. Toxicol.* 1986, **59**, 256-259
15. Clark, L. C. *Fed. Proc.* 1985, **44**(9), 2584-2589
16. Kazantzis, G. *EHP, Environ. Health Perspect.* 1981, **40**, 143-161
17. Lenga, R. E. *The Sigma-Aldrich Library of Chemical Safety Data* (2nd edition) (Sigma-Aldrich, Milwaukee, 1988)
18. *International Chemical Safety Cards: Selenium* (CEC, Luxembourg, 1990)
19. Isa, A.; Shirahige, K. *Kasai* 1983, **146**, 27-33

129. Selenium Dioxide

SELENIUM DIOXIDE
SeO_2

RISKS
Toxic by inhalation and if swallowed – Danger of cumulative effects (R23/25, R33)

SAFETY PRECAUTIONS
When using do not eat, drink or smoke – After contact with skin, wash immediately with plenty of water – If you feel unwell, seek medical advice (show label where possible) (S20/21, S28, S44)

IDENTIFIERS

SYNONYMS	selenium oxide (SeO_2); RCRA Waste No. U204
CHEMICAL ABSTRACTS No.	7446-08-4
NIOSH No.	VS 8575000
HAZCHEM CODE	not available
UN No.	not available

THRESHOLD LIMIT VALUES

US TLV (TWA)	0.2 mg/m^3 (as Se)
US TLV (STEL)	not available
UK EXPOSURE LIMITS (OES)	
Long-term (8 hr TWA value)	0.1 mg/m^3 (as Se)
Short-term (10 min TWA value)	not available
Germany	
MAK	0.1 mg/m^3
France	
VME	not available
VLE	not available
Sweden	
Short-term limit	not available
Level limit	0.1 mg/m^3 (as Se)

PHYSICAL PROPERTIES

Description White to slightly reddish, lustrous, crystalline powder or needles. Acidic taste, leaves a burning sensation. Its yellowish green vapour has a pungent, sour smell.

Boiling point	sublimes at 315°C
Melting point	sublimes at 340-350°C
Density	3.95 at 15°C
Vapour density	not available
Vapour pressure	1 mm Hg at 157°C
Flash point	not available
Explosive limits	not available
Autoignition temperature	not available

Solubility Soluble in water, methanol, ethanol, acetone, acetic acid or concentrated sulphuric acid.

PACKAGING AND TRANSPORTATION

Road transportation

hazard warning sign	not available
Hazchem code	not available

Sea transportation

IMDG page No.	not available
class	not available
label	not available
packaging group	not available

Air transportation

ICAO/IATA code	not available
class	not available
label	not available
packaging group	not available
packing instructions	
cargo	not available
passenger	not available
passenger aircraft max. quantity	not available
cargo aircraft max. quantity	not available

129. Selenium Dioxide

MANUFACTURE

Selenium oxide is produced by oxidising selenium with nitric acid followed by evaporation. It is also obtained by oxidising selenium with air or oxygen.

USES

The dioxide is the most widely used selenium compound. It is used in the production of other selenium compounds; as an oxidising agent for organics and pharmaceuticals; and as a reagent for alkaloids. It has also been used in etching and anti-corrosion coatings on magnesium.

CHEMICAL HAZARDS

An explosion occurred during attempted conversion of recovered selenium metal to the dioxide by heating in oxygen. This was attributed to organic impurities in the selenium. Organic impurities also lead to vigorous oxidation of recovered selenium by nitric acid, but this is a safe procedure (1). Such explosions have been wrongly attributed to the nitric acid procedure (2). Heating 1,3- bis(trichloromethyl)benzene with selenium dioxide to form the bis(acyl chloride) caused eruptions at higher temperatures and was too hazardous to continue (3). A mixture of cold phosphorus trichloride and selenium dioxide reaches red heat. Its hazardous reactions have been reviewed (4).

BIOLOGICAL HAZARDS

Selenium dioxide is formed when selenium is heated, hence exposure to selenium dioxide also occurs in industrial processes involving the element itself. The toxicity of selenium and its compounds has been reviewed (5).

Vapour Inhalation

Selenium dioxide is severely irritating to the nose and throat. High concentrations of selenium dioxide cause pulmonary oedema. Systemic effects of exposure include immediate sneezing, violent coughing and gagging, and headache, followed for several hours by blocked nose, dizziness, garlic-smelling breath, and a metallic taste. Other symptoms reported are dyspnoea, slight fever, nausea, vomiting, tiredness, irritability, bronchial spasms, pneumonia and unconsciousness. A generalised allergic body rash may occasionally occur. In most of these industrial incidents recovery occurred in a matter of days and there were no sequelae (5,6). Chronic exposure may cause rhinitis, nose bleeds, headache, weight loss, irritability, tiredness, weakness, sleeplessness, dyspnoea, painful extremities, gastrointestinal symptoms, and liver damage (5). The LC_{Lo} in rabbits (20 minutes) is 5890 mg/m^3, and lung oedema, and damage to the liver, kidneys and heart have been reported in animal studies (5).

Eye Contact

Selenium dioxide is severely irritating to the eyes, and causes a pink allergic-type reaction of the eyelids, which may become swollen and may be accompanied by conjunctivitis.

Skin Contact

Selenium dioxide causes very painful skin burns, the effects of which may not be apparent for 4 hours, and dermatitis (5,7). Its effects are similar to those of hydrofluoric acid. It causes severe pain if lodged under the fingernails (8).

Swallowing

No information is available on the oral toxicity of selenium dioxide.

CARCINOGENICITY

Fed at levels of 2 and 6 ppm in the drinking water to female mice, selenium dioxide inhibited mammary tumourigenesis without interfering with normal reproductive function and weight gain (9). It also appeared to have an antitumourigenic effect on B16 melanoma cells (10).

MUTAGENICITY

Selenium dioxide gave positive results in the rec assay with *Bacillus subtilis* (11), and in DNA repair tests in *Escherichia coli* (12). It induced sister chromatid exchanges in human whole blood cultures (13).

REPRODUCTIVE HAZARDS

Fed at levels of 2 and 6 ppm in the drinking water to female mice, selenium dioxide inhibited mammary tumourigenesis without interfering with normal reproductive function and weight gain (9). Intraperitoneal injection caused testicular degeneration and atrophy (14).

129. Selenium Dioxide

FIRST AID

Eyes Wash the eye with flowing water for 10 minutes.

Lungs Remove casualty from area of exposure. If unconscious, do not give anything to drink. Give artificial ventilation and chest compression or place in the recovery position as necessary. If conscious make the casualty lie or sit down quietly, give oxygen if available. Lung congestion may occur – a conscious casualty with breathing difficulties should be placed in a sitting position. Allergic asthma (wheezy breathing) may occur and immediate treatment is needed – Ventolin may be useful.

Mouth Do not make the casualty vomit. Treat unconscious casualties as for lungs but if conscious give 1 pint of water to drink. Immediately then give repeated drinks of water (1 cupful every 10 minutes.) Shock may result – if so do not give any drinks, and if conscious lie casualty flat with legs raised. Convulsions may occur and may cause unconsciousness.

Skin Remove contaminated clothing immediately, drench the affected area with running water for at least 10 minutes.

In all cases of exposure, the patient should be transferred to hospital as soon as possible.

HANDLING AND STORAGE

Selenium dioxide should be handled wearing an approved respirator, chemical-resistant gloves, safety goggles and other protective clothing. It should only be used in a chemical fume hood. It should be kept in a tightly closed container in a cool, dry place (6).

DISPOSAL

Ventilate the area, but isolate the material. Wear a laboratory coat, safety spectacles, compressed air breathing apparatus and rubber gloves. Cover with soda ash and mix well. Cautiously mop up with plenty of water and run to waste with water.

FIRE PRECAUTIONS

Selenium dioxide is not combustible and an extinguisher suitable to surrounding fire conditions should be used.

FURTHER READING

Bretherick, L. *Hazards in the Chemical Laboratory* (4th edition)

Bretherick, L. *Handbook of Reactive Chemical Hazards* (4th edition)

Sax, N. Irving *Dangerous Properties of Industrial Materials* (7th edition)

Encyclopaedia of Occupational Safety & Health

Patty's *Industrial Hygiene and Toxicology*

Kirk-Othmer *Encyclopedia of Chemical Technology*

National Fire Protection Association *Manual of Hazardous Reactions*

ACGIH *Documentation of TLVs and BEIs* (6th edition, 1986)

REFERENCES

1. Astin, S.; et al. *J. Chem. Soc.* 1933, 391
2. Watkins, C. R.; et al. *Chem. Rev.* 1945, **36**, 235
3. Rondevstedt, C. S. *J. Org. Chem.* 1976, **41**, 3574-3577
4. Leleu, J. *Cah. Notes Doc.* 1976, **83**, 281-294
5. *Environmental Health Criteria 58: Selenium* (WHO, Geneva, 1987)
6. Lenga, R. E. *The Sigma-Aldrich Library of Chemical Safety Data* (2nd edition) (Sigma-Aldrich, Milwaukee, 1988)
7. Cooper, W. C. *Symp.: Selenium Biomed., Int. Symp., 1st, Oregon State Univ.* 1966, 185-199
8. Glover, J. R. *Ind. Med. Surg.* 1970, **39**, 50
9. Medina, D.; Shepherd, F. *Cancer Lett.* 1980, **8**(3), 241-245
10. Hanada, K.; et al. *J. Dermatol.* 1986, **13**(1), 19-23
11. Kanematsu, N.; et al. *Mutat. Res.* 1980, **77**, 109-116

129. Selenium Dioxide

12. Yagi, T.; Nishioka, H. *Doshisha Daigaku Rikogaku Kenkyu Hokoku* 1977, **18**(2), 63-70
13. Ray, J. H.; et al. *Mutat. Res.* 1980, **78**(3), 261-266
14. Chowdhury, A. R.; Venkatakrishna-Bhatt, H. *Indian J. Physiol. Pharmacol.* 1983, **27**(3), 237-240

130. Selenium Disulphide

SELENIUM DISULPHIDE

SeS$_2$

RISKS
Toxic by inhalation and if swallowed – Danger of cumulative effects (R23/25, R33)

SAFETY PRECAUTIONS
When using do not eat, drink or smoke – After contact with skin, wash immediately with plenty of water – If you feel unwell, seek medical advice (show label where possible) (S20/21, S28, S44)

IDENTIFIERS

SYNONYMS Exsel; RCRA Waste No. U205; Selsun Blue; selenium sulphide (SeS$_2$)
CHEMICAL ABSTRACTS No. 7488-56-4
NIOSH No. .... VS 8925000
HAZCHEM CODE 2Z
UN No. 2657

THRESHOLD LIMIT VALUES

US TLV (TWA) .... 0.2 mg/m^3 (as Se)
US TLV (STEL) .... not available
UK EXPOSURE LIMITS (OES)
Long-term (8 hr TWA value) ..... 0.1 mg/m^3 (as Se)
Short-term (10 min TWA value) .... not available
Germany
MAK .... 0.1 mg/m^3
France
VME .... not available
VLE .... not available
Sweden
Short-term limit .... not available
Level limit .... 0.1 mg/m^3 (as Se)

PHYSICAL PROPERTIES

Description Red-yellow crystals.
Boiling point .... decomposes
Melting point .... < 100°C
Density .... not available
Vapour density .... not available
Vapour pressure .... not available
Flash point .... not available
Explosive limits .... not available
Autoignition temperature .... not available
Solubility .... not available.

PACKAGING AND TRANSPORTATION

Road transportation
hazard warning sign .... 2657 toxic substance
Hazchem code 2Z

Sea transportation
IMDG page No. 6249
class 6.1
label .... poison
packaging group II

Air transportation
ICAO/IATA code (UN No.) 2657
class 6.1
label .... poison
packaging group II
packing instructions
cargo 615
passenger 613
passenger aircraft max. quantity ... 25 kilograms
cargo aircraft max. quantity .... 100 kilograms

130. Selenium Disulphide

MANUFACTURE

Selenium disulphide can be prepared by reaction of hydrogen sulphide or alkali metal sulphides with selenious acid or its salts and excess acetic acid in a non-aqueous medium with a high dielectric constant. The precipitate formed is washed with ethanol or methanol (1).

USES

Selenium disulphide is used in shampoos as an antidandruff agent, and as an antifungal agent.

CHEMICAL HAZARDS

Selenium disulphide emits highly toxic fumes of sulphur oxides and selenium on heating to decomposition. It is incompatible with strong oxidising agents and reacts violently with silver(I) oxide (2).

BIOLOGICAL HAZARDS

Selenium disulphide is destructive to the eyes, skin and respiratory tract. It has caused tumours in laboratory animals, but results in mutagenicity assays are equivocal.

Vapour Inhalation

High concentrations of selenium disulphide are very destructive to tissues of the upper respiratory tract. It may cause coughing, dyspnoea, laryngitis, headache, nausea and vomiting (2).

Eye Contact

High concentrations of selenium disulphide are very destructive to the eyes (2).

Skin Contact

High concentrations of selenium disulphide are very destructive to the skin (2).

Swallowing

The oral LD_{50} in rats is 138 mg/kg.

CARCINOGENICITY

Selenium disulphide was positive in a two year rodent bioassay (given orally), and negative in the strain A mouse pulmonary tumour bioassay (3).

MUTAGENICITY

Selenium disulphide gave equivocal results when tested for induction of unscheduled DNA synthesis and S-phase synthesis in mouse and rat hepatocytes following *in vivo* treatment (4).

REPRODUCTIVE HAZARDS

No information is available specific to selenium disulphide. Dietary selenium appears to affect the fertility of livestock, and selenium dioxide has caused testicular damage in laboratory animals (5).

FIRST AID

Eyes Wash the eye with flowing water for 10 minutes.

Lungs Remove casualty from area of exposure. If unconscious, do not give anything to drink. Give artificial ventilation and chest compression or place in the recovery position as necessary. If conscious make the casualty lie or sit down quietly, give oxygen if available. Lung congestion may occur – a conscious casualty with breathing difficulties should be placed in a sitting position. Allergic asthma (wheezy breathing) may occur and immediate treatment is needed – Ventolin may be useful.

Mouth Do not make the casualty vomit. Treat unconscious casualties as for lungs but if conscious give 1 pint of water to drink. Shock may result – if so do not give any drinks, and if conscious lie casualty flat with legs raised. Convulsions may occur and may cause unconsciousness.

Skin Remove contaminated clothing immediately, drench the affected area with running water for at least 10 minutes.

In all cases of exposure, the patient should be transferred to hospital as soon as possible.

130. Selenium Disulphide

HANDLING AND STORAGE

Selenium disulphide should be handled wearing an approved respirator, chemical-resistant gloves, safety goggles and other protective clothing. It should only be used in a chemical fume hood. It should be kept in a tightly closed container, in a cool, dry place (2).

DISPOSAL

Ventilate the area, but isolate the material. Wear a laboratory coat, safety spectacles, compressed air breathing apparatus and rubber gloves. Mix powdered spillages with large quantities of sand then treat as normal refuse.

FIRE PRECAUTIONS

Selenium disulphide is not combustible and an extinguisher suitable to surrounding fire conditions should be used.

FURTHER READING

Bretherick, L. *Hazards in the Chemical Laboratory* (4th edition)

Bretherick, L. *Handbook of Reactive Chemical Hazards* (4th edition)

Sax, N. Irving *Dangerous Properties of Industrial Materials* (7th edition)

Encyclopaedia of Occupational Safety & Health

Patty's *Industrial Hygiene and Toxicology*

Kirk-Othmer *Encyclopedia of Chemical Technology*

National Fire Protection Association *Manual of Hazardous Reactions*

ACGIH Documentation of TLVs and BEIs (6th edition, 1986)

REFERENCES

1. Wolski, T.; et al. Pol. Pat. PL 106701 (1980)
2. Lenga, R. E. *The Sigma-Aldrich Library of Chemical Safety Data* (2nd edition) (Sigma-Aldrich, Milwaukee, 1988)
3. Maronpot, R. R.; et al. *Environ. Sci. Res.* 1983, **27**(short-term bioassays anal. complex environ.), 341-349
4. *Environ. Mol. Mutagen.* 1989, **14**(3), 155-164
5. *Environmental Health Criteria 58: Selenium* (WHO, Geneva, 1987)

131. Selenium Hexafluoride

SELENIUM HEXAFLUORIDE
SeF$_6$

PHYSICAL PROPERTIES

Description	Colourless gas.
Boiling point	-34.5°C
Melting point	-39°C (sublimes at -40.6°C)
Density	3.25 at -25°C
Vapour density	not available
Vapour pressure	651.2 torr at -48.7°C
Flash point	not available
Explosive limits	not available
Autoignition temperature	not available
Solubility	Insoluble in water.

RISKS
Toxic by inhalation and if swallowed – Danger of cumulative effects (R23/25, R33)

SAFETY PRECAUTIONS
When using do not eat, drink or smoke – After contact with skin, wash immediately with plenty of water – If you feel unwell, seek medical advice (show label where possible) (S20/21, S28, S44)

IDENTIFIERS

SYNONYMS	selenium fluoride (SeF$_6$)
CHEMICAL ABSTRACTS No.	7783-79-1
NIOSH No.	VS 9450000
HAZCHEM CODE	not available
UN No.	2194

THRESHOLD LIMIT VALUES

US TLV (TWA)	0.05 ppm (0.16 mg/m^3)
US TLV (STEL)	not available

UK EXPOSURE LIMITS (OES)
Long-term (8 hr TWA value) ..... 0.1 mg/m^3 (as Se)
Short-term (10 min TWA value) .... not available

Germany
MAK .... 0.1 mg/m^3

France
VME .... 0.05 ppm (0.2 mg/m^3)
VLE .... not available

Sweden
Short-term limit .... not available
Level limit .... 0.1 mg/m^3 (as Se)

PACKAGING AND TRANSPORTATION

Road transportation
hazard warning sign	2194 toxic gas
Hazchem code	not available

Sea transportation
IMDG page No.	2177
class	2;2.3
label	poison gas
packaging group	not available

Air transportation
ICAO/IATA code (UN No.)	2194
class	2;6.1
label	not available
packaging group	not available
packing instructions	
cargo	forbidden
passenger	forbidden
passenger aircraft max. quantity	forbidden
cargo aircraft max. quantity	forbidden

131. Selenium Hexafluoride

MANUFACTURE

Selenium hexafluoride is produced by passing gaseous fluorine over finely divided selenium in a copper vessel.

USES

Selenium hexafluoride is used as a gaseous electric insulator.

CHEMICAL HAZARDS

Selenium hexafluoride emits toxic fumes of fluoride when heated to decomposition.

BIOLOGICAL HAZARDS

Selenium hexafluoride has a high acute toxicity (1).

Vapour Inhalation

In animal studies, four hour exposures to 10 ppm were uniformly fatal, while exposure to 5 ppm caused pulmonary oedema (2).

Eye Contact

No information is available concerning the effects on the eyes.

Skin Contact

No information is available concerning the effects on the skin.

Swallowing

No information is available concerning the effects of ingestion.

CARCINOGENICITY

No information is available specific to selenium hexafluoride. Reported data on other inorganic selenium compounds suggests that the dioxide has an antitumour effect, and that the sulphide may be carcinogenic in rodents (2).

MUTAGENICITY

No information is available specific to selenium hexafluoride. Data from other inorganic selenium compounds suggests that the monosulphide is mutagenic in bacterial assays, and that the dioxide is mutagenic in bacterial and mammalian cell assays (2).

REPRODUCTIVE HAZARDS

No information is available specific to selenium hexafluoride. Dietary selenium appears to affect the fertility of livestock, and selenium dioxide has caused testicular damage in laboratory animals (2).

FIRST AID

Eyes Wash the eye with flowing water for 10 minutes, then irrigate with normal saline for at least 30 minutes. Do not use calcium gluconate gel.

Lungs Remove casualty from area of exposure. If unconscious do not give anything to drink, give artificial ventilation and chest compression or place in the recovery position as necessary. If conscious make the casualty lie or sit down quietly, give oxygen if available.

Mouth Do not make the casualty vomit. Treat unconscious casualties as for lungs but if conscious give milk or calcium gluconate by mouth.

Skin Remove contaminated clothing immediately, wearing protective gloves and clothes. If calcium gluconate gel is available, immediately rub into all affected areas and massage until all pain goes. If not, wash with soap and water for 30 minutes.

In all cases of exposure, the patient should be transferred to hospital as soon as possible.

131. Selenium Hexafluoride

HANDLING AND STORAGE

Selenium hexafluoride should be handled wearing an approved respirator, chemical-resistant gloves, safety goggles and other protective clothing. It should only be used in a chemical fume hood. It should be kept in a tightly closed container, in a cool, dry place.

DISPOSAL

Eliminate all sources of ignition and ventilate the area. Wear a laboratory coat, approved breathing apparatus, and heavy work gloves.

FIRE PRECAUTIONS

Selenium hexafluoride is not combustible and an extinguisher suitable to surrounding fire conditions should be used.

FURTHER READING

Bretherick, L. *Hazards in the Chemical Laboratory* (4th edition)

Bretherick, L. *Handbook of Reactive Chemical Hazards* (4th edition)

Sax, N. Irving *Dangerous Properties of Industrial Materials* (7th edition)

Encyclopaedia of Occupational Safety & Health

Patty's *Industrial Hygiene and Toxicology*

Kirk-Othmer *Encyclopedia of Chemical Technology*

National Fire Protection Association *Manual of Hazardous Reactions*

REFERENCES

1. *ACGIH Documentation of TLVs and BEIs* (6th edition, 1986)
2. *Environmental Health Criteria 58: Selenium* (WHO, Geneva, 1987)

SELENIUM HYDRIDE

H_2Se

RISKS
Toxic by inhalation and if swallowed – Danger of cumulative effects (R23/25, R33)

SAFETY PRECAUTIONS
When using do not eat, drink or smoke – After contact with skin, wash immediately with plenty of water – If you feel unwell, seek medical advice (show label where possible) (S20/21, S28, S44)

IDENTIFIERS

SYNONYMS dihydrogen selenide; Electronic E-2; hydrogen selenide; hydrogen selenide (H_2Se); selane; selenium dihydride

CHEMICAL ABSTRACTS No.	7783-07-5
NIOSH No.	MX 1050000
HAZCHEM CODE	not available
UN No.	2202

THRESHOLD LIMIT VALUES

US TLV (TWA) .... 0.05 ppm (0.16 mg/m^3)
US TLV (STEL) .... not available
UK EXPOSURE LIMITS (OES)
Long-term (8 hr TWA value)   0.05 ppm (0.2 mg/m^3)
Short-term (10 min TWA value) .... not available

Germany
MAK .... not available

France
VME .... 0.02 ppm (0.08 mg/m^3)
VLE .... not available

Sweden
Short-term limit .... 0.05 ppm (0.2 mg/m^3)
Level limit .... 0.01 ppm (0.03 mg/m^3)

132. Selenium Hydride

PHYSICAL PROPERTIES

Description Colourless, flammable gas with a disagreeable smell.

Boiling point	-41.4°C
Melting point	-64°C
Density	3.614g (gas); 2.12 at -42°C (liquid)
Vapour density	not available
Vapour pressure	10 atm at 23.4°C
Flash point	not available
Explosive limits	not available
Autoignition temperature	not available

Solubility Soluble in carbonyl chloride and carbon disulphide.

PACKAGING AND TRANSPORTATION

Road transportation
hazard warning sign ....
.... 2202 toxic gas; flammable gas
Hazchem code .... not available

Sea transportation
IMDG page No.	2151
class	2;2.3
label	flammable gas; poison gas
packaging group	not available

Air transportation
ICAO/IATA code (UN No.)	2202
class	not available
label	not available
packaging group	not available
packing instructions	
cargo	not available
passenger	not available
passenger aircraft max. quantity	not available
cargo aircraft max. quantity	not available

132. Selenium Hydride

MANUFACTURE

Selenium hydride is produced by action of acids or water on a few metal selenides, mainly aluminium or iron selenides. It can also be produced by passing hydrogen and selenium vapour over pumice at 177°C, which gives a 58% yield.

USES

Selenium hydride has potential applications in the electronics industry.

CHEMICAL HAZARDS

Selenium hydride ignites when fuming nitric acid is dripped onto it (1). It reacts very rapidly with hydrogen peroxide. It is a dangerous fire hazard on exposure to heat or flame and forms explosive mixtures with air.

BIOLOGICAL HAZARDS

Selenium hydride is highly toxic and irritating to the eyes and mucous membranes. The most characteristic symptom of selenium hydride poisoning is a garlicky odour of the breath (2). It may cause damage to the liver, lungs and spleen.

Vapour Inhalation

Selenium hydride is a severe respiratory tract irritant and is highly toxic (3). Gastrointestinal complaints, dental caries, conjunctivitis, nail deformities and garlicky breath were reported by a woman following repeated exposure to selenium hydride (2). It may also cause tight chestedness, nausea, dizziness, violent sneezing, coughing, running nose, nausea, vomiting, tiredness, lung oedema, bronchitis, bronchopneumonia and dyspnoea. The development of severe hyperglycaemia has also been reported (4). Chronic exposure causes nausea, vomiting, dizziness, tiredness and a metallic taste in the mouth. In animal studies repeated exposures to 0.3 ppm are fatal due to pneumonitis. At 0.3 ppm the odour of selenium hydride is readily detected but does not cause irritation. Levels of 1.5 ppm or above cause marked nasal irritation. Olfactory fatigue occurs rapidly, hence its odour and irritation do not offer adequate warning.

Eye Contact

At levels of 1.5 ppm or above, selenium hydride is strongly irritating to the eyes, causing lacrimation and conjunctivitis.

Skin Contact

Selenium hydride is a skin irritant, and may cause a bluish-red erythema of the skin (4).

Swallowing

Not applicable.

CARCINOGENICITY

No information is available specific to selenium hydride. Reported data on other inorganic selenium compounds suggests that the dioxide has an antitumour effect, and that the sulphide may be carcinogenic in rodents (4).

MUTAGENICITY

No information is available specific to selenium hydride. Data from other inorganic selenium compounds suggests that the monosulphide is mutagenic in bacterial assays, and that the dioxide is mutagenic in bacterial and mammalian cell assays (4).

REPRODUCTIVE HAZARDS

No information is available specific to selenium hydride. Dietary selenium appears to affect the fertility of livestock, and selenium dioxide has caused testicular damage in laboratory animals (4).

FIRST AID

Eyes Wash the eye with flowing water for 10 minutes.

Lungs Remove casualty from area of exposure. If unconscious, do not give anything to drink. Give artificial ventilation and chest compression or place in the recovery position as necessary. If conscious make the casualty lie or sit down quietly, give oxygen if available. Lung congestion may occur – a conscious casualty with breathing difficulties should be placed in a sitting position. Allergic asthma (wheezy breathing) may occur and immediate treatment is needed – Ventolin may be useful.

132. Selenium Hydride

Mouth Do not make the casualty vomit. Treat unconscious casualties as for lungs but if conscious give 1 pint of water to drink. Shock may result – if so do not give any drinks, and if conscious lie casualty flat with legs raised. Convulsions may occur and may cause unconsciousness.

Skin Remove contaminated clothing immediately, drench the affected area with running water for at least 10 minutes.

In all cases of exposure, the patient should be transferred to hospital as soon as possible.

HANDLING AND STORAGE

Selenium hydride should be handled wearing an approved respirator, chemical-resistant gloves, safety goggles and other protective clothing. It should only be used in a chemical fume hood. It should be kept in a tightly closed container, in a cool, dry place.

DISPOSAL

Eliminate all sources of ignition and ventilate the area. Wear a laboratory coat, approved breathing apparatus, and heavy work gloves.

FIRE PRECAUTIONS

Selenium hydride is not combustible and an extinguisher suitable to surrounding fire conditions should be used.

FURTHER READING

Bretherick, L. *Hazards in the Chemical Laboratory* (4th edition)

Bretherick, L. *Handbook of Reactive Chemical Hazards* (4th edition)

Sax, N. Irving *Dangerous Properties of Industrial Materials* (7th edition)

Encyclopaedia of Occupational Safety & Health

Patty's *Industrial Hygiene and Toxicology*

Kirk-Othmer *Encyclopedia of Chemical Technology*

National Fire Protection Association *Manual of Hazardous Reactions*

ACGIH *Documentation of TLVs and BEIs* (6th edition, 1986)

REFERENCES

1. Hofmann, A. W. *Ber.* 1870, **3**, 658-660
2. Alderman, L. C.; Bergin, J. J. *Arch. Environ. Health* 1986, **41**(6), 354-358
3. Cooper, W. C. *Symp.: Selenium Biomed., Int. Symp., 1st, Oregon State Univ.* 1966, 185-199
4. *Environmental Health Criteria 58: Selenium* (WHO, Geneva, 1987)

133. Selenium Monochloride

SELENIUM MONOCHLORIDE
Se_2Cl_2

RISKS
Toxic by inhalation and if swallowed – Danger of cumulative effects (R23/25, R33)

SAFETY PRECAUTIONS
When using do not eat, drink or smoke – After contact with skin, wash immediately with plenty of water – If you feel unwell, seek medical advice (show label where possible) (S20/21, S28, S44)

IDENTIFIERS

SYNONYMS	diselenium dichloride (Se_2Cl_2)
CHEMICAL ABSTRACTS No.	10025-68-0
NIOSH No.	not available
HAZCHEM CODE	not available
UN No.	not available

THRESHOLD LIMIT VALUES

US TLV (TWA)	0.2 mg/m^3 (as Se)
US TLV (STEL)	not available
UK EXPOSURE LIMITS (OES)	
Long-term (8 hr TWA value)	0.1 mg/m^3 (as Se)
Short-term (10 min TWA value)	not available
Germany	
MAK	0.1 mg/m^3
France	
VME	not available
VLE	not available
Sweden	
Short-term limit	not available
Level limit	0.1 mg/m^3 (as Se)

PHYSICAL PROPERTIES

Description	Deep red, oily liquid.
Boiling point	decomposes at 127°C at 733 mm Hg
Melting point	-85°C
Density	2.7741 at 25°C
Vapour density	not available
Vapour pressure	not available
Flash point	not available
Explosive limits	not available
Autoignition temperature	not available

Solubility Soluble in chloroform, benzene, carbon tetrachloride, carbon disulphide or fuming sulphuric acid. Decomposes in water.

PACKAGING AND TRANSPORTATION

Road transportation

hazard warning sign	not available
Hazchem code	not available

Sea transportation

IMDG page No.	not available
class	not available
label	not available
packaging group	not available

Air transportation

ICAO/IATA code	not available
class	not available
label	not available
packaging group	not available
packing instructions	
cargo	not available
passenger	not available
passenger aircraft max. quantity	not available
cargo aircraft max. quantity	not available

133. Selenium Monochloride

MANUFACTURE

Selenium monochloride is produced by reduction of selenium tetrachloride by sulphur and selenium; or by the action of hydrogen chloride on a solution of selenium in fuming sulphuric acid. It may also be produced by the action of phosphorus pentachloride on selenium, phosphorus or antimony selenides, selenium dioxide or selenium oxychloride.

USES

No information is available on the uses of selenium monochloride.

CHEMICAL HAZARDS

The black compound formed by reaction of selenium monochloride with trimethylsilyl azide in acetonitrile exploded at approximately 100°C (1). Selenium monochloride reacts explosively with potassium. It reacts very violently with potassium oxide, and violently with sodium peroxide. It reacts vigorously with sodium on heating, emitting heat and light. It reacts incandescently with aluminium above 80°C (2).

BIOLOGICAL HAZARDS

The chronic effects of most forms of selenium include depression, tiredness, dizziness, garlic-smelling breath, excess dental caries, dermatitis, gastrointestinal disturbances, partial hair loss and loss of fingernails.

Vapour Inhalation

No information is available concerning the effects of inhalation.

Eye Contact

No information is available concerning the effects on the eyes.

Skin Contact

No information is available concerning the effects on the skin.

Swallowing

No information is available concerning the effects of ingestion.

CARCINOGENICITY

No information is available specific to selenium monochloride. Reported data on other inorganic selenium compounds suggests that the dioxide has an antitumour effect, and that the sulphide may be carcinogenic in rodents (2).

MUTAGENICITY

No information is available specific to selenium monochloride. Data from other inorganic selenium compounds suggests that the monosulphide is mutagenic in bacterial assays, and that the dioxide is mutagenic in bacterial and mammalian cell assays (2).

REPRODUCTIVE HAZARDS

No information is available specific to selenium monochloride. Dietary selenium appears to affect the fertility of livestock, and selenium dioxide has caused testicular damage in laboratory animals (2).

FIRST AID

Eyes Wash the eye with flowing water for 10 minutes.

Lungs Remove casualty from area of exposure. If unconscious, do not give anything to drink. Give artificial ventilation and chest compression or place in the recovery position as necessary. If conscious make the casualty lie or sit down quietly, give oxygen if available. Lung congestion may occur – a conscious casualty with breathing difficulties should be placed in a sitting position. Allergic asthma (wheezy breathing) may occur and immediate treatment is needed – Ventolin may be useful.

Mouth Do not make the casualty vomit. Treat unconscious casualties as for lungs but if conscious give 1 pint of water to drink. Immediately then give repeated drinks of water (1 cupful every 10 minutes.) Shock may result – if so do not give any drinks, and if conscious lie casualty flat with legs raised. Convulsions may occur and may cause unconsciousness.

Skin Remove contaminated clothing immediately, drench the affected area with running water for at least 10 minutes.

133. Selenium Monochloride

In all cases of exposure, the patient should be transferred to hospital as soon as possible.

HANDLING AND STORAGE

Selenium monochloride should be handled wearing an approved respirator, chemical-resistant gloves, safety goggles and other protective clothing. It should only be used in a chemical fume hood. It should be kept in a tightly closed container, in a cool, dry place.

DISPOSAL

Ventilate the area, but isolate the material. Wear a laboratory coat, safety spectacles, compressed air breathing apparatus and rubber gloves. Mix powdered spillages with large quantities of sand and arrange for disposal by a licenced cintractor.

FIRE PRECAUTIONS

Selenium monochloride is not combustible and an extinguisher suitable to surrounding fire conditions should be used.

FURTHER READING

Bretherick, L. *Hazards in the Chemical Laboratory* (4th edition)

Bretherick, L. *Handbook of Reactive Chemical Hazards* (4th edition)

Sax, N. Irving *Dangerous Properties of Industrial Materials* (7th edition)

Encyclopaedia of Occupational Safety & Health

Patty's *Industrial Hygiene and Toxicology*

Kirk-Othmer *Encyclopedia of Chemical Technology*

National Fire Protection Association *Manual of Hazardous Reactions*

ACGIH Documentation of TLVs and BEIs (6th edition, 1986)

REFERENCES

1. Kennett, F. A.; et al. *J. Chem. Soc., Dalton Trans.* 1982, **5**, 851-857
2. Lenher, V.; et al. *J. Amer. Chem. Soc.* 1926, **48**, 1553

134. Selenium Monosulphide

SELENIUM MONOSULPHIDE
SeS

RISKS
Toxic by inhalation and if swallowed – Danger of cumulative effects (R23/25, R33)

SAFETY PRECAUTIONS
When using do not eat, drink or smoke – After contact with skin, wash immediately with plenty of water – If you feel unwell, seek medical advice (show label where possible) (S20/21, S28, S44)

IDENTIFIERS

SYNONYMS   NCI-C50033; selenuim sulfide; selenium sulphide; sulfur selenide

CHEMICAL ABSTRACTS No.	7446-34-6
NIOSH No.	YT 0525000
HAZCHEM CODE	not available
UN No.	not available

THRESHOLD LIMIT VALUES

US TLV (TWA)	0.2 mg/m^3 (as Se)
US TLV (STEL)	not available

UK EXPOSURE LIMITS (OES)
Long-term (8 hr TWA value) ..... 0.1 mg/m^3 (as Se)
Short-term (10 min TWA value) .... not available

Germany
MAK .... 0.1 mg/m^3

France
VME .... not available
VLE .... not available

Sweden
Short-term limit .... not available
Level limit .... 0.1 mg/m^3 (as Se)

PHYSICAL PROPERTIES

Description   Orange-yellow tablets or powder.

Boiling point	decomposes at 118-119°C
Melting point	111.03°C
Density	3.056 at 0°C
Vapour density	not available
Vapour pressure	not available
Flash point	not available
Explosive limits	not available
Autoignition temperature	not available

Solubility   Barely soluble in water, acetone or dimethyl sulphoxide. Soluble in carbon disulphide.

PACKAGING AND TRANSPORTATION

Road transportation

hazard warning sign	not available
Hazchem code	not available

Sea transportation

IMDG page No.	not available
class	not available
label	not available
packaging group	not available

Air transportation

ICAO/IATA code	not available
class	not available
label	not available
packaging group	not available

packing instructions
cargo .... not available
passenger .... not available

passenger aircraft max. quantity	not available
cargo aircraft max. quantity	not available

134. Selenium Monosulphide

MANUFACTURE

No information is available concerning the manufacture of selenium monosulphide.

USES

Selenium monosulphide is used as a treatment for seborrheic dermatitis, dandruff, eczemas, dermatomycoses and non-specific dermatoses.

CHEMICAL HAZARDS

Selenium monosulphide emits toxic fumes of sulphur oxides and selenium when heated to decomposition (1).

BIOLOGICAL HAZARDS

Selenium monosulphide may cause hair loss and discolouration, dental caries, garlicky-smelling breath, dizziness, pallor, depression, gastrointestinal disturbances and nausea (1). The toxicity of selenium and its compounds has been reviewed (2).

Vapour Inhalation

No information is available concerning the effects of inhalation.

Eye Contact

Selenium monosulphide is a severe eye irritant.

Skin Contact

Selenium monosulphide is a severe skin irritant, and may be absorbed through the skin (1).

Swallowing

Selenium compounds can be readily absorbed from the gut (2). The oral LD_{50} in mice is 370 mg/kg.

CARCINOGENICITY

In an NCI carcinogenesis bioassay there was clear evidence of carcinogenicity in mice and rats following administration by gavage (3). It induced lung and liver tumours in female mice (4). Dermal studies on mice by the NTP have been interpreted as showing no carcinogenic effect (5).

MUTAGENICITY

Selenium monosulphide gave positive results in *Salmonella typhimurium* (3).

REPRODUCTIVE HAZARDS

No information is available specific to selenium monosulphide. Dietary selenium appears to affect the fertility of livestock, and selenium dioxide has caused testicular damage in laboratory animals (2).

FIRST AID

Eyes Wash the eye with flowing water for 10 minutes.

Lungs Remove casualty from area of exposure. If unconscious, do not give anything to drink. Give artificial ventilation and chest compression or place in the recovery position as necessary. If conscious make the casualty lie or sit down quietly, give oxygen if available. Lung congestion may occur – a conscious casualty with breathing difficulties should be placed in a sitting position. Allergic asthma (wheezy breathing) may occur and immediate treatment is needed – Ventolin may be useful.

Mouth Do not make the casualty vomit. Treat unconscious casualties as for lungs but if conscious give 1 pint of water to drink. Shock may result – if so do not give any drinks, and if conscious lie casualty flat with legs raised. Convulsions may occur and may cause unconsciousness.

Skin Remove contaminated clothing immediately, drench the affected area with running water for at least 10 minutes.

In all cases of exposure, the patient should be transferred to hospital as soon as possible.

134. Selenium Monosulphide

HANDLING AND STORAGE

Selenium monosulphide should be handled wearing an approved respirator, chemical-resistant gloves, safety goggles and other protective clothing. It should only be used in a chemical fume hood. It should be kept in a tightly closed container, in a refrigerator or cool, dry place (2).

DISPOSAL

Ventilate the area, but isolate the material. Wear a laboratory coat, safety spectacles, compressed air breathing apparatus and rubber gloves. Mix powdered spillages with large quantities of sand then treat as normal refuse.

FIRE PRECAUTIONS

Selenium monosulphide is not combustible and an extinguisher suitable to surrounding fire conditions should be used.

FURTHER READING

Bretherick, L. *Hazards in the Chemical Laboratory* (4th edition)

Bretherick, L. *Handbook of Reactive Chemical Hazards* (4th edition)

Sax, N. Irving *Dangerous Properties of Industrial Materials* (7th edition)

Encyclopaedia of Occupational Safety & Health

Patty's *Industrial Hygiene and Toxicology*

Kirk-Othmer *Encyclopedia of Chemical Technology*

National Fire Protection Association *Manual of Hazardous Reactions*

ACGIH Documentation of TLVs and BEIs (6th edition, 1986)

REFERENCES

1. Keith, L. H.; Walters, D. B. (editors) *Compendium of Safety Data Sheets for Research and Industrial Chemicals* (VCH, Deerfield Park, 1987)
2. *Environmental Health Criteria 58: Selenium* (WHO, Geneva, 1987)
3. *Natl. Cancer Inst. Tech. Rep. NCI-CG-TR-194* (NCI, Bethesda, 1980)
4. Ashby, J.; Tennant, R. W. *Mutat. Res.* 1991, **257**, 229-306
5. Haseman, J. K.; et al. *J. Toxicol. Environ. Health* 1984, **14**, 621-639

135. Selenium Tetrachloride

SELENIUM TETRACHLORIDE
SeCl$_4$

RISKS
Toxic by inhalation and if swallowed – Danger of cumulative effects (R23/25, R33)

SAFETY PRECAUTIONS
When using do not eat, drink or smoke – After contact with skin, wash immediately with plenty of water – If you feel unwell, seek medical advice (show label where possible) (S20/21, S28, S44)

IDENTIFIERS

SYNONYMS selenium chloride (SeCl$_4$); tetrachloroselenium

CHEMICAL ABSTRACTS No.	10026-03-6
NIOSH No.	YS 7875000
HAZCHEM CODE	not available
UN No.	not available

THRESHOLD LIMIT VALUES

US TLV (TWA) .... 0.2 mg/m^3 (as Se)
US TLV (STEL) .... not available

UK EXPOSURE LIMITS (OES)
Long-term (8 hr TWA value) ...... 0.1 mg/m^3 (as Se)
Short-term (10 min TWA value) .... not available

Germany
MAK .... 0.1 mg/m^3

France
VME .... not available
VLE .... not available

Sweden
Short-term limit .... not available
Level limit .... 0.1 mg/m^3 (as Se)

PHYSICAL PROPERTIES

Description	Cubic white-yellow deliquescent crystals.
Boiling point	decomposes at 288°C
Melting point	300°C
Density	2.6
Vapour density	not available
Vapour pressure	1 mm Hg at 74°C
Flash point	not available
Explosive limits	not available
Autoignition temperature	not available

Solubility Decomposes in water and moist air. Insoluble in liquid bromine. Decomposed by dry ammonia.

PACKAGING AND TRANSPORTATION

Road transportation
hazard warning sign	not available
Hazchem code	not available

Sea transportation
IMDG page No.	not available
class	not available
label	not available
packaging group	not available

Air transportation
ICAO/IATA code	not available
class	not available
label	not available
packaging group	not available
packing instructions	
cargo	not available
passenger	not available
passenger aircraft max. quantity	not available
cargo aircraft max. quantity	not available

135. Selenium Tetrachloride

MANUFACTURE

Selenium tetrachloride, which is formed in hydrochloric acid, is extended with an oxygen-containing organic solvent or alkylamine solvent, and then evaporated or distilled at normal or reduced pressures (1).

USES

Selenium tetrachloride is used in the manufacture of selenium monochloride.

CHEMICAL HAZARDS

Selenium tetrachloride emits highly toxic fumes of sulphur oxides and selenium on heating to decomposition.

BIOLOGICAL HAZARDS

The subcutaneous LD_{50} in guinea pigs is 19 mg/kg. Repeated administration of 5-60 mg/kg selenium tetrachloride caused depletion of liver glycogen but no histopathological changes in the liver, heart, spleen, or kidneys. The LD_{50} (route unspecified) in mice, guinea pigs and rabbits was 57, 78.7 and 100 mg/kg respectively (2).

Vapour Inhalation

No information is available concerning the effects of inhalation.

Eye Contact

No information is available concerning the effects on the eyes.

Skin Contact

No information is available concerning the effects on the skin.

Swallowing

No information is available concerning the effects of ingestion.

CARCINOGENICITY

Antineoplastic effects have been reported (3).

MUTAGENICITY

No information is available specific to selenium tetrachloride. Data from other inorganic selenium compounds suggests that the monosulphide is mutagenic in bacterial assays, and that the dioxide is mutagenic in bacterial and mammalian cell assays (4).

REPRODUCTIVE HAZARDS

No information is available specific to selenium tetrachloride. Dietary selenium appears to affect the fertility of livestock, and selenium dioxide has caused testicular damage in laboratory animals (4).

FIRST AID

Eyes Wash the eye with flowing water for 10 minutes.

Lungs Remove casualty from area of exposure. If unconscious, do not give anything to drink. Give artificial ventilation and chest compression or place in the recovery position as necessary. If conscious make the casualty lie or sit down quietly, give oxygen if available. Lung congestion may occur – a conscious casualty with breathing difficulties should be placed in a sitting position. Allergic asthma (wheezy breathing) may occur and immediate treatment is needed – Ventolin may be useful.

Mouth Do not make the casualty vomit. Treat unconscious casualties as for lungs but if conscious give 1 pint of water to drink. Shock may result – if so do not give any drinks, and if conscious lie casualty flat with legs raised. Convulsions may occur and may cause unconsciousness.

Skin Remove contaminated clothing immediately, drench the affected area with running water for at least 10 minutes.

In all cases of exposure, the patient should be transferred to hospital as soon as possible.

135. Selenium Tetrachloride

HANDLING AND STORAGE

Selenium tetrachloride should be handled wearing an approved respirator, chemical-resistant gloves, safety goggles and other protective clothing. It should only be used in a chemical fume hood. It should be kept in a tightly closed container, in a cool, dry place.

DISPOSAL

Ventilate the area, but isolate the material. Wear a laboratory coat, safety spectacles, compressed air breathing apparatus and rubber gloves. Mix powdered spillages with large quantities of sand and arrange for disposal by a licenced cintractor.

FIRE PRECAUTIONS

Selenium tetrachloride is not combustible and an extinguisher suitable to surrounding fire conditions should be used.

FURTHER READING

Bretherick, L. *Hazards in the Chemical Laboratory* (4th edition)

Bretherick, L. *Handbook of Reactive Chemical Hazards* (4th edition)

Sax, N. Irving *Dangerous Properties of Industrial Materials* (7th edition)

Encyclopaedia of Occupational Safety & Health

Patty's *Industrial Hygiene and Toxicology*

Kirk-Othmer *Encyclopedia of Chemical Technology*

National Fire Protection Association *Manual of Hazardous Reactions*

ACGIH Documentation of TLVs and BEIs (6th edition, 1986)

REFERENCES

1. Nanjo, M.; et al. Jpn. Kokai Tokkyo Koho JP 01042302 (1989)
2. Fidel'skaya, R. I.; et al. *Sb. Rab.-Leningr. Vet. Inst.* 1977, **50**, 5-60
3. Suzuki, M.; et al. Eur. Pat. Appl. EP 182,317 (Cl. A61K31/095) (1986)
4. *Environmental Health Criteria 58: Selenium* (WHO, Geneva, 1987)

136. Strontium Chromate

STRONTIUM CHROMATE

SrCrO$_4$

RISKS
May cause cancer – Harmful if swallowed (R45, R22)

SAFETY PRECAUTIONS
Avoid exposure-obtain special instructions before use – If you feel unwell, seek medical advice (show label where possible) (S53, S44)

IDENTIFIERS

SYNONYMS chromic acid (H$_2$CrO$_4$), strontium salt (1:1); C.I. Pigment Yellow 32; Deep Lemon Yellow; Strontium chromate (1:1); strontium chromate I 2170; strontium chromate A; strontium chromate (Sr Cr O$_4$); stromtium chromate (VI); strontium chromate X 2396; strontium yellow

CHEMICAL ABSTRACTS No.	7789-06-2
NIOSH No.	GB 3240000
HAZCHEM CODE	not available
UN No.	not available

THRESHOLD LIMIT VALUES

US TLV (TWA)	not available
US TLV (STEL)	not available

UK EXPOSURE LIMITS (OES)
Long-term (8 hr TWA value)	not available
Short-term (10 min TWA value)	not available

Germany
MAK	not available

France
VME	not available
VLE	not available

Sweden
Short-term limit	not available
Level limit	0.02 mg/m^3 (as Cr)

PHYSICAL PROPERTIES

Description	Monoclinic, yellow crystals.
Boiling point	not available
Melting point	not available
Density	3.895 at 15°C
Vapour density	not available
Vapour pressure	not available
Flash point	not available
Explosive limits	not available
Autoignition temperature	not available

Solubility Soluble in 840 parts cold water, about 5 parts boiling water. Freely soluble in dilute hydrochloric, nitric or acetic acids.

PACKAGING AND TRANSPORTATION

Road transportation
hazard warning sign	not available
Hazchem code	not available

Sea transportation
IMDG page No.	not available
class	not available
label	not available
packaging group	not available

Air transportation
ICAO/IATA code	not available
class	not available
label	not available
packaging group	not available
packing instructions	
cargo	not available
passenger	not available
passenger aircraft max. quantity	not available
cargo aircraft max. quantity	not available

136. Strontium Chromate

MANUFACTURE

Strontium chromate is prepared from sodium chromate and strontium chloride.

USES

Strontium chromate is used as a corrosion inhibitor in metal protective coatings, in polyvinyl chloride resins, to give primrose yellows, and to control sulphate concentration of solutions in electrochemical processes. The German Technical Rule for Dangerous Substances TRG 602 concluded it was not yet possible to impose an overall ban on strontium chromate anticorrosive pigments until adequate substitutes are available (1).

CHEMICAL HAZARDS

No information is available concerning the chemical hazards of strontium chromate.

BIOLOGICAL HAZARDS

Strontium chromate is an eye, skin and respiratory irritant, and may cause chrome ulcers and skin sores.

Vapour Inhalation

Strontium chromate is a respiratory irritant, and inhalation may cause lung damage (2).

Eye Contact

Strontium chromate is an eye irritant, and may cause eye damage (2).

Skin Contact

Strontium chromate is irritating to the skin, causing dermatitis and skin ulcers. Skin and nasal lesions occurred in strontium chromate production workers (3).

Swallowing

Oral administration of strontium chromate caused dystrophy of the kidneys and paresis of the digestive system in rats (4). The lethal dose of hexavalent chromium compounds is estimated as between 1.5 g and 16 g (2).

CARCINOGENICITY

A review of experimental animal data provided strong evidence for the carcinogenicity of strontium chromate (5). Intramuscular implantation of strontium chromate resulted in implantation-site tumours in 15 out of 33 rats, and tumours at the site of intrapleural implantation in 17 out of 28 rats (6). The IARC evaluation is that there is sufficient evidence for the carcinogenicity of strontium chromate in humans and animals (6,7). The UK Industrial Injuries Advisory Council has extended the list of jobs covered for lung cancer compensation to include workers exposed to strontium chromate (8,9). The carcinogenicity of strontium chromate has been reviewed (10).

MUTAGENICITY

Chromium(VI) compounds including strontium chromate were inactive or scarcely active in the *Salmonella*/microsome test when dissolved in water. Mutagenicity was increased when solubilised by 0.5N sodium hydroxide or nitrilotriacetic acid trisodium salt (NTA). It was directly clastogenic in the sister chromatid exchange assay in Chinese hamster ovary cells, and NTA significantly increased chromosome damaging activity (11).

REPRODUCTIVE HAZARDS

No information is available concerning the reproductive hazards of strontium chromate. Hexavalent chromium (as chromium trioxide) is teratogenic in hamsters (12,13).

FIRST AID

Eyes Wash the eye with plenty of water or normal saline for at least 20 minutes.

Lungs Wearing appropriate respiratory protection, remove casualty from area of exposure. If unconscious, do not give anything to drink, give artificial ventilation and chest compression or place in the recovery position as necessary. If conscious make the casualty lie or sit down quietly, give oxygen if available.

Mouth Do not make the casualty vomit. Treat unconscious casualties as for lungs but if conscious give a glass or two of milk or water to drink immediately.

136. Strontium Chromate

Skin Remove contaminated clothing immediately, wash the affected area with soap and copious amounts of water. Absorption through the skin may cause symptoms similar to those of inhalation.

In all cases of exposure, the patient should be transferred to hospital as soon as possible.

HANDLING AND STORAGE

Strontium chromate should be handled wearing an approved respirator, rubber gloves, boots, safety goggles and other protective clothing. All equipment should be fully decontaminated or disposed of after use (2). Guidelines for safe handling of chromate pigments have been published (14). It should be kept in a tightly closed container, in a well ventilated area away from heat and water (2).

DISPOSAL

Wearing a laboratory coat or overalls, safety glasses (with a polycarbonate visor), rubber gloves, suitable safety boots and breathing apparatus if the spill is large and in a confined area, cover with a reducing agent, eg. sodium metabisulphite, sodium thiosulphate or a ferrous salt (do not use carbon or sulphur), mix and spray with water. Transfer the slurry to a large container of water, neutralise with soda ash, and run to waste with excess cold running water. For large spills absorb onto sand and arrange for removal by a licenced contractor.

FIRE PRECAUTIONS

Fires involving strontium chromate should be extinguished using water spray, carbon dioxide, dry chemical powder, alcohol foam or polymer foam (2).

FURTHER READING

Bretherick, L. *Hazards in the Chemical Laboratory* (4th edition)

Bretherick, L. *Handbook of Reactive Chemical Hazards* (4th edition)

Sax, N. Irving *Dangerous Properties of Industrial Materials* (7th edition)

Encyclopaedia of Occupational Safety & Health

Patty's *Industrial Hygiene and Toxicology*

Kirk-Othmer *Encyclopedia of Chemical Technology*

National Fire Protection Association *Manual of Hazardous Reactions*

ACGIH Documentation of TLVs and BEIs (6th edition, 1986)

REFERENCES

1. *Bundesarbeitsblatt* 1988, **5**, no pages given
2. *Dangerous Prop. Ind. Mater. Rep.* 1981, **1**(7), 74-76
3. McAughey, J. J.; et al. *Sci. Total Environ.* 1988, **71**(3), 317-322
4. Kochanov, M. M. *Toksikol. Vysokomol. Mater. Khim. Syr'ya Ikh Sin., Gos. Nauch.-Issled. Inst. Polim. Plast. Mass.* 1966, 264-273
5. Langard, S. *Sci. Total Environ.* 1988, **71**(3), 341-350
6. IARC *Monographs on the evaluation of the carcinogenic risk of chemicals to humans* 1973, **2**, 100
7. IARC *Monographs on the evaluation of the carcinogenic risk of chemicals to humans* 1980, **23**, 205
8. *Hazards* 1987, **14**, 5
9. *Occup. Saf. Health* 1987, **17**(5), 4
10. *The toxicology of chemicals. 1.Carcinogenicity* (CEC, Luxembourg, 1989)
11. Venier, P.; et al. *Mutat. Res.* 1985, **156**(3), 219-228
12. Gale, T. F.; Bunch, J. D. *Teratology* 1979, **19**, 81-86
13. Gale, T. F. *Environ. Res.* 1982, **29**, 196-203
14. *Guidelines on the safe handling of chromate pigments and chromate containing pigments* (Paintmakers Association, 1988)

137. Sulphur Dioxide

SULPHUR DIOXIDE
SO_2

RISKS
Toxic by inhalation – Irritating to eyes and respiratory system (R23, R36/37)

SAFETY PRECAUTIONS
Keep container tightly closed and in a well ventilated place – If you feel unwell, seek medical advice (show label where possible) (S7/9, S44)

IDENTIFIERS

SYNONYMS bisulphite; fermenticide liquid; sulphurous acid anhydride; sulphurous oxide; sulphurous anhydride; sulphur oxide (SO_2)

CHEMICAL ABSTRACTS No.	7446-09-5
NIOSH No.	WS 4550000
HAZCHEM CODE	2RE
UN No.	1079

THRESHOLD LIMIT VALUES

US TLV (TWA)	2 ppm (5.2 mg/m^3)
US TLV (STEL)	5 ppm (13 mg/m^3)

UK EXPOSURE LIMITS (OES)
Long-term (8 hr TWA value) .... 2 ppm (5 mg/m^3)
Short-term (10 min TWA value) 5 ppm (13 mg/m^3)

Germany
MAK .... 2 ppm (5 mg/m^3)

France
VME .... not available
VLE .... not available

Sweden
Ceiling limit .... 5 ppm (13 mg/m^3)
Level limit .... 2 ppm (5 mg/m^3)

PHYSICAL PROPERTIES

Description Colourless gas or liquid with a pungent smell.

Boiling point	-10°C
Melting point	-72.7°C
Density	1.434 at 0°C (liquid)
Vapour density	2.264 at 0°C
Vapour pressure	2538 mm Hg at 21.1°C
Flash point	not available
Explosive limits	not available
Autoignition temperature	not available

Solubility Soluble in water, alcohol, methanol, chloroform or ether.

PACKAGING AND TRANSPORTATION

Road transportation

hazard warning sign	not available
Hazchem code	2RE

Sea transportation

IMDG page No.	2179
class	2;2.3
label	poison gas
packaging group	not available

Air transportation

ICAO/IATA code (UN No.)	1079
class	2;6.1
label	gas poisonous
packaging group	not available
packing instructions	
cargo	200
passenger	forbidden
passenger aircraft max. quantity	forbidden
cargo aircraft max. quantity	25 kilograms

137. Sulphur Dioxide

MANUFACTURE

Sulphur dioxide is prepared by combustion of Frasch-process sulphur and sulphur recovered from natural gas or oil-refinery gases, or by burning pyrite.

USES

Sulphur dioxide is used in sulphuric acid manufacture, for sulphite pulping in the pulp and paper industry, and as an intermediate for bleaching agents. It is used as a fumigant, preservative and bleach for food and drink products. It is also used in water treatment, in the oil industry, mineral technology, kaolin processing, and as a catalyst and solvent.

CHEMICAL HAZARDS

Sulphur dioxide may react explosively with sodium hydride (unless diluted with hydrogen) (1); silver azide; diethylzinc; fluorine; propene and lithium nitrate; while solutions of sulphur dioxide in ethanol or ether may explode on contact with potassium chlorate. Contact of sulphur dioxide and potassium chlorate above 60°C causes flashing. Violent or vigorous reactions occur with sodium. Heated barium peroxide reacts incandescently in sulphur dioxide. Metal acetylides, caesium oxide, iron(II) oxide, tin oxide and lead(IV) oxide ignite and incandesce in sulphur dioxide. Finely divided chromium incandesces in sulphur dioxide and pyrophoric manganese burns brightly on heating in sulphur dioxide. Caesium azide ignites in sulphur dioxide (2). Contact with bromine pentafluoride or chlorine trifluoride at ambient or slightly elevated temperature is violently, ignition often occurring. Generation of sulphur dioxide from reaction of copper and sulphuric acid is too hazardous for school experiments (3). Sulphur dioxide accelerated self-heating of peat (4). Polymeric tubing failed below 2 bar when used for sulphur dioxide (5). Sulphur dioxide is covered by CIMAH; quantity limits were reduced to 500 tonnes in 1986 (6).

BIOLOGICAL HAZARDS

Sulphur dioxide is an irritant gas. The principal effects are on the upper respiratory tract and bronchi, and it may cause oedema of the lungs and glottis, and respiratory paralysis (7).

Vapour Inhalation

Levels of 0.3-1 ppm are detectable. 6-12 ppm are immediately irritating to the nose and throat (7), although upper respiratory tract irritation has been reported at 2-5 ppm. 20 ppm is markedly irritant and causes choking and sneezing. 50 ppm rapidly causes nose and throat irritation, cough and nasal discharge. Dyspnoea, substernal pain and tearing are also commonly reported (8,9). 400-500 ppm is immediately dangerous to life. High levels may cause bronchitis, asthma-and 'flu-like symptoms, bronchopneumonia, asphyxia and death. Bronchial hyperreactivity is a frequent sequela and may persist for several years (10). Sulphur dioxide appears to have to have a low degree of chronic toxicity. The effects of inhalation have been reviewed (11).

Eye Contact

Sulphur dioxide causes conjunctival inflammation (7). Levels of 20 ppm are irritating to the eyes.

Skin Contact

10,000 ppm causes irritation to moist areas of the skin within a few minutes (7). A case of severe burns has been reported following the bursting of a cylinder (12).

Swallowing

Not applicable.

CARCINOGENICITY

NIOSH found no data to suggest sulphur dioxide was a primary carcinogen (13). An association between occupational sulphur dioxide exposure and lung cancer deaths has been suggested (14).

MUTAGENICITY

Some *in vitro* assays, such as the Ames test, suggest that sulphur dioxide may be weakly mutagenic (15). Positive results are reported in *Saccharomyces cerevisiae* (16). Increased frequency of chromosome aberrations and sister chromatid exchanges was reported in lymphocyte cultures of workers chronically exposed to sulphur dioxide (17).

REPRODUCTIVE HAZARDS

No information is available concerning the reproductive hazards of sulphur dioxide.

137. Sulphur Dioxide

FIRST AID

Eyes Wash the eye with flowing water for 10 minutes.

Lungs Remove casualty from area of exposure. If unconscious, do not give anything to drink. Give artificial ventilation and chest compression or place in the recovery position as necessary. If conscious make the casualty lie or sit down quietly, give oxygen if available. Lung congestion may occur – a conscious casualty with breathing difficulties should be placed in a sitting position. Allergic asthma (wheezy breathing) may occur and immediate treatment is needed – Ventolin may be useful.

Mouth None.

Skin Remove contaminated clothing immediately, drench the affected area with running water for a least 10 minutes.

In all cases of exposure, the patient should be transferred to hospital as soon as possible.

HANDLING AND STORAGE

Sulphur dioxide should be handled wearing an approved respirator in non-ventilated areas, rubber gloves, and safety goggles. It should only be used in a chemical fume hood. Cylinders should be properly secured and cylinder temperature should not exceed 52°C, and equipment rated for cylinder pressure used. The cylinder valve should be closed when not in use or empty (18). It should be kept in a fireproof area, ventilated along the floor, and away from ammonia, acrolein, metals, metal oxides, halogens, and inter-halogens (19). Problems of long term storage, following explosion of sulphur dioxide drums, have been discussed (20). Further information on handling is available (21,22).

DISPOSAL

Eliminate all sources of ignition. Wear a laboratory coat, safety spectacles and rubber gloves and breathing apparatus. Ventilate the area to dispel gas. If a gas valve is leaking, EITHER, bubble through calcium hypochlorite solution, including a trap in the line to prevent suckback, OR place gas cylinder in ventilated fume cupboard and allow to empty.

FIRE PRECAUTIONS

Water spray should be used to keep cylinders cool, and cylinders should be moved away from fire if it is safe to do so.

FURTHER READING

Bretherick, L. *Hazards in the Chemical Laboratory* (4th edition)

Bretherick, L. *Handbook of Reactive Chemical Hazards* (4th edition)

Sax, N. Irving *Dangerous Properties of Industrial Materials* (7th edition)

Encyclopaedia of Occupational Safety & Health

Patty's *Industrial Hygiene and Toxicology*

Kirk-Othmer *Encyclopedia of Chemical Technology*

National Fire Protection Association *Manual of Hazardous Reactions*

ACGIH Documentation of TLVs and BEIs (6th edition, 1986)

REFERENCES

1. Moissan, H. *Compt. Rend.* 1902, **135**, 647
2. Kennet, F. A.; et al. *J. Chem. Soc., Dalton Trans.* 1982, 853
3. Campbell, D. A. *School Sci. Rev.* 1939, **20**(80), 631
4. Byrne, P. J. *Proc. Int. Peat Congr., 7th* 1984, **4**, 130-145
5. *MCA Case History No. 1044*
6. *Saf. Pract.* 1986, **4**(4), 45
7. *Dangerous Prop. Ind. Mater. Rep.* 1981, **1**, 78-79
8. Savic, M.; et al. *Int. Arch. Occup. Environ. Health* 1987, **59**(5), 513-518
9. Osterman, J. W.; et al. *Br. J. Ind. Med.* 1989, **46**(9), 629-635
10. Harkonen, H.; et al. *Am. Rev. Respir. Dis.* 1983, **128**(5), 890-893

137. Sulphur Dioxide

11. Jaeger, M. J. *Res. Top. Physiol.* 1982, **5**(Air Pollut. Physiol. Eff.), 81-105
12. *Saf. Manage. Veiligheidsbestuur* 1991, **17**(3), 47
13. *Criteria for a Recommended Standard, Occupational Exposure to Sulphur Dioxide* (U.S. Dept. Health, Educ., Welfare, Public Health Service, 1974)
14. Bond, G. G.; et al. *Am. J. Epidemiol.* 1986, **124**(1), 53-66
15. Shapiro, R. *Mutat. Res.* 1977, **39**, 149-176
16. Guerra, D.; et al. *Experientia* 1981, **37**(7), 691-693
17. Meng, Z.; Zhang, L. *Mutat. Res.* 1990, **241**(1), 15-20
18. Lenga, R. E. *The Sigma-Aldrich Library of Chemical Safety Data* (2nd edition) (Sigma-Aldrich, Milwaukee, 1988)
19. *International Chemical Safety Cards: Sulphur dioxide* (CEC, Luxembourg, 1990)
20. *Health Saf. Work* 1985, **7**(10), 10
21. *Handling chemicals safely. Instruction bulletin 70-9001* (Fischer & Porter)
22. *Australian Standard AS 2508 Safe storage and handling information cards for hazardous materials. 2.010. Sulphur dioxide (liquefied)* (Standards Assoc., Australia, Sydney, 1984)

138. 1,1,2,2-Tetrabromoethane

1,1,2,2-TETRA-BROMOETHANE

Br$_2$CHCHBr$_2$

RISKS
Very toxic by inhalation – Irritating to eyes (R26, R36)

SAFETY PRECAUTIONS
Keep locked up – Avoid contact with skin – Take off immediately all contaminated clothing – In case of accident or if you feel unwell, seek medical advice immediately (show label where possible) (S1, S24, S27, S45)

IDENTIFIERS

SYNONYMS acetylene tetrabromide; ethane,1,1,2,2-tetrabromo-; Muthmann's liquid; TBE; tetrabomoacetylene; 1,1,2,2-tetrabromoethylene

CHEMICAL ABSTRACTS No.	79-27-6
NIOSH No.	KI 8225000
HAZCHEM CODE	2Z
UN No.	2504

THRESHOLD LIMIT VALUES

US TLV (TWA)	not available
US TLV (STEL)	not available

UK EXPOSURE LIMITS (OES)
Long-term (8 hr TWA value) .... 0.5 ppm (7 mg/m^3)
Short-term (10 min TWA value) .... not available

Germany
MAK .... 1 ppm (14 mg/m^3)

France
VME .... 1 ppm (15 mg/m^3)
VLE .... not available

Sweden
Short-term limit .... not available
Level limit .... not available

PHYSICAL PROPERTIES

Description Colourless or yellow liquid, smelling of camphor and iodoform.

Boiling point	151°C at 54 mm Hg
Melting point	0°C
Density	2.9638 at 20°C
Vapour density	11.9
Vapour pressure	0.04 torr at 24°C
Flash point	not available
Explosive limits	not available
Autoignition temperature	335°C

Solubility Insoluble in water. Miscible with alcohol, chloroform, ether, aniline or glacial acetic acid.

PACKAGING AND TRANSPORTATION

Road transportation

hazard warning sign	2504 harmful substance
Hazchem code	2Z

Sea transportation

IMDG page No.	6263
class	6.1
label	harmful stow away from foodstuffs; marine pollutant
packaging group	III

Air transportation

ICAO/IATA code (UN No.)	2504
class	6.1
label	keep away from food
packaging group	III
packing instructions	
cargo	618
passenger	611
passenger aircraft max. quantity	60 litres
cargo aircraft max. quantity	220 litres

138. 1,1,2,2-Tetrabromoethane

MANUFACTURE

1,1,2,2-Tetrabromoethane is manufactured by bromination of acetylene.

USES

1,1,2,2-Tetrabromoethane is used as a gauge fluid, for balancing equipment, for ore separation, in microscopy and as a solvent.

CHEMICAL HAZARDS

1,1,2,2-Tetrabromoethane is incompatible with strong bases, strong oxidising agents and magnesium. When heated to decomposition it produces toxic fumes of carbon monoxide and dioxide, carbonyl bromide and bromide.

BIOLOGICAL HAZARDS

1,1,2,2-Tetrabromoethane is a central nervous system depressant, and large doses may cause narcosis, coma and eventually death. It is toxic to the liver and kidneys.

Vapour Inhalation

1,1,2,2-Tetrabromoethane is irritating to the mucous membranes and upper respiratory tract. Exposure of chemists by inhalation of vapours and skin absorption resulted in headaches, anorexia, stomach ache, heartburn, and 'near fatal' liver injury (1). Rabbits exposed for up to 2.5 hours and rats for up to 3 hours to near saturated atmospheres of 1,1,2,2-tetrabromoethane survived without any deaths. Guinea pigs exposed for 30 minutes survived, 50% died after 1 hour, and all succumbed to 1.5 hour exposures. In the animals that died liver and kidney injuries were observed. Animals exposed to a saturated atmosphere for 15 minutes/day for 47-92 days suffered no treatment related effects (2). Exposure of rats, guinea pigs, rabbits and monkeys to 14 ppm 1,1,2,2-tetrabromoethane 7 hours/day, 5 days/week for 100-106 days showed increased liver weight and histopathological changes to lungs and liver in all animals. Slight histopathological changes to the lungs and liver were observed at 4 ppm, and at 1.1 ppm all animals appeared normal (3). The LC_{50} for rats exposed for 4 hours was 549 mg/m^3 (4). The odour threshold of 1,1,2,2-tetrabromoethane is considered insufficient to be a good warning property, although it may be sufficient to prevent serious acute exposure.

Eye Contact

1,1,2,2-Tetrabromoethane is irritating to the eyes. 100 mg applied to the eyes of rabbits caused mild irritation (3).

Skin Contact

1,1,2,2-Tetrabromoethane is irritating to the skin. 500 mg applied to rabbit skin for 24 hours caused moderate irritation (3). The LD_{50} in rats by skin exposure is 5250 mg/kg (4).

Swallowing

1,1,2,2-Tetrabromoethane is poisonous by ingestion. The oral LD_{50} in rabbits and guinea pigs is approximately 400 mg/kg (2). The oral LD_{50} was 1100 mg/kg in rats and 269 mg/kg in mice (4). Rats survived exposure to 600 mg/kg but succumbed to 1600 mg/kg (3).

CARCINOGENICITY

1,1,2,2-Tetrabromoethane induced a statistically significant incidence of forestomach papillomas when given to mice by repeated skin application (5).

MUTAGENICITY

1,1,2,2-Tetrabromoethane was negative in the *Salmonella typhimurium* assay, but positive in a DNA polymerase deficient *Escherichia coli* assay (6,7).

REPRODUCTIVE HAZARDS

No information is available concerning the reproductive hazards of 1,1,2,2-tetrabromoethane.

FIRST AID

Eyes Wash the eye with flowing water for 10 minutes.

Lungs Remove casualty from area of exposure. If unconscious, do not give anything to drink. Give artificial ventilation and chest compression or place in the recovery position as necessary. If conscious make the casualty lie or sit down

138. 1,1,2,2-Tetrabromoethane

quietly, give oxygen if available. Lung congestion may occur – a conscious casualty with breathing difficulties should be placed in a sitting position. Convulsions may occur and may cause unconsciousness. Shock may result – if so do not give any drinks, and if conscious lie casualty flat with legs raised.

Mouth Do not make the casualty vomit. Treat unconscious casualties as for lungs, but if conscious give 1 pint of water to drink.

Skin Remove contaminated clothing immediately, wash the affected area with soap and copious amounts of water. Absorption through the skin may cause symptoms similar to those of inhalation.

In all cases of exposure, the patient should be transferred to hospital as soon as possible.

HANDLING AND STORAGE

1,1,2,2-Tetrabromoethane should be handled wearing an approved respirator, chemical-resistant gloves, safety goggles and other protective clothing. It should only be used in a chemical fume hood. It should be kept in a tightly closed container, and stored in a cool, dry place.

DISPOSAL

Eliminate all sources of ignition and ventilate the area. Wearing a laboratory coat, approved breathing apparatus, heavy work gloves.

FIRE PRECAUTIONS

1,1,2,2-Tetrabromoethane is non-combustible and fires should be extinguished using media appropriate to the surrounding conditions.

FURTHER READING

Sax, N. Irving *Dangerous Properties of Industrial Materials* (7th edition)

Encyclopedia of Occupational Safety & Health

Patty's *Industrial Hygiene and Toxicology*

REFERENCES

1. Van Itaaften, A. B. *Am. Ind. Hyg. Assoc. J.* 1969, **30**, 251
2. Gray, M. G. *Arch. Ind. Hyg. Occup. Med.* 1950, **2**, 407
3. Hollingsworth, R. L.; et al. *Am. Ind. Hyg. Assoc. J.* 1963, **24**, 28
4. Izmerov, N. F.; et al. *Toxicometric Parameters of Industrial Toxic Chemicals Under Single Exposure* (Centre of International Projects, Moscow, 1982) p. 107
5. Van Duuren, B. L.; et al. *JNCI, J. Natl. Cancer Institute* 1979, **637**(6), 1433-1439
6. Rosenkranz, H. S.; et al. *Mutat. Res.* 1976, **41**, 61-70
7. Rosenkranz, H. S.; et al. *Environ. Health Perspect.* 1977, **21**, 79-84
8. Brem, H.; et al. *Cancer Res.* 1974, **34**(10), 2576-2579

139. 1,1,2,2-Tetrachloroethane

1,1,2,2-TETRA-CHLOROETHANE

$Cl_2CHCHCl_2$

RISKS
Very toxic by inhalation and in contact with skin (R26/27)

SAFETY PRECAUTIONS
Keep out of reach of children – In case of insufficient ventilation, wear suitable respiratory equipment – In case of accident or if you feel unwell, seek medical advice immediately (show label where possible) (S2, S38, S45)

IDENTIFIERS

SYNONYMS   acetylene tetrachloride; Bonoform; Cellon; ethane, 1,1,2,2-tetrachloro-; NCI-CO3554; RCRA Waste No. U209; 1,1,2,2-tetrachlorethane; tetrachloroethane; *s*- tetrachloroethane; *sym*-tetrochloroethane; Westron

CHEMICAL ABSTRACTS No.	79-34-5
NIOSH No.	KI 8575000
HAZCHEM CODE	2XE
UN No.	1702

THRESHOLD LIMIT VALUES

US TLV (TWA)	1 ppm (6.9 mg/m^3)
US TLV (STEL)	not available

UK EXPOSURE LIMITS (OES)
Long-term (8 hr TWA value) .... not available
Short-term (10 min TWA value) .... not available

Germany
MAK .... 1 ppm (7 mg/m^3)

France
VME .... 1 ppm (7 mg/m^3)
VLE .... 5 ppm (35 mg/m^3)

Sweden
Short-term limit .... not available
Level limit .... not available

PHYSICAL PROPERTIES

Description   Heavy, colourless, mobile liquid, chloroform-like odour.

Boiling point	146.6°C
Melting point	-43.8°C
Density	1.600 at 20°C
Vapour density	not available
Vapour pressure	6 torr at 25°C
Flash point	not available
Explosive limits	not available
Autoignition temperature	not available

Solubility   Very sparingly soluble in water. Miscible with methanol, ethanol, benzene, ether, petroleum ether, carbon tetrachloride, chloroform or carbon disulphide

PACKAGING AND TRANSPORTATION

Road transportation

hazard warning sign	1702 toxic substance
Hazchem code	2XE

Sea transportation

IMDG page No.	6263
class	6.1
label	poison; marine pollutant
packaging group	II

Air transportation

ICAO/IATA code (UN No.)	1702
class	6.1
label	poison
packaging group	II
packing instructions	
cargo	612
passenger	610
passenger aircraft max. quantity	5 litres
cargo aircraft max. quantity	60 litres

139. 1,1,2,2-Tetrachloroethane

MANUFACTURE

1,1,2,2-Tetrachloroethane is manufactured commercially by the reaction of acetylene with chlorine. It is also produced by the chlorination of 1,2-dichloroethylene and the 2 stage chlorination of 1,2-dichloroethane.

USES

1,1,2,2-Tetrachloroethane is used as an intermediate in the manufacture of trichloroethylene from acetylene; as a solvent; in the manufacture of paint, varnish, and rust remover; in soil sterilization, weed killer and insecticides; determination of theobromine in cacao; and immersion fluid in crystallography.

CHEMICAL HAZARDS

1,1,2,2-Tetrachloroethane is not in itself flammable or explosive. However, it is capable of forming explosive mixtures with potassium, sodium and nitrogen tetroxide. Heating in contact with potassium hydroxide produces chloro-or dichloroacetylene gas, which ignites spontaneously in air. 1,1,2,2-Tetrachloroethane also reacts violently with 2,4-dinitrophenyl disulphide, nitrates, sodium-potassium alloy and bromoform. It is incompatible with strong oxidising agents and strong bases. The presence of moisture causes hydrolysis, with the evolution of hydrogen chloride. This process is gradual at room temperature but becomes rapid above 110°C. Adjacent metals may be corroded. Heating the compound to decomposition produces toxic fumes of hydrogen chloride and carbon monoxide.

BIOLOGICAL HAZARDS

1,1,2,2,-Tetrachloroethane is highly toxic by inhalation, ingestion and absorption through the skin. There are many case reports of human poisoning, some fatal (1,2). Acute poisoning causes central nervous system depression, which may be severe enough to cause death from respiratory failure. Other symptoms include drowsiness, dizziness, hallucinations, narcosis, loss of consciousness, tremors, nausea, and vomiting. Autopsy after fatal poisoning has revealed atrophy and cirrhosis of the liver, fatty degeneration of the kidney and heart, and oedema of the brain. Exposure to lower concentrations initially produces vague symptoms of gastrointestinal complaints and nervous system disorders such as tremor and headaches (3). Chronic exposure causes jaundice, liver enlargement and possibly liver failure. Its toxicity has been reviewed (4,5).

Vapour Inhalation

1,1,2,2-Tetrachloroethane is irritating to the mucous membranes and respiratory tract, and may produce coughing, wheezing and a burning sensation. It is toxic by inhalation as described above. It is reported lethal to rats at concentration as low as 1000 ppm (6).

Eye Contact

1,1,2,2-Tetrachloroethane is an eye irritant, causing lachrymation and conjunctivitis.

Skin Contact

1,1,2,2,-Tetrachloroethane is readily absorbed through the skin. External effects are irritation and inflammation of exposed skin, and dry, scaly dermatitis.

Swallowing

Toxic by ingestion. There may be abnormal salivation and irritation of the mouth and throat. The oral LD_{50} in rats is 800 mg/kg.

CARCINOGENICITY

The IARC evaluation is that there is limited evidence for the carcinogenicity of 1,1,2,2-tetrachloroethane in experimental animals (7). An oral NCI carcinogenesis bioassay found it to be a liver carcinogen in mice, but gave no conclusive evidence in rats (8). The effects of human exposure have been well documented since the 1920s without evidence that it is carcinogenic (9).

MUTAGENICITY

1,1,2,2-Tetrachloroethane is mutagenic in *Salmonella typhimurium* (7,10) and *Escherichia coli* (10). In Chinese hamster ovary cells it produced sister chromatid exchange but not chromosome aberrations (11).

REPRODUCTIVE HAZARDS

This compound has teratogenic effects in mice (12). No information is available on any human reproductive effects.

139. 1,1,2,2-Tetrachloroethane

FIRST AID

Eyes Wash the eye with flowing water for 10 minutes.

Lungs Remove casualty from area of exposure. If unconscious, do not give anything to drink. Give artificial ventilation and chest compression or place in the recovery position as necessary. If conscious make the casualty lie or sit down quietly, give oxygen if available. Lung congestion may occur – a conscious casualty with breathing difficulties should be placed in a sitting position. Shock may result – if so do not give any drinks, and if conscious, lie casualty flat with legs raised. Convulsions may occur and may cause unconsciousness.

Mouth Do not make the casualty vomit. Treat unconscious casualties as for lungs, but if conscious give 1 pint of water to drink. Convulsions may occur and may cause unconsciousness. Shock may result – do not give any drinks, and if conscious lie casualty flat with legs raised.

Skin Remove contaminated clothing immediately. Wash the affected area with soap and copious amounts of water. Absorption through the skin may cause symptoms similar to lung congestion.

In all cases of exposure, the patient should be transferred to hospital as soon as possible.

HANDLING AND STORAGE

1,1,2,2-Tetrachloroethane should be handled wearing an approved respirator, PVA or Viton (NOT rubber) gloves (13), safety goggles and other protective clothing. Avoid getting it on clothing. It should only be used in a chemical fume hood. It should be kept in a tightly closed container, in a cool place and protected from moisture. Glove permeability studies are available (14).

DISPOSAL

Eliminate all sources of ignition, ventilate the area and wear a laboratory coat or overalls, rubber gloves, approved compressed air breathing apparatus and safety boots. Absorb small spills onto paper towels, remove to a safe open air site and allow to evaporate in a metal tray. Wash spillage site with detergent. For large spills, absorb onto vermiculite-sodium carbonate mixture (90-10) or sand-soda ash (90-10) and mix carefully. EITHER transport in dry buckets to a safe open air area for atmospheric evaporation OR shovel into paper boxes and incinerate.

FIRE PRECAUTIONS

Fires involving 1,1,2,2-tetrachloroethane should be extinguished using water spray, carbon dioxide, dry chemical powder, alcohol foam or polymer foam.

FURTHER READING

Bretherick, L. *Handbook of Reactive Chemical Hazards* (4th edition)

Sax, N. Irving *Dangerous Properties of Industrial Materials* (7th edition)

ACGIH Documentation of TLVs and BEIs (6th edition, 1986)

Lenga, R. E. *The Sigma-Aldrich Library of Chemical Safety Data* (2nd edition) (Sigma-Aldrich, Milwaukee, 1988)

Dangerous Prop. Ind. Mater. Report 1985, **5**(4), 10-30

Exposure and Risk Assessment for 1,1,2,2-Tetrachloroethane (Arthur D. Little, Cambridge, MA, 1985)

REFERENCES

1. Wilson, R. H.; Brumly, D. R. *Ind. Med.* 1944, **13**, 320
2. Von Oettingen, W. F. *J. Ind. Hyg. Tox.* 1937, **19**, 349
3. Lobo-Medorca, R. *Brit. J. Ind. Med.* 1963, **20**, 50
4. Pebay-Peyroula, F.; Nicaise, A. M. *J. Eur. Toxicol.* 1970, **3**(5), 300-308
5. Parker, J. C.; et al. *Am. Ind. Hyg. Assoc. J.* 1979, **40**(3), A46,A48,A50,A52,A60
6. Smyth, H. F. Jr. *Am. Ind. Hyg. Assoc. Q.* 1956, **17**, 129
7. *IARC Monographs on the Evaluation of the Carcinogenic Risk of Chemicals to Humans* 1979, **20**, 477-489
8. *Natl. Cancer Inst. Carcinogenesis Tech. Report* 1978, **27** (NCI, Bethesda, MD)
9. *AGCIH Documentation of TLVs and BEIs* (6th edition, 1986)

139. 1,1,2,2-Tetrachloroethane

10. Rosenkranz, H. S. *Environ. Health Perspect.* 1977, **21**, 79-84
11. Galloway, S. M.; et al. *Environ. Mol. Mutagen.* 1987, **10**(Suppl. 10), 1-175
12. Schmidt, R. *Biol. Rundschau* 1976, **14**, 220-223
13. Forsberg, K.; Mansdorf, S. Z. *Quick Selection Guide to Chemical Proctective Clothing* (Van Nostrand Reinhold, New York, 1989)
14. Sansone, E. B.; et al. *J. Am. Ind. Hyg. Assoc.* 1978, **39**(2), 169-174

2,3,4,6-TETRA-CHLOROPHENOL

140. 2,3,4,6-Tetrachlorophenol

RISKS
Toxic if swallowed – Irritating to eyes and skin (R25, R36/38)

SAFETY PRECAUTIONS
In case of contact with eyes, rinse immediately with plenty of water and seek medical advice – After contact with skin, wash immediately with plenty of water – Wear suitable gloves – If you feel unwell, seek medical advice (show label where possible) (S26, S28, S37, S44)

IDENTIFIERS
SYNONYMS Dowicide 6; phenol, 2,3,4,6-tetrachloro-; TCP; 2,4,5,6-tetrachlorophenol

CHEMICAL ABSTRACTS No.	58-90-2
NIOSH No.	SM 9275000
HAZCHEM CODE	2X
UN No.	2020

THRESHOLD LIMIT VALUES
US TLV (TWA)	not available
US TLV (STEL)	not available

UK EXPOSURE LIMITS (OES)
Long-term (8 hr TWA value)	not available
Short-term (10 min TWA value)	not available

Germany
MAK	not available

France
VME	not available
VLE	not available

Sweden
Short-term limit	1.5 mg/m^3
Level limit	0.5 mg/m^3

PHYSICAL PROPERTIES
Description	Needles.
Boiling point	150°C at 15 mm Hg
Melting point	70°C
Density	1.839 at 25°C
Vapour density	not available
Vapour pressure	1 mm Hg at 100°C
Flash point	not available
Explosive limits	not available
Autoignition temperature	not available

Solubility Practically insoluble in water. Soluble in ether and benzene. Very soluble in dimethyl sulphoxide, ethanol and acetone.

PACKAGING AND TRANSPORTATION

Road transportation
hazard warning sign	2020 harmful substance
Hazchem code	2X

Sea transportation
IMDG page No.	6107
class	6.1
label	harmful stow away from foodstuffs; marine pollutant
packaging group	III

Air transportation
ICAO/IATA code (UN No.)	2020
class	6.1
label	keep away from food
packaging group	III
packing instructions	
cargo	619
passenger	619
passenger aircraft max. quantity	100 kilograms
cargo aircraft max. quantity	200 kilograms

140. 2,3,4,6-Tetrachlorophenol

MANUFACTURE

Tetrachorophenols are produced by the chlorination of phenol using potassium tellurate as a catalyst. The isomers produced are separated using ion-exchange resin systems.

USES

2,3,4,6-Tetrachlorophenol is the most widely used isomer, and is used as a disinfectant, antimicrobial agent and a preservative for wood, latex and leather. Tetrachlorophenols are present in small quantities in pentachlorophenol.

CHEMICAL HAZARDS

Tetrachlorophenols are incompatible with acid chlorides, acid anhydrides and oxidising agents. When heated to decomposition they emit toxic fumes of carbon oxides and hydrogen chloride.

BIOLOGICAL HAZARDS

Tetrachlorophenols are toxic by inhalation, ingestion and skin absorption. They may be able to stimulate tissue oxygen metabolism and include hyperpyrexia (1). Chronic exposure to a mixture of penta-and tetrachlorophenols is reported to have cause aplastic anaemia in an adult (1).

Vapour Inhalation

Inhalation of tetrachlorophenol dust is irritating to the mucous membranes of the respiratory tract.

Eye Contact

2,3,4,6-Tetrachlorophenol is reported to cause conjunctivitis and slight to moderate corneal injuries (1).

Skin Contact

Solid and 10% aqueous suspensions of 2,3,4,6-tetrachlorophenol are not primary skin irritants, but may cause an acneform dermatitis on repeated exposure. When dissolved in organic solvents, tetrachlorophenol may be absorbed through the skin in toxic amounts (1). The LD_{50} when applied to the skin of rabbits was 250 mg/kg.

Swallowing

The oral LD_{50} for 2,3,4,5-tetrachlorophenol was 140 mg/kg in rats and 250 mg/kg in guinea pigs. The oral LD_{50} in mice was 400 mg/kg for the 2,3,4,5-isomer and 109 mg/kg for the 2,3,5,6-isomer.

CARCINOGENICITY

The IARC evaluation is that there is limited evidence for the carcinogenicity of occupational exposure to chlorophenols to humans (2). 2,3,4,6-Tetrachlorophenol is an experimental carcinogen. 100 mg/kg injected subcutaneously into mice caused carcinogenic effects (3).

MUTAGENICITY

2,3,4,6-Tetrachlorophenol was not mutagenic to *Salmonella* (4). It was mutagenic in a Chinese hamster V79 cell assay (5) and was cytotoxic, but non-mutagenic in another assay using Chinese hamster V79 cells (6).

REPRODUCTIVE HAZARDS

2,3,4,6-Tetrachlorophenol caused teratogenic effects when orally administered to pregnant rats on days 6-15 of gestation (7).

FIRST AID

Eyes Wash the eye with flowing water for 10 minutes.

Lungs Remove casualty from area of exposure. If unconscious, do not give anything to drink. Give artificial ventilation and chest compression or place in the recovery position as necessary. If conscious make the casualty lie or sit down quietly, give oxygen if available. Lung congestion may occur – a conscious casualty with breathing difficulties should be given oxygen. Convulsions may occur and may cause unconsciousness. Shock may result – do not give any drinks, if conscious lie casualty flat with legs raised. Further artificial ventilation may be needed.

Mouth Do not make the casualty vomit. Treat unconscious casualties as for lungs but if conscious give 1 pint of water to drink.

140. 2,3,4,6-Tetrachlorophenol

Skin Remove all contaminated clothing immediately, wash affected areas with soap and copious amounts of water. Absorption through the skin may cause symptoms similar to those of inhalation.

In all cases of exposure, the patient should be transferred to hospital as soon as possible.

HANDLING AND STORAGE

2,3,4,6-Tetrachlorophenol should be handled wearing an approved respirator, chemical-resistant gloves, safety goggles and other protective clothing. It should only be used in a chemical fume hood. It should be kept in a tightly closed container (8).

DISPOSAL

Eliminate all sources of ignition and ventilate the area. Wear a laboratory coat, safety spectacles, butyl rubber gloves, and approved self-contained breathing apparatus. While stirring constantly, add contaminated material to approximately 30 times its weight of a solution prepared by dissolving 1 part sodium sulphide in 6 parts water.

FIRE PRECAUTIONS

Fires involving tetrachlorophenols should be extinguished using carbon dioxide, dry chemical powder, alcohol foam or polymer foam.

FURTHER READING

Sax, N. Irving *Dangerous Properties of Industrial Materials* (7th edition)

Kirk-Othmer *Encyclopedia of Chemical Technology*

REFERENCES

1. Gosselin, R. E.; et al. *Clinical Toxicology of Commercial Products* (5th edition) (Williams & Wilkins, Baltimore, 1984)
2. IARC *Monographs on the evaluation of the carcinogenic risk of chemicals to humans* 1986, **41**, 319-356
3. *Natl. Tech. Info. Service PB223-159* (NTIS, Springfield)
4. DeMarine, D. M.; et al. *Environ. Mol. Mutagen.* 1990, **15**(1), 1-9
5. Hattula, M. L.; Knuutines, J. *Chemosphere* 1985, **14**(10), 1617-1625
6. Jansson. K.; Jansson, V. *Mutat. Res.* 1986, **171**(2-3), 165-168
7. *Toxicol. Appl. Pharmacol.* 1974, **28**, 146
8. Lenga, R. E. *The Sigma-Aldrich Library of Chemical Safety Data* (2nd edition) (Sigma-Aldrich, Milwaukee, 1988)

141. Thioglycolic Acid

THIOGLYCOLIC ACID

$HSCH_2CO_2H$

RISKS
Toxic by inhalation, in contact with skin and if swallowed – Causes burns (R23/24/25, R34)

SAFETY PRECAUTIONS
Keep out of reach of children – Avoid contact with eyes – Take off immediately all contaminated clothing – After contact with skin, wash immediately with plenty of water (S2, S25, S27, S28)

IDENTIFIERS

SYNONYMS acetic acid, mercepto-; mercaptoacetic acid; β-mercatoacetic acid; 2-mercaptoacetic acid; 2-mercaptoethonoic acid; 2-thioglycolic acid; thiovanic acid; Mercaptoacetate

CHEMICAL ABSTRACTS No.	68-11-1
NIOSH No.	AI 5950000
HAZCHEM CODE	2X
UN No.	1940

THRESHOLD LIMIT VALUES

US TLV (TWA)	1 ppm (3.8 mg/m^3)
US TLV (STEL)	not available

UK EXPOSURE LIMITS (OES)
Long-term (8 hr TWA value) .... 1 ppm (5 mg/m^3)
Short-term (10 min TWA value) .... not available

Germany
MAK .... not available

France
VME .... not available
VLE .... not available

Sweden
Short-term limit .... not available
Level limit .... not available

PHYSICAL PROPERTIES

Description Liquid with a strong odour. Readily oxidized by air.

Boiling point	108°C at 15 mm Hg
Melting point	-16.5°C
Density	1.3253 at 20°C
Vapour density	not available
Vapour pressure	1 mm Hg at 60°C
Flash point	125°C
Explosive limits	lel 5.9%
Autoignition temperature	350°C

Solubility Miscible with water, alcohol, ether, chloroform, benzene and many other organic solvents.

PACKAGING AND TRANSPORTATION

Road transportation

hazard warning sign	1940 corrosive substance
Hazchem code	2X

Sea transportation

IMDG page No.	8235
class	8
label	corrosive
packaging group	II

Air transportation

ICAO/IATA code (UN No.)	1940
class	8
label	corrosive
packaging group	II
packing instructions	
cargo	813
passenger	809
passenger aircraft max. quantity	1 litre
cargo aircraft max. quantity	30 litres

141. Thioglycolic Acid

MANUFACTURE

Thioglycolic acid is produced by reaction of sodium or potassium chloroacetate with alkali-metal hydrosulphide. Thioglycolic acid is obtained from the reaction mixture by acidification, and purified by vacuum distillation (or directly converted into derivatives).

USES

The salts and esters of thioglycolic acid are used in hair-waving and depilatory preparations. It is also used in the manufacture of pharmaceuticals, thioglycolates, and as a vinyl stabiliser.

CHEMICAL HAZARDS

Thioglycolic acid is incompatible with strong oxidising agents (1).

BIOLOGICAL HAZARDS

Thioglycolic acid is a corrosive irritant, and is poisonous if swallowed, absorbed through the skin, and by intravenous or intraperitoneal injection.

Vapour Inhalation

No observable adverse effects occurred in rats exposed to 620 ppm. The LC_{50} in mice (6 hours) is 256 mg/l.

Eye Contact

Thioglycolic acid is a corrosive irritant, and caused severe pain, conjunctival inflammation, corneal opacity and severe iritis in rabbits. No improvement was noted after 14 days, and the effect was the same even with immediate washing after exposure (2).

Skin Contact

Thioglycolic acid is a corrosive irritant, and causes severe skin burns and blistering (3). Necrosis, oedema and hyperaemia have been reported in single application patch tests in rabbits. Fatalities have been reported in guinea pigs treated with a 10% solution at less than 5 ml/kg. The LD_{50} in rabbits was 848 mg/kg (10% solution). Symptoms included weakness, breathing difficulties and convulsions (2).

Swallowing

The oral LD_{50} in rats has been reported as 114 mg/kg (1), although an LD_{50} of below 50 mg/kg for the undiluted material has been recorded (2). Liver injury and possible gut irritation were seen on autopsy.

CARCINOGENICITY

Sodium thioglycolate was noncarcinogenic in lifetime skin painting studies in mice and rabbits (4).

MUTAGENICITY

No information is available concerning the mutagenicity of thioglycolic acid.

REPRODUCTIVE HAZARDS

No information is available concerning the reproductive hazards of thioglycolic acid.

FIRST AID

Eyes Wash the eye with flowing water for 10 minutes.

Lungs Remove casualty from area of exposure. If unconscious, do not give anything to drink, give artificial ventilation and chest compression or place in the recovery position as necessary. If conscious make the casualty lie or sit down quietly, give oxygen if available. Lung congestion may occur – a conscious casualty with breathing difficulties should be placed in a sitting position.

Mouth Do not make the casualty vomit. Treat unconscious casualties as for lungs but if conscious give 1 pint of water to drink immediately; give repeated drinks of water (1 cupful every 10 minutes).

Skin Remove contaminated clothing immediately, drench the affected area with running water for at least 10 minutes.

In all cases of exposure, the patient should be transferred to hospital as soon as possible.

141. Thioglycolic Acid

HANDLING AND STORAGE

Thioglycolic acid should be handled in the same way as strong acids, wearing rubber or impervious plastic gloves and boots, safety goggles, and a neoprene apron when working with over gram amounts of neat acid. It should only be used in a chemical fume hood. It should be kept refrigerated in a tightly closed container (1).

DISPOSAL

Eliminate all sources of ignition, ventilate the area and wear a laboratory coat or overalls, butyl rubber gloves, and approved self-contained breathing apparatus or all purpose canister respirator.
EITHER oxidise with an aqueous solution (10-20% w/v) of either calcium or sodium hypochlorite and stir vigorously. Neutralise with 6M ammonium hydroxide, test with litmus paper (red to blue), allow to stand for 2 hours and run to waste down outside drain with x1000 volume of running cold water.
OR Cover spill with solid calcium hypochlorite, mix thoroughly, scoop into buckets and allow to stand overnight in a fume cupboard. Neutralise with 6M ammonium hydroxide], testing before and after with litmus. Wash down outside drain with excess running water.
Wash spillage site with a strong soap solution to which 10% solid calcium hypochlorite or sodium hypochlorite has been added.

FIRE PRECAUTIONS

Fires involving thioglycolic acid should be extinguished using water spray, carbon dioxide, dry chemical powder, alcohol foam or polymer foam.

FURTHER READING

Sax, N. Irving *Dangerous Properties of Industrial Materials* (7th edition)

Encyclopedia of Occupational Safety & Health

Patty's *Industrial Hygiene and Toxicology*

Kirk-Othmer *Encyclopedia of Chemical Technology*

National Fire Protection Association *Manual of Hazardous Reactions*

REFERENCES

1. Lenga, R. E. *The Sigma-Aldrich Library of Chemical Safety Data* (2nd edition) (Sigma-Aldrich, Milwaukee, 1988)
2. *ACGIH Documentation of TLVs and BEIs* (6th edition, 1986) p. 571
3. Gosselin, R. E.; et al. *Clinical Toxicology of Commercial Products* (5th edition) (Williams & Wilkins, Baltimore, 1984)
4. Stenback, F. G.; et al. *Food Chem. Toxicol.* 1977, **15**, 601

142. o-Tolidine

o-TOLIDINE

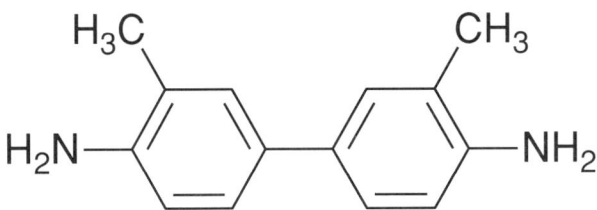

RISKS
May cause cancer – Harmful if swallowed (R45, R22)

SAFETY PRECAUTIONS
Avoid exposure-obtain special instructions before use – If you feel unwell, seek medical advice (show label where possible) (S53, S44)

IDENTIFIERS

SYNONYMS   benzidine, 3,3'-dimethyl-, bianisidine; C.I. 37230; C.I. Azoic Diazo Component 113; 3,3'-dimethylbenzidine; 3,3'-dimethyl-4,4'-biphenyl-diamine; 3,3'dimethlbiphenyl-4,4'diamine, 3,3'- dimethyl-(1, 1'-biphenyl)4,4'-diamine; 3,3'-dimethyl-4-4'-diphenyl-diamine,4,4'-DI-o-toluidine; (1-1'-biphenyl-4-4'diamine, 3, 3'- dimethyl; diaminoditolyl, 4,4'-diamino-3-3'dimethyl (1,1'-biphenyl); 4,4'-diamino-3,3'-dimethyl-diphenyl; Fast Dark Blue Base R; RCRA Waste No. U095; o-tolidin, 2-tolidin, 2tolidino,tolidine; 2-tolidine; o,o'-tolidine, 3, 3'-tolidine, 4,4'-BI-o-toluidine

CHEMICAL ABSTRACTS No.	119-93-7
NIOSH No.	DD 1225000
HAZCHEM CODE	not available
UN No.	not available

THRESHOLD LIMIT VALUES

US TLV (TWA)	not available
US TLV (STEL)	not available

UK EXPOSURE LIMITS (OES)
Long-term (8 hr TWA value)	not available
Short-term (10 min TWA value)	not available

Germany
MAK	not available

France
VME	not available
VLE	not available

Sweden
Short-term limit	not available
Level limit	not available

PHYSICAL PROPERTIES

Description   White to reddish crystals.

Boiling point	415°C
Melting point	129-131°C
Density	1.0
Vapour density	not available
Vapour pressure	not available
Flash point	not available
Explosive limits	not available
Autoignition temperature	not available

Solubility   Very slightly soluble in water. Soluble in alcohol, ether or acetic acid.

PACKAGING AND TRANSPORTATION

Road transportation
hazard warning sign	not available
Hazchem code	not available

Sea transportation
IMDG page No.	not available
class	not available
label	not available
packaging group	not available

Air transportation
ICAO/IATA code	not available
class	not available
label	not available
packaging group	not available
packing instructions	
cargo	not available
passenger	not available
passenger aircraft max. quantity	not available
cargo aircraft max. quantity	not available

142. o-Tolidine

MANUFACTURE

o-Tolidine is manufactured by alkaline reduction of o-nitrotoluene using zinc as the reductant to form o-hydrazotoluene, which is rearranged to o-tolidine by boiling with hydrochloric acid.

USES

o-Tolidine is used in the manufacture of dyestuffs and pigments. It is used as a laboratory reagent for the detection of gold and free chlorine in water and air.

CHEMICAL HAZARDS

o-Tolidine is not explosive or violently reactive. It may decompose on heating to give toxic fumes of carbon and nitrogen oxides.

BIOLOGICAL HAZARDS

o-Tolidine is harmful by inhalation, ingestion and skin contact. Symptoms of exposure to the dihydrochloride include cyanosis, air hunger, nausea, vomiting, low blood pressure and convulsions (1). The LD_{Lo} in mice and rats administered o-tolidine intraperitoneally was 125 mg/kg.

Vapour Inhalation

o-Tolidine may cause sneezing and dizziness if inhaled.

Eye Contact

No data is available.

Skin Contact

o-Tolidine is readily absorbed through the skin.

Swallowing

o-Tolidine is moderately toxic by ingestion. The oral LD_{50} in rats was 404 mg/kg and the LD_{Lo} in dogs was 600 mg/kg.

CARCINOGENICITY

Exposure of dye worker to mixtures of biphenyl amine compounds which includes otolidine has lead to bladder cancer, and o-tolidine is structurally related to benzidine, a suspected bladder carcinogen. o-Tolidine administered in drinking water caused lung neoplasms in male, but no female mice at the highest dose given (140 ppm) (2). When administered orally in oil to female rats it was a weak inducer of mammary carcinomas (3). It did not induce tumours in hamsters fed commercial o-tolidine in the diet (4,5). When administered by subcuteneous injection to rats it induced tumours of the external auditory canal (6) and zymbals glands (7). In a series of six tests designed to predict carcinogenicity, o-tolidine was positive in two (Ames test and endoplasmic reticulum degranulation test) and negative in four (cell transformation test, sebaceous gland suppression test, tetrazolium-reduction test, and tissue implant test) (8).

MUTAGENICITY

o-Tolidine induced sister chromatid exchanges and chromosome aberrations in Chinese hamster ovary cells (9). It induced DNA repair in rat and hamster hepatocytes (10) and unscheduled DNA synthesis in HeLa cells (11). It was mutagenic in *Salmonella typhimurium* (12,13). It was weakly mutagenic in rabbit lymphocytes (waalkens). Urine from rats treated with o-tolidine was of greater mutagenicity in *Salmonella typhimurim* then the substance itself (14).

REPRODUCTIVE HAZARDS

o-Tolidine was not foetotoxic to rats when injected subcuteneously on day 7 of gestation (15).

FIRST AID

Eyes Wash the eye with flowing water for 10 minutes.

Lungs Remove casualty from area of exposure. If unconscious, do not give anything to drink. Give artificial ventilation and chest compression or place in the recovery position as necessary. If conscious make the casualty lie or sit down quietly, give oxygen if available. Convulsions may occur and may cause unconsciousness. Shock may result – if so do not give any drinks, and if conscious, lie casualty flat with legs raised.

142. o-Tolidine

Mouth Do not make the casualty vomit. Treat unconscious casualties as for lungs, but if conscious give 1 pint of water to drink.

Skin Remove contaminated clothing immediately, wash the affected area with soap and copious amounts of water. Absorption through the skin may cause symptoms similar to those of inhalation.

In all cases of exposure, the patient should be transferred to hospital as soon as possible.

HANDLING AND STORAGE

o-Tolidine should be handled wearing an approved respirator, chemical-resistant gloves, safety goggles and other protective clothing. It should only be used in chemical fume hood. It should be kept in a tightly closed container. Safe handling has been discussed (16).

DISPOSAL

Wear a laboratory coat, safety spectacles, butyl rubber gloves and suitable safety shoes, and have an approved self-contained breathing apparatus or canister respirator available. Absorb small liquid spills onto paper towels, evaporate in an iron pan in a fume cupboard, add crumpled paper and burn carefully. Brush small solid spills onto paper, put in an iron pan, cover with crumpled paper and burn carefully in a safe place outside. For large spills,
EITHER cover with sand and mix carefully. Shovel into containers, disperse in an excess solution of dilute hydrochloric acid, mix well and leave to stand for 24 hours stirring occasionally. Carefully decant acid extract into the drains, diluting with a large volume of cold tap water. Wash the sand thoroughly with cold water.
OR cover with a sand-soda ash mix (90-10), mix well, shovel into cardboard boxes, pack with crumpled paper and incinerate. Further information on disposal is available (17).

FIRE PRECAUTIONS

Fires involving o-tolidine should be extinguished using water spray, carbon dioxide, dry chemical powder, alcohol foam or polymer foam.

FURTHER READING

Sax, N. Irving *Dangerous Properties of Industrial Materials* (7th edition)

Kirk-Othmer *Encyclopedia of Chemical Technology*

National Fire Protection Association *Manual of Hazardous Reactions*

ACGIH *Documentation of TLVs and BEIs* (6th edition, 1986)

IARC Monographs on the evaluation of the carcinogenic risk of chemicals to humans 1972, **17**, 87-91

REFERENCES

1. *Dangerous Prop. Ind. Mater. Rep.* 1985, **5**(3), 75-77
2. Schieferstein, G. J.; et al. *Food. Chem. Toxicol.* 1989, **27**(12), 801-806
3. Giswold, D. P.; et al. *Cancer Res.* 1968, **28**(5), 924-933
4. Saffioti, U.; et al. in: Deichman, Io.; Lampe, K. F. (Eds.) *Bladder Cancer; a symposium* (Aesculapius Publishing Co., Birmingham, Al, 1967) pp. 129
5. Sellakimur, A. R.; et al. *Proc. Am. Assoc. Cancer Res.* 1969, **10**, 78
6. Spitz, S.; et al. *Cancer (Philad.)* 1950, **3** 789
7. Pliss, G. B.; Zabezhinsky, M. A. *JNCI, J. Natl. Cancer Inst.* 1970, **45**, 283
8. Purchase, I. H. F.; et al. *Br. J. Cancer* 1978, **37** 873
9. Galloway, S. M.; et al. *Environ. Mol. Mutagen.* 1987, **10** (Suppl. 10), 1-175
10. Kornburst, D. J.; Barfknecht, T. R. *Mutat. Res.* 1984, **136**, 255-266
11. Mortin, C. N.; et al. *Cancer Res.* 1978, **38**, 2621
12. Waalkens D. H.; et al. Mutat. Res. 1981, **89**, 197-202
13. Krishna, G.; et al. *J. Toxicol. Environ. Mutagen.* 1984, **6**(2), 145-151
14. Tanaka, K. I.; et al. *Mutat. Res. 1980,* **79**, 173-176

142. o-Tolidine

15. Ema, M.; et al. *Nippon Yakirigaku Zasshi* 1984, **83**(5), 459-465
16. *Lab. Pract.* 1982, **31**(11), 1090-1092
17. *Am. Ind. Hyg. Assoc. J.* 1985, **46**(4), 187-191

143. Toluene-2,4-Diisocyanate

TOLUENE-2,4-DIISOCYANATE

RISKS
Very toxic by inhalation – Irritating to eyes, respiratory system and skin – May cause sensitisation by inhalation (R26, R36/37/38, R42)

SAFETY PRECAUTIONS
In case of contact with eyes, rinse immediately with plenty of water and seek medical advice – After contact with skin, wash immediately with plenty of water – In case of insufficient ventilation, wear suitable respiratory equipment – In case of accident or if you feel unwell, seek medical advice immediately (show label where possible) (S26, S28, S38, S45)

IDENTIFIERS

SYNONYMS benzene,2,4-diisocyanato-1-methyl; isocyanic acid, 4-methyl-*m*-phenylene ester; 2,4-TDI; 2,4-toluene diisocyanate; 2,4-tolylene diisocyanate; Desmodur T80; diisocyanatotoluene; Hylene TM; isocyanic acid, methylphenylene ester; 4-methyl-phenylene diisocyante; 4-methyl-phenylene isocyanate; Mondur TDS; Nacconate IOO; NCI-C50533; Niax TDI; RCRA Waste No. U223; Rubuinte TDI 80/20; 2,4-toluenediisocyanate; toluylene-2, 4-diisocyanate; 2,4-tolyenediisocyanate

CHEMICAL ABSTRACTS No.	584-84-9
NIOSH No.	CZ 6300000
HAZCHEM CODE	2XE
UN No.	2078

THRESHOLD LIMIT VALUES

US TLV (TWA)	0.005 ppm (0.036 mg/m^3)
US TLV (STEL)	0.002 ppm (0.14 mg/m^3)

UK EXPOSURE LIMITS (MEL)
Long-term (8 hr TWA value) .... 0.02 ppm (0.07 mg/m^3) (as -NCO)
Short-term (10 min TWA value) .... 0.07 mg/m^3 (as -NCO)

Germany
MAK .... 0.01 ppm (0.07 mg/m^3)

France
VME .... 0.01 ppm (0.08 mg/m^3)
VLE .... 0.02 ppm (0.16 mg/m^3)

Sweden
Ceiling limit .... 0.01 ppm (0.07 mg/m^3)
Level limit .... 0.005 ppm (0.04 mg/m^3)

PHYSICAL PROPERTIES

Description Clear, faintly yellow liquid with a sharp pungent odour.

Boiling point	251°C
Melting point	19.5-21.5°C
Density	1.2244 at 20°C
Vapour density	6.0
Vapour pressure	0.01 mm Hg at 20°C
Flash point	132°C (open cup)
Explosive limits	0.9% vol – 9.5% vol
Autoignition temperature	not available

Solubility Miscible with alcohol (decomposes) and most organic solvents.

PACKAGING AND TRANSPORTATION

Road transportation

hazard warning sign	2078 toxic substance
Hazchem code	2XE

Sea transportation

IMDG page No.	6269
class	6.1
label	poison
packaging group	II

Air transportation

ICAO/IATA code (UN No.)	2078
class	6.1
label	poison
packaging group	II
packing instructions	
cargo	611
passenger	609
passenger aircraft max. quantity	5 litres
cargo aircraft max. quantity	60 litres

143. Toluene-2,4-Diisocyanate

MANUFACTURE

Toluene diisocyanate is manufactured by reaction of toluenediamine and phosgene, and is usually formulated as the 80:20 mixture of the 2,4- and 2,6-isomers (1).

USES

Toluene diisocyanate is used in flexible polyurethane foams (eg. for furnishings), polyurethane coatings, varnishes, paints and elastomers.

CHEMICAL HAZARDS

Toluene 2,4-diisocyanate may polymerise exothermically in contact with bases or more than traces of acyl chlorides (2). Slow absorption of water vapour causing deposition of urea and liberation of carbon dioxide and hence pressure, may cause bursting of polythene containers of toluene 2,4-diisocyanate after prolonged storage.

BIOLOGICAL HAZARDS

The toxicology of toluene diisocyanate has been reviewed (3).

Vapour Inhalation

Toluene 2,4-diisocyanate is a strong respiratory irritant and sensitiser, and causes asthma attacks, loss of lung function, chronic restrictive lung disease, hypersensitivity pneumonitis and chronic bronchitis (4). Fatalities have been reported. Respiratory symptoms such as breathlessness, wheezing, and cough may persist in sensitised workers no longer exposed to toluene diisocyanate (5), and recur in sensitised workers on re-exposure, even after many years (6). Sensitisation may occur on first exposure or may be developed after exposure over days, months or years, and may occur after high level or chronic low level exposures. Long term sequelae of severe exposures include headache, loss of memory and concentration, confusion and personality changes. A chronic-like syndrome with coughing, wheezing and tight-chestedness may occur after repeated low level exposures. The LC_{50} (4 hours) in rats is 14 ppm (7). Sensitisation, reduced respiratory rate, pneumonitis, tracheitis, bronchitis, necrotic rhinitis and fibrosis in the bronchiole walls, and effects on the liver, kidney and gastrointestinal tract have been reported in animal studies. The effects on the lung have been reviewed (8,9), as have exposure effects in the polyurethane industry (10).

Eye Contact

Toluene diisocyanate is irritating to the eyes at levels above 0.35 mg/m^3 (11), and causes lacrimation.

Skin Contact

Respiratory hypersensitivity occurred in guinea pigs after dermal application. It may cause inflammation and allergic dermatitis (12).

Swallowing

The oral LD_{50} in rats is 5800 mg/kg (7). Bronchopneumonia has been reported in rats following administration by gavage. Symptoms may include abdominal distress, nausea and vomiting.

CARCINOGENICITY

The IARC evaluation states that there is "sufficient evidence for the carcinogenicity of toluene diisocyanate to experimental animals" but "inadequate evidence" in humans (1). In studies on rats and mice with the commercial grade (80:20) there was no evidence of carcinogenicity after inhalation exposure, but increased tumour incidence after oral administration (11). Experimental evidence for the NIOSH recommendation that toluene diisocyanate be regarded as a potential occupational carcinogen has been discussed (13).

MUTAGENICITY

Toluene 2,4-diisocyanate is not mutagenic in *Salmonella typhimurium* TA1535, TA1538, TA98 or TA100 with metabolic activation. The commercial grade (80:20) gave positive results in the presence of phenobarbital induced rat liver S9. Positive results in the Ames test have been reported (14). There was no increase in micronuclei in bone marrow of rats and mice exposed by inhalation to the commercial grade (80:20) (1), and negative results have been reported in *in vitro* mammalian cell assays (11). In tests on Chinese hamster ovary cells *in vitro* the 2, 4-isomer gave negative results for chromosome aberrations, negative results for sister chromatid exchanges with S9 and equivocal results for sister chromatid exchanges without S9. The 2,6-isomer gave positive results for sister chromatid exchanges and chromosome aberrations without S9, and negative results with S9 (15). The breakdown products of toluene diisocyanate, 2,4- and 2,6-diaminotoluene, are mutagenic in *S. typhimurium* with metabolic activation, and in cultured mammalian cells (1).

143. Toluene-2,4-Diisocyanate

REPRODUCTIVE HAZARDS

No information is available concerning the reproductive hazards of toluene 2,4-diisocyanate (1).

FIRST AID

Eyes Wash the eye with flowing water for 10 minutes.

Lungs Remove casualty from area of exposure. If unconscious, do not give anything to drink, give artificial ventilation and chest compression or place in the recovery position as necessary. If conscious make the casualty lie or sit down quietly, give oxygen if available. Lung congestion may occur – a conscious casualty with breathing difficulties should be placed in a sitting position. Allergic asthma (wheezy breathing) may occur and immediate treatment is needed – Ventolin may be useful.

Mouth Do not make the casualty vomit. Treat unconscious casualties as for lungs but if conscious give 1 pint of water to drink immediately.

Skin Remove contaminated clothing immediately, drench the affected area with running water for at least 10 minutes.

In all cases of exposure, the patient should be transferred to hospital as soon as possible.

HANDLING AND STORAGE

Toluene 2,4-diisocyanate should be handled wearing an approved respirator, chemical-resistant gloves (NOT natural rubber, neoprene or polyethylene (16)), safety goggles and other protective clothing. It should only be used in a chemical fume hood. It should be kept in a tightly closed container (7), in a cool, dry, well-ventilated place, away from oxidisers or areas where there is an acute fire hazard, and protected from physical damage. If stored in tanks, it should be blanketed with nitrogen. Outside or detached storage is preferred. Guidelines for its handling in the paint and ink industries are available (17), as is information on safe storage (18).

DISPOSAL

Eliminate all sources of ignition, ventilate the area and wearing self-contained breathing apparatus, a chemical resistant suit, gloves, and PVC safety boots, spread calcium hypochlorite over the (liquid) spill (or disperse excess calcium hypochlorite solution onto it). Mix and mop into polythene buckets, stand 24 hours and run to waste diluting x1000 with running cold water (see discharge limits from water authorities).

For solid spills either: a) sweep up, place in a large volume of water, add excess sodium hypochlorite, leave to stand 24 hours, then run to waste diluting x1000 with running water; or b) scoop into a large container, make alkaline with 2% sodium hydroxide solution and stir well. Add excess ferrous sulphate solution, mix well, stand for 2-3 hours and run to waste diluting x1000 with running water.

Large spills need specialist help – contact fire brigade.

FIRE PRECAUTIONS

Fires involving toluene 2,4-diisocyanate should be extinguished using carbon dioxide or dry chemical powder.

FURTHER READING

Bretherick, L. *Hazards in the Chemical Laboratory* (4th edition)

Bretherick, L. *Handbook of Reactive Chemical Hazards* (4th edition)

Sax, N. Irving *Dangerous Properties of Industrial Materials* (7th edition)

Encyclopedia of Occupational Safety & Health

Patty's *Industrial Hygiene and Toxicology*

Kirk-Othmer *Encyclopedia of Chemical Technology*

National Fire Protection Association *Manual of Hazardous Reactions*

ACGIH *Documentation of TLVs and BEIs* (6th edition, 1986)

Chemical Infogram. Toluene-2,4-diisocyanate (2,4-TDI) (Can. Cent. Occup. Health Saf., Hamilton, 1988)

De Craecker, W. *Promosafe* 1987, **14**(1), 52-54

Hazard databank. Sheet No. 65. Toluene diisocyanate *Saf. Pract.* 1985, **3**(5), 32-35

143. Toluene-2,4-Diisocyanate

REFERENCES

1. IARC *Monographs on the evaluation of the carcinogenic risk of chemicals to humans* 1986, **39**, 287-323
2. *MCA SD-73,* 1971
3. Woolrich, P. F. *Am. Ind. Hyg. Assoc. J.* 1982, **43**(2), 89-97
4. Baur, X. *Allergologie* 1986, **9**(11), 487-496
5. Luo, J. C. J.; et al. *Br. J. Ind. Med.* 1990, **47**(4), 239-241
6. Banks, D. E.; Rando, R. J. *Thorax* 1988, **43**(8), 660-662
7. Lenga, R. E. *The Sigma-Aldrich Library of Chemical Safety Data* (2nd edition) (Sigma-Aldrich, Milwaukee, 1988)
8. Karol, M. H. *CRC, Crit. Rev. Toxicol.* 1986, **16**(4), 349-379
9. Kay, S. *Food Chem. Toxicol.* 1985, **23**(3), 411-413
10. Burrows, G. E. *Cell. Poly.* 1983, **2**(3), 205-212
11. *Environmental Health Criteria 75: Toluene Diisocyanates* (WHO, Geneva, 1987)
12. Ohsawa, T. *J. Tokyo Womens Med. Coll.* 1983, **53**(3), 237-246
13. *Occup. Health Bull.* 1990, (25), 2
14. Andersen, M.; et al. *Scand. J. Work, Environ. Health* 1982, **8**(1), 80-81
15. Gulati, D. K.; et al. *Environ. Mol. Mutagen.* 1989, **13**(2), 133-193
16. Forsberg, K.; Mansdorf, S. Z. *Quick selection guide to chemical protective clothing* (Van Nostrand Reinhold, New York, 1989)
17. Scott, I. C. *Radiat. Curing* 1987, **14**(2), 20-22
18. Australian Standard AS 2508 Safe storage and handling information cards for hazardous materials. 6.010. Toluene diisocyanate (TDI) (Standards Assoc., Australia, Sydney, 1983)

o-TOLUIDINE

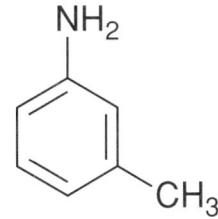

RISKS
Toxic by inhalation, in contact with skin and if swallowed – Danger of cumulative effects (R23/24/25, R33)

SAFETY PRECAUTIONS
After contact with skin, wash immediately with plenty of water – Wear protective clothing and gloves – If you feel unwell, seek medical advice (show label where possible) (S28, S36/37, S44)

IDENTIFIERS

SYNONYMS benzenamine, 2-methyl; *o*-aminotoluene; 2-aminotoluene; *o*-methylaniline; 2-methylaniline; *o*- methylbenzenamine; 2-methylbenzenamine; 2-toluidine; *o*- tolylamine; 1-amino-2-methylbenzene; 2 amino-1-methylbenzene

CHEMICAL ABSTRACTS No.	95-53-4
NIOSH No.	XU 2975000
HAZCHEM CODE	3X
UN No.	1708

THRESHOLD LIMIT VALUES

US TLV (TWA)	2 ppm (8.8 mg/m^3)
US TLV (STEL)	not available

UK EXPOSURE LIMITS (OES)
Long-term (8 hr TWA value)   2 ppm (9 mg/m^3)
Short-term (10 min TWA value)   5 ppm (22 mg/m^3)

Germany
MAK   not available

France
VME   2 ppm (9 mg/m^3)
VLE   not available

Sweden
Short-term limit   not available
Level limit   not available

PHYSICAL PROPERTIES

Description	Yellow liquid. Darkens in air.
Boiling point	200-202°C
Melting point	-14.7°C (β form)
Density	1.004 at 20°C
Vapour density	3.69
Vapour pressure	1 mm Hg at 44°C
Flash point	85°C (closed cup)
Explosive limits	lel 1.5%
Autoignition temperature	481.7°C

Solubility Slightly soluble in water. Very soluble in alcohol, ether or dilute acids.

PACKAGING AND TRANSPORTATION

Road transportation
hazard warning sign	1708 toxic substance
Hazchem code	3X

Sea transportation
IMDG page No.	6270
class	6.1
label	poison
packaging group	II

Air transportation
ICAO/IATA code (UN No.)	1708
class	6.1
label	poison
packaging group	II

packing instructions
cargo
liquid   611
solid   615
passenger
liquid   609
soild   613

passenger aircraft max. quantity
liquid   5 litres
solid   25 kilograms

cargo aircraft max. quantity
liquid   60 litres
solid   100 kilograms

144. o-Toluidine

MANUFACTURE

o-Toluidine is prepared by catalytic hydrogenation or iron reduction of 1-methyl-2-nitrobenzene with alcoholic ammonium sulphide (1). Reduction of crude nitrotoluene produces a mixture of o- and p-toluidine.

USES

o-Toluidine is used in the manufacture of dyes, rubber chemicals (accelerators), agrochemicals and pharmaceuticals. It is also used as a reagent for gold.

CHEMICAL HAZARDS

o-Toluidine is combustible. It gives off flammable vapours on heating, and vapours may form explosive mixtures in air. It is hypergolic with red fuming nitric acid. 2-Toluenediazonium salts are unstable. In an incident in which an operator was found dead under a reaction vessel where o-aminoazotoluene had been prepared from o-toluidine by reaction with sodium nitrate and sulphuric acid, death was found to be due to the presence of methanol in the system which reacted with sodium nitrate and sulphuric acid to give methyl nitrate (2).

BIOLOGICAL HAZARDS

Toxicity is similar to that of aniline; o-toluidine forms methaemoglobin and causes anaemia, reticulocytosis and haematuria. Signs of poisoning include headache, breathing difficulties, kidney and bladder irritation and psychological disturbances. Following the NIOSH alert (3) it is recommended that exposure be reduced to the lowest feasible concentration (4).

Vapour Inhalation

o-Toluidine is absorbed via the respiratory tract.

Eye Contact

o-Toluidine is strongly irritating to the eyes of rabbits (5).

Skin Contact

o-Toluidine is moderately irritating to the skin of rabbits (5). o-Toluidine is absorbed through the skin, though not as readily as m- or p-toluidine (6). The dermal LD_{50} in rabbits is 3250 mg/kg (7).

Swallowing

The oral LD_{50} in rats is 670 mg/kg (6). o-Toluidine is less toxic than the p-isomer in rats after single or repeated intragastric doses (8). Intragastric administration to rats caused cyanosis, spleen congestion, and hypercellularity in bone marrow (9). Oral administration to rats caused changes in the bladder epithelium and a low incidence of papillomas (10).

CARCINOGENICITY

The IARC evaluation is that there is "sufficient evidence for the carcinogenicity of o-toluidine hydrochloride in experimental animals", and that "o-toluidine should be regarded, for practical purposes, as if it presented a carcinogenic risk to humans" (1). NIOSH concluded that direct contact with aniline and o- toluidine was a major factor in increased incidence of bladder tumours (11) and issued an alert (3).

MUTAGENICITY

o-Toluidine is not mutagenic to *Salmonella typhimurium* or *Escherichia coli* (12,13). Both negative (14) and positive results (15) have been reported in mouse lymphoma cells. In mutagenicity tests on urine from rats orally dosed with o-toluidine using *S. typhimurium* in the presence of metabolic activation from PCB treated rats, both positive (16) and negative (17) results have been reported. In tests for sister chromatid exchanges and chromosome aberrations in Chinese hamster ovary cells it gave positive (15,18,19) and negative (20) results. Positive (21) and negative (12) results have been reported in tests for unscheduled DNA synthesis in rat hepatocytes. It induced increases in UDS with S9 in HeLa cells (22). It did not cause single strand breaks in DNA of V29 Chinese hamster lung fibroblasts (23), but did so in liver and kidney cells of mice (14). It did not induce sister chromatid exchanges in human lymphocytes *in vitro* (24).

REPRODUCTIVE HAZARDS

No information is available concerning the reproductive hazards of o-toluidine.

144. *o*-Toluidine

FIRST AID

Eyes Wash the eye with flowing water for 10 minutes.

Lungs Remove casualty from area of exposure. If unconscious, do not give anything to drink. Give artificial ventilation and chest compression or place in the recovery position as necessary. If conscious make the casualty lie or sit down quietly, give oxygen if available. Convulsions may occur and may cause unconsciousness. Shock may result – if so do not give any drinks, and if conscious, lie casualty flat with legs raised.

Mouth Do not make the casualty vomit. Treat unconscious casualties as for lungs, but if conscious give 1 pint of water to drink.

Skin Remove contaminated clothing immediately, wash the affected area with soap and copious amounts of water. Absorption through the skin may cause symptoms similar to those of inhalation.

In all cases of exposure, the patient should be transferred to hospital as soon as possible.

HANDLING AND STORAGE

o-Toluidine should be handled wearing an approved respirator, chemical-resistant gloves (NOT polyethylene) (25), safety goggles and other protective clothing. It should only be used in a chemical fume hood (6). Following the NIOSH alert, certain manufacturers using *o*-toluidine supply full-face respirators, Tyvek suits and rubber boots to workers (26). It should be kept in a tightly closed container under nitrogen, in a cool, dry, well-ventilated place, away from other storage and fire hazards, and protected against physical damage. Outside or detached storage is preferred.

DISPOSAL

Wear a laboratory coat, safety spectacles, butyl rubber gloves and suitable safety shoes, and have an approved self-contained breathing apparatus or canister respirator available. Absorb small liquid spills onto paper towels, evaporate in an iron pan in a fume cupboard, add crumpled paper and burn carefully. Brush small solid spills onto paper, put in an iron pan, cover with crumpled paper and burn carefully in a safe place outside. For large spills,
EITHER cover with sand and mix carefully. Shovel into containers, disperse in an excess solution of dilute hydrochloric acid, mix well and leave to stand for 24 hours stirring occasionally. Carefully decant acid extract into the drains, diluting with a large volume of cold tap water. Wash the sand thoroughly with cold water.
OR cover with a sand-soda ash mix (90-10), mix well, shovel into cardboard boxes, pack with crumpled paper and incinerate.

FIRE PRECAUTIONS

Fires involving *o*-toluidine should be extinguished using water spray, carbon dioxide, dry chemical powder, alcohol foam or polymer foam.

FURTHER READING

Bretherick, L. *Hazards in the Chemical Laboratory* (4th edition)

Bretherick, L. *Handbook of Reactive Chemical Hazards* (4th edition)

Sax, N. Irving *Dangerous Properties of Industrial Materials* (7th edition)

Encyclopedia of Occupational Safety & Health

Patty's *Industrial Hygiene and Toxicology*

Kirk-Othmer *Encyclopedia of Chemical Technology*

National Fire Protection Association *Manual of Hazardous Reactions*

ACGIH Documentation of TLVs and BEIs (6th edition, 1986)

Dangerous Prop. Ind. Mater. Rep. 1982, **2**(1),

Toxicology card 197. *o*-Toluidine *Cah. Notes Doc.* 1984, (115), 259-262

REFERENCES

1. IARC *Monographs on the evaluation of the carcinogenic risk of chemicals to humans* 1982, **27**, 155-175

2. *Sichere Chemarbeit.* 1986, **38**(2), 18-19

3. *NIOSH Alert: request for assistance in preventing bladder cancer from exposure to o-toluidine and aniline* (NIOSH, Cincinnati, 1991)

144. o-Toluidine

4. *Chem. Week* 10 Apr. 1991, 6
5. Smyth, H. F.; et al. *Am. Ind. Hyg. Assoc. J.* 1962, **23**, 95-102
6. Senczuk, W.; Rucinaka, H. *Bromatol. Chem. Toksykol.* 1984, **17**(2), 109-112
7. Lenga, R. E. *The Sigma-Aldrich Library of Chemical Safety Data* (2nd edition) (Sigma-Aldrich, Milwaukee, 1988)
8. Senczuk, W.; Rucinaka, H. *Bromatol. Chem. Toksykol.* 1984, **17**(1), 51-52
9. Short, C. R.; et al. *Fundam. Appl. Toxicol.* 1983, **3**, 285-292
10. Ekman, B.; Stroembeck, J. P. *Acta Physiol. Scand.* 1947, **14**, 43-50
11. *Eur. Chem. News* 1991, **56**(1467), 35
12. Thompson, C. Z.; et al. *Environ. Mol. Mutagen.* 1983, **5**, 803-811
13. Rexroat, M. A.; Probst, G. S. *Prog. Mutat. Res.* 1985, **5**(Eval. Short-term Test Carcinog.), 201-212
14. Cesarone, C. F.; et al. *Arch. Toxicol., Suppl.* 1982, **5**, 355-359
15. *Annual Plan* (National Toxicol. Prog., Washington, D.C., 1984)
16. Tanaka, K. J.; et al. *Mutat. Res.* 1980, **79**, 173-176
17. Perry, P. E.; Thompson, E. J. in: DeSerres, F.; Hoffman, G. R. *Prog. Mutat. Res.* vol. 1 (Elsevier, Amsterdam, 1981) pp. 560-569
18. Gulati, D. K.; et al. *Prog. Mutat. Res.* 1985, **5**(Eval. Short-term Test Carcinog.), 413-426
19. Palitti, F.; et al. *Prog. Mutat. Res.* 1985, **5**(Eval. Short-term Test Carcinog.), 443-450
20. Natarajan, A. T.; et al. *Prog. Mutat. Res.* 1985, **5**(Eval. Short-term Test Carcinog.), 433-437
21. Kornbrust, D. J.; Barfknecht, T. R. *Mutat. Res.* 1984, **136**, 255-266
22. Barrett, R. H. *Prog. Mutat. Res.* 1985, **5**(Eval. Short-term Test Carcinog.), 347-352
23. Zimmer, D.; et al. *Mutat. Res.* 1980, **77**, 317-326
24. Obe, G.; et al. *Prog. Mutat. Res.* 1985, **5**(Eval. Short-term Test Carcinog.), 439-442
25. Forsberg, K.; Mansdorf, S. Z. *Quick selection guide to chemical protective clothing* (Van Nostrand Reinhold, New York, 1989)
26. *Chem. Week* 17 Apr. 1991, 7

p-TOLUIDINE

RISKS
Toxic by inhalation, in contact with skin and if swallowed – Danger of cumulative effects (R23/24/25, R33)

SAFETY PRECAUTIONS
After contact with skin, wash immediately with plenty of water – Wear protective clothing and gloves – If you feel unwell, seek medical advice (show label where possible) (S28, S36/37, S44)

IDENTIFIERS

SYNONYMS 4-amino-1-methylbenzene; benzanenamine, 4- methyl-; p-aminotoluene; 4-aminotoluene; C.I. 37107; C.I. Azoic Coupling Component 107; *p*-methylaniline; 4-Methylaniline; *p*-methylbenzenamine; 4-methylbenzenamine; Naphthol AS-KG; Naphthol AS-KGLL; 4-toluidine; *p*-tolylamine

CHEMICAL ABSTRACTS No.	106-49-0
NIOSH No.	XU 3150000
HAZCHEM CODE	3X
UN No.	1708

THRESHOLD LIMIT VALUES

US TLV (TWA)	2 ppm (8.8 mg/m^3)
US TLV (STEL)	not available
UK EXPOSURE LIMITS (OES)	
Long-term (8 hr TWA value)	not available
Short-term (10 min TWA value)	not available
Germany	
MAK	not available
France	
VME	not available
VLE	not available
Sweden	
Short-term limit	not available
Level limit	not available

PHYSICAL PROPERTIES

Description White crystals.

Boiling point	200°C
Melting point	44°C
Density	1.046 at 20°C
Vapour density	3.9
Vapour pressure	1 mm Hg at 42°C
Flash point	86°C (closed cup)
Explosive limits	1.1%-6.6%
Autoignition temperature	482°C

Solubility Slightly soluble in water. Very soluble in alcohol or ether.

PACKAGING AND TRANSPORTATION

Road transportation
hazard warning sign	1708 toxic substance
Hazchem code	3X

Sea transportation
IMDG page No.	6270
class	6.1
label	poison
packaging group	II

Air transportation
ICAO/IATA code (UN No.)	1708
class	6.1
label	poison
packaging group	II

packing instructions
cargo	
liquid	611
solid	615
passenger	
liquid	609
solid	613

passenger aircraft max. quantity
liquid	5 litres
solid	25 kilograms

cargo aircraft max. quantity
liquid	60 litres
solid	100 kilograms

145. p-Toluidine

MANUFACTURE

p-Toluidine is produced by iron reduction or catalytic hydrogenation of p-nitrotoluene.

USES

p-Toluidine is used as an intermediate in dye manufacture and as a reagent for lignin, nitrite and phloroglucinol.

CHEMICAL HAZARDS

p-Toluidine is a volatile solid. It gives off flammable vapours on heating, and vapours may form explosive mixtures in air. It can react vigorously with oxidisers. 4-Toluenediazonium salts are unstable.

BIOLOGICAL HAZARDS

Toxicity is similar to that of aniline; it forms methaemoglobin (1) and causes haematuria. Its toxicity has not been studied as extensively as the o-isomer, however its industrial hazards have been reviewed (2).

Vapour Inhalation

Exposure to 640 mg/m^3 caused nasal irritation in rats. It is absorbed via the respiratory tract.

Eye Contact

p-Toluidine is a severe eye irritant. Exposure to 640 mg/m^3 caused eye irritation in rats.

Skin Contact

p-Toluidine is a moderate to strong skin irritant in guinea pigs, and causes sensitisation (3). p- and m-Toluidine are more easily absorbed through the skin than o-toluidine (4). The dermal LD_{50} in rabbits is 890 mg/kg.

Swallowing

The oral LD_{50} in rats and mice is 656 mg/kg and 794 mg/kg respectively. p-Toluidine is more toxic than the o-isomer in rats after single or repeated intragastric doses (5).

CARCINOGENICITY

p-Toluidine produced liver tumours in mice but not rats after oral administration (6); its carcinogenicity appears to be less than that of the o-isomer.

MUTAGENICITY

Data on mutagenicity are conflicting. It is not mutagenic in *Salmonella typhimurium* or *Escherichia coli* (7). It inhibited unscheduled DNA synthesis in rat hepatocytes with metabolic activation (7). It did not cause single strand DNA breaks in V79 Chinese hamster lung cells *in vitro* (8) but did so in liver and kidney cells of mice following intraperitoneal injection (9).

REPRODUCTIVE HAZARDS

No information is available concerning the reproductive hazards of p-toluidine.

FIRST AID

Eyes Wash the eye with flowing water for 10 minutes.

Lungs Remove casualty from area of exposure. If unconscious, do not give anything to drink. Give artificial ventilation and chest compression or place in the recovery position as necessary. If conscious make the casualty lie or sit down quietly, give oxygen if available. Convulsions may occur and may cause unconsciousness. Shock may result – if so do not give any drinks, and if conscious, lie casualty flat with legs raised.

Mouth Do not make the casualty vomit. Treat unconscious casualties as for lungs, but if conscious give 1 pint of water to drink.

Skin Remove contaminated clothing immediately, wash the affected area with soap and copious amounts of water. Absorption through the skin may cause symptoms similar to those of inhalation.

In all cases of exposure, the patient should be transferred to hospital as soon as possible.

145. p-Toluidine

HANDLING AND STORAGE

p-Toluidine should be handled wearing an approved respirator, chemical-resistant gloves, safety goggles and other protective clothing. It should only be used in a chemical fume hood (10). It should be kept in a tightly closed container, in a cool, dry, well- ventilated place, away from other storage, heat or open flame, and protected against physical damage. Outside or detached storage is preferred.

DISPOSAL

Wear a laboratory coat, safety spectacles, butyl rubber gloves and suitable safety shoes, and have an approved self-contained breathing apparatus or canister respirator available. Absorb small liquid spills onto paper towels, evaporate in an iron pan in a fume cupboard, add crumpled paper and burn carefully. Brush small solid spills onto paper, put in an iron pan, cover with crumpled paper and burn carefully in a safe place outside. For large spills,
EITHER cover with sand and mix carefully. Shovel into containers, disperse in an excess solution of dilute hydrochloric acid, mix well and leave to stand for 24 hours stirring occasionally. Carefully decant acid extract into the drains, diluting with a large volume of cold tap water. Wash the sand thoroughly with cold water.
OR cover with a sand-soda ash mix (90-10), mix well, shovel into cardboard boxes, pack with crumpled paper and incinerate.

FIRE PRECAUTIONS

Fires involving *p*-toluidine should be extinguished using water spray, carbon dioxide, dry chemical powder, alcohol foam or polymer foam.

FURTHER READING

Bretherick, L. *Hazards in the Chemical Laboratory* (4th edition)

Bretherick, L. *Handbook of Reactive Chemical Hazards* (4th edition)

Sax, N. Irving *Dangerous Properties of Industrial Materials* (7th edition)

Encyclopedia of Occupational Safety & Health

Patty's *Industrial Hygiene and Toxicology*

Kirk-Othmer *Encyclopedia of Chemical Technology*

National Fire Protection Association *Manual of Hazardous Reactions*

ACGIH Documentation of TLVs and BEIs (6th edition, 1986)

REFERENCES

1. Smith, R. P.; et al. *Biochem. Pharmacol.* 1967, **16**(2), 317-328
2. Ikeda, M.; et al. *Sumitomo. Sangyo. Eisei.* 1985, **21**, 131-151
3. Kleniewska, D.; Maibach, H. *Dermatosen, Beruf, Umwelt* 1980, **28**, 11-13
4. Senczuk, W.; Rucinaka, H. *Bromatol. Chem. Toksykol.* 1984, **17**(2), 109-112
5. Senczuk, W.; Rucinaka, H. *Bromatol. Chem. Toksykol.* 1984, **17**(1), 51-52
6. Weisburger, E. K.; et al. *J. Environ. Pathol. Toxicol.* 1978, **2**, 325-356
7. Thompson, C. Z.; et al. *Environ. Mol. Mutagen.* 1983, **5**, 803-811
8. Zimmer, D.; et al. *Mutat. Res.* 1980, **77**, 317-326
9. Cesarone, C. F.; et al. *Arch. Toxicol., Suppl.* 1982, **5**, 355-359
10. Lenga, R. E. *The Sigma-Aldrich Library of Chemical Safety Data* (2nd edition) (Sigma-Aldrich, Milwaukee, 1988)

146. Trichloroacetonitrile

TRICHLORO-ACETONITRILE

Cl_3CCN

RISKS
Toxic by inhalation, in contact with skin and if swallowed (R23/24/25)

SAFETY PRECAUTIONS
If you feel unwell, seek medical advice (show label where possible) (S44)

IDENTIFIERS

SYNONYMS acetonitrile, trichloro-; cyanotrichloromethane; trichloromethyl cyanide; Tritox

CHEMICAL ABSTRACTS No.	545-06-2
NIOSH No.	AM 2450000
HAZCHEM CODE	not available
UN No.	not available

THRESHOLD LIMIT VALUES

US TLV (TWA)	not available
US TLV (STEL)	not available
UK EXPOSURE LIMITS (OES)	
Long-term (8 hr TWA value)	not available
Short-term (10 min TWA value)	not available
Germany	
MAK	not available
France	
VME	not available
VLE	not available
Sweden	
Short-term limit	not available
Level limit	not available

PHYSICAL PROPERTIES

Description Crystals, odour of chloral and hydrogen cyanide.

Boiling point	83-84°C
Melting point	61°C
Density	1.4403 at 25°C
Vapour density	not available
Vapour pressure	58 mm Hg at 20°C
Flash point	none
Explosive limits	not available
Autoignition temperature	not available
Solubility	not available.

PACKAGING AND TRANSPORTATION

Road transportation

hazard warning sign	not available
Hazchem code	not available

Sea transportation

IMDG page No.	not available
class	not available
label	not available
packaging group	not available

Air transportation

ICAO/IATA code	not available
class	not available
label	not available
packaging group	not available
packing instructions	
cargo	not available
passenger	not available
passenger aircraft max. quantity	not available
cargo aircraft max. quantity	not available

146. Trichloroacetonitrile

MANUFACTURE

Trichloroacetonitrile may be prepared by reaction of methylnitrile, hydrogen chloride and chlorine; from aqueous ammonia and ethyl trichloroacetate; and by reaction of trichloroacetamide with phosphorus pentoxide.

USES

Trichloroacetonitrile is used as a fumigant for stored products (1).

CHEMICAL HAZARDS

Attempted preparation of 5-trichloromethyltetrazole from trichloroacetonitrile, ammonium chloride and sodium azide, by a published method (2) but at lower initial temperature, gave an oily product after vacuum distillation, which exploded on sampling with a pipette. This may have been due to formation of an azidomethyltetrazole (3). An alternative suggestion that isomerisation of 5-trichloromethyltetrazole was responsible has been discounted (4). 5-Trichloromethyltetrazole is, in fact, relatively stable (5). Trichloroacetonitrile is incompatible with strong acids, strong bases and strong reducing or oxidising agents. It may emit carbon monoxide, carbon dioxide, nitrogen oxides and hydrogen chloride when heated to decomposition (6).

BIOLOGICAL HAZARDS

Trichloroacetonitrile is a strong eye and skin irritant, and is poisonous if swallowed or administered intravenously. It is very hazardous by inhalation.

Vapour Inhalation

Inhalation is very damaging to the upper respiratory tract and bronchi, and may be fatal. In animal studies survivors suffered severe degenerative heart, liver and kidney lesions (1).

Eye Contact

Trichloroacetonitrile is a strong eye irritant, causing lacrimation.

Skin Contact

Trichloroacetonitrile is a strong skin irritant and vesicant, causing fissures and petechia (small, round, red spots due to intradermal or submucous haemorrhage) (1). The dermal LD_{50} in rabbits is 900 mg/kg (6).

Swallowing

The oral LD_{50} in rats is 250 mg/kg (6). Death is preceded by ataxia and convulsions in rabbits (1).

CARCINOGENICITY

Trichloroacetonitrile's genotoxicity may indicate a potential for carcinogenic activity (7,8).

MUTAGENICITY

Trichloroacetonitrile is mutagenic in *Salmonella typhimurium* TA1535, TA1537, TA98 and TA100 with or without metabolic activation (9). It produced DNA strand breaks in cultured human lymphoblastic cells (7,8).

REPRODUCTIVE HAZARDS

In developmental toxicity studies in rats, trichloroacetonitrile reduced fertility, birth weight and post natal growth, and increased early implantation failure (10).

FIRST AID

Eyes Wash the eye with flowing water.

Lungs Remove casualty from area of exposure, wearing protective clothing and approved breathing apparatus if necessary. If unconscious, do not give anything to drink. If breathing has stopped DO NOT use mouth to mouth or mouth to nose ventilation but use a resuscitation bag and mask instead. If breathing, place in the recovery position. Break two amyl nitrate capsules open under the casualty's nose so that the vapour is inhaled. If Kelo-cyanor (dicobalt edetate) and personnel trained in its use are available and you are certain the casualty has been poisoned by cyanide, give it immediately. NOTE: KELO-CYANOR IS EXTREMELY DANGEROUS WHEN GIVEN TO ANYONE NOT SUFFERING FROM CYANIDE POISONING. If conscious make the casualty lie or sit down quietly. Oxygen may be beneficial.

146. Trichloroacetonitrile

Mouth Do not make the casualty vomit. Treat as for inhalation. If conscious give 1 pint of water immediately.

Skin Remove all contaminated clothing, wash affected areas with soap and copious amounts of water. Absorption through the skin may result in symptoms similar to those of ingestion and inhalation.

In all cases of exposure, the patient should be transferred to hospital as soon as possible.

HANDLING AND STORAGE

Trichloroacetonitrile should be handled wearing an approved respirator, PVA gloves (11), safety goggles and other protective clothing. It should only be used in a chemical fume hood It should be kept refrigerated, in a tightly closed container (6).

DISPOSAL

Eliminate all sources of ignition, ventilate the area and wearing self-contained breathing apparatus, a chemical resistant suit, gloves, and PVC safety boots, spread calcium hypochlorite over the (liquid) spill (or disperse excess calcium hypochlorite solution onto it). Mix and mop into polythene buckets, stand 24 hours and run to waste diluting x1000 with running cold water (see discharge limits from water authorities).

For solid spills either: a) sweep up, place in a large volume of water, add excess sodium hypochlorite, leave to stand 24 hours, then run to waste diluting x1000 with running water; or b) scoop into a large container, make alkaline with 2% sodium hydroxide solution and stir well. Add excess ferrous sulphate solution, mix well, stand for 2-3 hours and run to waste diluting x1000 with running water.

Large spills need specialist help – contact fire brigade.

FIRE PRECAUTIONS

Fires involving trichloroacetonitrile should be extinguished using water spray, carbon dioxide, dry chemical powder, alcohol foam or polymer foam.

FURTHER READING

Bretherick, L. *Hazards in the Chemical Laboratory* (4th edition)

Bretherick, L. *Handbook of Reactive Chemical Hazards* (4th edition)

Sax, N. Irving *Dangerous Properties of Industrial Materials* (7th edition)

Encyclopedia of Occupational Safety & Health

Patty's *Industrial Hygiene and Toxicology*

Kirk-Othmer *Encyclopedia of Chemical Technology*

National Fire Protection Association *Manual of Hazardous Reactions*

ACGIH Documentation of TLVs and BEIs (6th edition, 1986)

REFERENCES

1. Gosselin, R. E.; et al. *Clinical Toxicology of Commercial Products* (5th edition) (Williams & Wilkins, Baltimore, 1984)
2. Finnegan, W. G.; et al. *J. Am. Chem. Soc.* 1958, **80**, 3908
3. Howe, R. K.; et al. *Chem. Eng. News* 1983, **61**(3), 4
4. Burke, L. A. *Chem. Eng. News* 1983, **61**(17), 2
5. Beck, W.; et al. *Chem. Eng. News* 1984, **62**(10), 39
6. Lenga, R. E. *The Sigma-Aldrich Library of Chemical Safety Data* (2nd edition) (Sigma-Aldrich, Milwaukee, 1988)
7. Daniel, F. B.; et al. *Fundam. Appl. Toxciol.* 1986, **6**(3), 447-453
8. *NTIS Report* 1984, EPA/600/D-84/200 Order No. PB84-246230
9. Mortelmans, K.; et al. *Environ. Mol. Mutagen.* 1986, **8**(suppl. 7), 1-119
10. Smith, M. K.; et al. *Toxicology* 1987, **46**(1), 83-93
11. Forsberg, K.; Mansdorf, S. Z. *Quick selection guide to chemical protective clothing* (Van Nostrand Reinhold, New York, 1989)

TRICHLOROANILINE

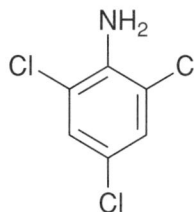

RISKS
Toxic by inhalation, in contact with skin and if swallowed – Danger of cumulative effects (R23/24/25, R33)

SAFETY PRECAUTIONS
After contact with skin, wash immediately with plenty of water – Wear protective clothing and gloves – If you feel unwell, seek medical advice (show label where possible) (S28, S36/37, S44)

IDENTIFIERS

SYNONYMS   aniline, 2,4,6-trichloro-; benzenamine, 2,4,6- trichloro; *sym*-trichloroaniline; 2,4,6-trichloroaniline; 2,4,6- trichlorobenzenamine

CHEMICAL ABSTRACTS No.	634-93-5
NIOSH No.	BZ 0250000
HAZCHEM CODE	not available
UN No.	not available

THRESHOLD LIMIT VALUES

US TLV (TWA)	not available
US TLV (STEL)	not available
UK EXPOSURE LIMITS (OES)	
Long-term (8 hr TWA value)	not available
Short-term (10 min TWA value)	not available
Germany	
MAK	not available
France	
VME	not available
VLE	not available
Sweden	
Short-term limit	not available
Level limit	not available

PHYSICAL PROPERTIES

Description   Needles from liquid.

Boiling point	262°C at 46 mm Hg
Melting point	77.5-78.5 °C
Density	not available
Vapour density	not available
Vapour pressure	not available
Flash point	not available
Explosive limits	not available
Autoignition temperature	not available

Solubility   Insoluble in phosphoric acid. Soluble in alcohol or ether.

PACKAGING AND TRANSPORTATION

Road transportation

hazard warning sign	not available
Hazchem code	not available

Sea transportation

IMDG page No.	not available
class	not available
label	not available
packaging group	not available

Air transportation

ICAO/IATA code	not available
class	not available
label	not available
packaging group	not available
packing instructions	
cargo	not available
passenger	not available
passenger aircraft max. quantity	not available
cargo aircraft max. quantity	not available

147. Trichloroaniline

MANUFACTURE

2,4,6-Trichloroaniline may be prepared by chlorination of aniline derivatives such as the hydrochlorides (1,2).

USES

2,4,6-Trichloroaniline is used in photographic materials (1).

CHEMICAL HAZARDS

2,4,6-Trichloroaniline is incompatible with acids, acid chlorides, acid anhydrides, chloroformates and strong oxidisers. It may emit toxic fumes of carbon monoxide, carbon dioxide, nitrogen oxides and hydrogen chloride when heated to decomposition (3). The energy of decomposition (in the range 220-400°C) of 2,3,4-trichloroaniline (CAS Registry No. 634-67-3) is measured as 0.21 kJ/g (4).

BIOLOGICAL HAZARDS

2,4,6-Trichloroaniline may be irritating to the eyes, skin and upper respiratory tract, and may be a methaemoglobin former (5).

Vapour Inhalation

2,4,6-Trichloroaniline may be irritating to the upper respiratory tract (3).

Eye Contact

2,4,6-Trichloroaniline may be irritating to the eyes (3).

Skin Contact

2,4,6-Trichloroaniline may be irritating to the skin (3).

Swallowing

The oral LD_{50} in rats is 3850 mg/kg (3).

CARCINOGENICITY

2,4,6-Trichloroaniline was inactive in rats but induced a significant increase in vascular tumours in male mice (6).

MUTAGENICITY

2,4,6-Trichloroaniline is mutagenic in *Salmonella typhimurium* TA98, TA100, TA1538 and *Escherichia coli* in the preincubation method (7).

REPRODUCTIVE HAZARDS

No information is available concerning the reproductive hazards of 2,4,6-trichloroaniline.

FIRST AID

Eyes Wash the eye with flowing water for 10 minutes.

Lungs Remove casualty from area of exposure. If unconscious, do not give anything to drink. Give artificial ventilation and chest compression or place in the recovery position as necessary. If conscious make the casualty lie or sit down quietly, give oxygen if available. Convulsions may occur and may cause unconsciousness. Shock may result – if so do not give any drinks, and if conscious, lie casualty flat with legs raised.

Mouth Do not make the casualty vomit. Treat unconscious casualties as for lungs, but if conscious give 1 pint of water to drink.

Skin Remove contaminated clothing immediately, wash the affected area with soap and copious amounts of water. Absorption through the skin may cause symptoms similar to those of inhalation.

In all cases of exposure, the patient should be transferred to hospital as soon as possible.

HANDLING AND STORAGE

2,4,6-Trichloroaniline should be handled wearing an approved respirator, chemical-resistant gloves, safety goggles and other protective clothing. Mechanical exhaust is required. It should be kept in a closed container, in a cool dry place (3).

147. Trichloroaniline

DISPOSAL

Wear a laboratory coat, safety spectacles, butyl rubber gloves and suitable safety shoes, and have an approved self-contained breathing apparatus or canister respirator available. Absorb small liquid spills onto paper towels, evaporate in an iron pan in a fume cupboard, add crumpled paper and burn carefully. Brush small solid spills onto paper, put in an iron pan, cover with crumpled paper and burn carefully in a safe place outside. For large spills,
EITHER cover with sand and mix carefully. Shovel into containers, disperse in an excess solution of dilute hydrochloric acid, mix well and leave to stand for 24 hours stirring occasionally. Carefully decant acid extract into the drains, diluting with a large volume of cold tap water. Wash the sand thoroughly with cold water.
OR cover with a sand-soda ash mix (90-10), mix well, shovel into cardboard boxes, pack with crumpled paper and incinerate.

FIRE PRECAUTIONS

Fires involving 2,4,6-trichloroaniline should be extinguished using water spray, carbon dioxide, dry chemical powder, alcohol foam or polymer foam.

FURTHER READING

Bretherick, L. *Hazards in the Chemical Laboratory* (4th edition)

Sax, N. Irving *Dangerous Properties of Industrial Materials* (7th edition)

Encyclopedia of Occupational Safety & Health

Patty's *Industrial Hygiene and Toxicology*

Kirk-Othmer *Encyclopedia of Chemical Technology*

National Fire Protection Association *Manual of Hazardous Reactions*

ACGIH Documentation of TLVs and BEIs (6th edition, 1986)

REFERENCES

1. Winning, E.; et al. Ger. (East) DD 226,878 (Cl. C07C87/60) (1987, to VEB Film. Wolfen. Fotochem. Kombin.)
2. Werner, F.; et al. Ger. Offen. DE 3,200,069 (Cl. C07C87/60) (1983, to Bayer, A.-G.)
3. Lenga, R. E. *The Sigma-Aldrich Library of Chemical Safety Data* (2nd edition) (Sigma-Aldrich, Milwaukee, 1988)
4. Bretherick, L. *Handbook of Reactive Chemical Hazards* (4th edition) p. 582
5. McLean, S.; et al. *J. Pharm. Pharmacol.* 1969, **21**(7), 441-450
6. Weisburger, E. k.; et al. *J. Environ. Pathol. Toxicol.* 1978, **2**, 325-356
7. Shimizu, H.; Takemura, N. *Occup. Health Chem. Ind. Proc. Int. Congr., 11th* 1983 (Publ. 1984), 497-506

148. Trichloronitromethane

TRICHLORO-NITROMETHANE
Cl_3CNO_2

RISKS
Very toxic by inhalation, in contact with skin and if swallowed – Irritating to eyes, respiratory system and skin (R26/27/28, R36/37/38)

SAFETY PRECAUTIONS
In case of contact with eyes, rinse immediately with plenty of water and seek medical advice – Wear suitable protective clothing – In case of accident or if you feel unwell, seek medical advice immediately (show label where possible) (S26, S36, S45)

IDENTIFIERS

SYNONYMS Acquinite; chloropicrin; Dolochlor; G 25; Larvacide; methane, trichloronitro-; Microlysin; nitrochloroform; nitrotrichloromethane; Picride; Picfume; PS

CHEMICAL ABSTRACTS No.	76-06-2
NIOSH No.	PB 6300000
HAZCHEM CODE	2XE
UN No.	1580

THRESHOLD LIMIT VALUES

US TLV (TWA)	0.1 ppm (0.67 mg/m^3)
US TLV (STEL)	not available

UK EXPOSURE LIMITS (OES)
Long-term (8 hr TWA value) ..... 0.1 ppm (0.7 mg/m^3)
Short-term (10 min TWA value) ... 0.3 ppm (2 mg/m^3)

Germany
MAK ..... 0.1 ppm (0.7 mg/m^3)

France
VME ..... 0.1 ppm (0.7 mg/m^3)
VLE ..... not available

Sweden
Short-term limit ..... not available
Level limit ..... not available

PHYSICAL PROPERTIES

Description Slightly oily, colourless liquid with intense penetrating odour.

Boiling point	112.3°C at 766 mm Hg
Melting point	-64°C
Density	1.651 at 22.8°C
Vapour density	6.69
Vapour pressure	40 mm Hg at 33.8°C
Flash point	not available
Explosive limits	not available
Autoignition temperature	not available

Solubility Practical insoluble in water. Soluble in alcohol, ether, acetone, dimethyl sulphoxide, benzene or carbon disulphide.

PACKAGING AND TRANSPORTATION

Road transportation
hazard warning sign	1580 toxic substance
Hazchem code	2XE

Sea transportation
IMDG page No.	6108
class	6.1
label	poison
packaging group	I

Air transportation
ICAO/IATA code (UN No.)	1580
class	6.1
label	not available
packaging group	not available
packing instructions	
cargo	forbidden
passenger	forbidden
passenger aircraft max. quantity	forbidden
cargo aircraft max. quantity	forbidden

148. Trichloronitromethane

MANUFACTURE

Trichloronitromethane is manufactured from nitromethane and alkaline hypochlorite.

USES

Trichloronitromethane is used in the synthesis of methyl violet, as a fumigant for stored products, as a soil insecticide and war gas.

CHEMICAL HAZARDS

Bulk containers of trichloronitromethane above a critical volume can be shocked into detonation (1). It reacts violently with excess aniline at 145°C (2), or alcoholic sodium hydroxide (3). The temperature should not be allowed to fall much below 50°C during addition of trichloronitromethane in methanol to sodium methoxide solution to avoid a violent and hazardous exotherm (4). An explosion was reported of an insecticdal mixture of trichloronitromethane and 3-bromopropyne (5). Photochemical transformation to phosgene has been reported.

BIOLOGICAL HAZARDS

Its toxicity is greater than that of chlorine, but less than that of phosgene. Cases of poisoning have been reported (6). It is a strong irritant. Effects include congestion, haemorrhage, oedema and infiltration of lung tissue, followed by necrosis of kidneys, liver and skeletal muscle. Vertigo, fatigue, hypotension, abdominal pain and diarrhoea may also occur. Higher levels cause vomiting, bronchitis, pulmonary oedema and death.

Vapour Inhalation

Trichloronitromethane is severely irritating to the upper respiratory tract. 340 ppm is lethal in one minute; 110 ppm in 20 minutes. Toxic effects are outline above.

Eye Contact

Trichloronitromethane causes severe eye irritation and is a strong lacrimator, having a tear gas-like effect. 0.3-0.37 ppm cause painful eye irritation in 3 to 30 seconds. 15 ppm cannot be tolerated for over a minute even by accustomed individuals.

Skin Contact

Trichloronitromethane is a strong skin irritant.

Swallowing

The oral LD_{50} in rats is 250 mg/kg, and causes gastrointestinal tract irritation.

CARCINOGENICITY

There was no evidence of carcinogenicity in mice given trichloronitromethane by gavage.

MUTAGENICITY

Trichloronitromethane gave negative results in *Salmonella typhimurium* TA1535, TA1537, TA1538, a weakly positive result in TA98 and was mutagenic to TA100 with metabolic activation. It gave a weak positive result in *Escherichia coli* (7).

REPRODUCTIVE HAZARDS

No information is available concerning the reproductive effects of trichloronitromethane.

FIRST AID

Eyes Wash the eye with flowing water for 10 minutes.

Lungs Remove casualty from area of exposure. If unconscious, do not give anything to drink. Give artificial ventilation and chest compression or place in the recovery position as necessary. If conscious make the casualty lie or sit down quietly, give oxygen if available. Lung congestion may occur – a conscious casualty with breathing difficulties should be placed in a sitting position.

Mouth Do not make the casualty vomit. Treat unconscious casualties as for lungs but if conscious give 1 pint of water to drink immediately then give repeated drinks of water (1 cupful every 10 minutes.)

148. Trichloronitromethane

Skin Remove all contaminated clothing, wash affected areas with soap and copious amounts of water. Absorption through the skin may result in symptoms similar to those of ingestion and inhalation.

In all cases of exposure, the patient should be transferred to hospital as soon as possible.

HANDLING AND STORAGE

Trichloronitromethane should be handled wearing an approved respirator, chemical-resistant gloves, safety goggles and other protective clothing. It should be kept in a tightly closed container. It is inert to normal gas mask chemicals (8).

DISPOSAL

Eliminate all sources of ignition, ventilate the area and wear a laboratory coat or overalls, rubber gloves, approved compressed air breathing apparatus and safety boots. Absorb small spills onto paper towels, remove to a safe open air site and allow to evaporate in a metal tray. Wash spillage site with detergent. For large spills, absorb onto vermiculite-sodium carbonate mixture (90-10) or sand-soda ash (90-10) and mix carefully. EITHER transport in dry buckets to a safe open air area for atmospheric evaporation OR shovel into paper boxes and incinerate.

FIRE PRECAUTIONS

Trichloronitromethane is not flammable at normal temperatures; use an extinguisher suitable to surrounding fire conditions (8). An emergency procedure guide is available (9).

FURTHER READING

Bretherick, L. *Hazards in the Chemical Laboratory* (4th edition)

Bretherick, L. *Handbook of Reactive Chemical Hazards* (4th edition)

Sax, N. Irving *Dangerous Properties of Industrial Materials* (7th edition)

Encyclopedia of Occupational Safety & Health

Patty's *Industrial Hygiene and Toxicology*

Kirk-Othmer *Encyclopedia of Chemical Technology*

National Fire Protection Association *Manual of Hazardous Reactions*

ACGIH *Documentation of TLVs and BEIs* (6th edition, 1986)

REFERENCES

1. Anon. *Chem. Eng. News* 1972, **50**(38), 13
2. Jackson, K. E. *Chem. Rev.* 1934, **14**, 269
3. Scholtz, S. *Explosivstoffe* 1963, **11**, 159, 181
4. Ramsey, B. G.; et al. *J. Amer. Chem. Soc.* 1966, **88**, 3059
5. *BCISC Quart. Safety Summ.* 1968, **39**, 12
6. Okada, E.; et al. *Nippon Naika Gakkai Zasshi* 1970, **59**(11), 1214-1221
7. Moriya, M.; et al. *Mutat. Res.* 1983, **116**(1), 185-216
8. *Dangerous Prop. Ind. Mater. Rep.* 1982, **2**(2), 17-18
9. *Australian Standard AS 1678. Emergency procedure guide – transport. 6.0.013. Chloropicrin* (Standards Assoc. Australia, Sydney, 1984)

TRIPHENYLTIN ACETATE

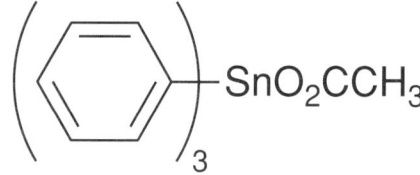

RISKS
Toxic by inhalation, in contact with skin and if swallowed (R23/24/25)

SAFETY PRECAUTIONS
Keep out of reach of children – Keep away from food, drink and animal feeding stuffs – If you feel unwell, seek medical advice (show label where possible) (S2, S13, S44)

IDENTIFIERS

SYNONYMS   acetatotriphenylstannane; acetoxytriphenylstannane; (acetyloxy)tripenylstannane; acetoxytriphenyltin; Batasan; Brestan; Brestan 60; ENT 25208; Fenoloyo acetate; fentin acetate; GC 6936; Liromatin; Lirostanol; phentin acetate; stannane, (acetyloxy) triphenyl-; stannane, acetoxytriphenyl- ; Suzu; Tinestan; tin triphenyl acetate; TPTA; triphenylaceto stannane; Tubotin; VP 19-40

CHEMICAL ABSTRACTS No.	900-95-8
NIOSH No.	WH 6650000
HAZCHEM CODE	not available
UN No.	2786

THRESHOLD LIMIT VALUES

US TLV (TWA)	not available
US TLV (STEL)	not available

UK EXPOSURE LIMITS (OES)

Long-term (8 hr TWA value)	not available
Short-term (10 min TWA value)	not available

Germany

MAK	not available

France

VME	not available
VLE	not available

Sweden

Short-term limit	not available
Level limit	not available

PHYSICAL PROPERTIES

Description   Practically insoluble, crystalline solid.

Boiling point	not available
Melting point	120°C
Density	not available
Vapour density	not available
Vapour pressure	not available
Flash point	not available
Explosive limits	not available
Autoignition temperature	not available

Solubility   Practically insoluble.

PACKAGING AND TRANSPORTATION

Road transportation

hazard warning sign	2786 2786
Hazchem code	not available

Sea transportation

IMDG page No.	6221
class	6.1
label	marine pollutant;harmful stow away from foodstuffs
packaging group	III

Air transportation

ICAO/IATA code (UN No.)	2786
class	not available
label	not available
packaging group	not available

packing instructions

cargo	not available
passenger	not available
passenger aircraft max. quantity	not available
cargo aircraft max. quantity	not available

149. Triphenyltin Acetate

MANUFACTURE

Triphenyltin acetate can be prepared by treating triphenyltin hydroxide or anhydride with organic acids or phenols (1). It can also be obtained by reacting triphenyltin-113 chloride with sodium acetate (2).

USES

Triphenyltin acetate is used as fungicide, molluscicide and algicide.

CHEMICAL HAZARDS

Triphenyltin acetate produces acrid smoke and fumes of carbon monoxide, carbon dioxide, tin and tin oxides when heated to decomposition.

BIOLOGICAL HAZARDS

Triphenyltin acetate is toxic through ingestion and contact with the skin, and via intravenous, subcutaneous and intraperitoneal administration. The common symptoms are severe headache, nausea and vomiting and abdominal pain. Dizziness, transient loss of consciousness, convulsions, photophobia and blurred vision may also occur (3), and there may be glycosuria and hyperglycaemia. Liver enlargement and mild liver dysfunction have been reported in cases of occupational exposure, but as other pesticides were also involved it is no certain that triphenyltin acetate was the cause. There have been no reports of fatal poisoning.

Vapour Inhalation

The symptoms described above have occurred in cases of respiratory exposure to triphenyltin acetate during crop spraying.

Eye Contact

Irritation of the mucous membranes occurred in workers bagging a fungicide product containing 20% triphenyltin acetate.

Skin Contact

Triphenyltin acetate is toxic when absorbed through the skin. There are reports of skin irritation, acute urticarial eruption and genital oedema (4).

Swallowing

Th oral LD_{50} is 140 mg/kg for rats and 30 mg/kg for rabbits.

CARCINOGENICITY

No information is available on the carcinogenicity of triphenyltin acetate.

MUTAGENICITY

No information is available on the mutagenicity of triphenyltin acetate.

REPRODUCTIVE HAZARDS

Triphenyltin acetate is an experimental teratogen (5).

FIRST AID

Eyes Wash the eye with flowing water for 10 minutes.

Lungs Remove casualty from area of exposure. If unconscious, do not give anything to drink. Give artificial ventilation and chest compression or place in the recovery position as necessary. If conscious make the casualty lie or sit down quietly, give oxygen if available. Convulsions may occur and may cause unconsciousness.

Mouth Do not make the casualty vomit. Treat unconscious casualties as for lungs but if conscious give 1 pint of water to drink.

Skin Remove contaminated clothing immediately, drench the affected area with running water for at least 10 minutes.

In all cases of exposure, the patient should be transferred to hospital as soon as possible.

149. Triphenyltin Acetate

HANDLING AND STORAGE

Triphenyltin acetate should be handled wearing an approved respirator, chemical-resistant gloves, safety goggles and other protective clothing. It should only be used in a chemical fume hood. It should be kept in a tightly closed container.

DISPOSAL

Eliminate all sources of ignition and ventilate the area. Wearing a laboratory coat or overalls, safety glasses and gloves, and absorb the spill onto paper towels and allow to evaporate in a fume-cupboard. For large spills, absorb onto sand or vermiculite, and remove in buckets for atmospheric evaporation in a safe open area. Ideally, waste should be burned in an incinerator with afterburner.

FIRE PRECAUTIONS

Fires involving triphenyltin acetate should be extinguished using water spray, carbon dioxide, dry chemical powder, alcohol foam or polymer foam.

FURTHER READING

Sax, N. Irving *Dangerous Properties of Industrial Materials* (7th edition)

Patty's *Industrial Hygiene and Toxicology*

Kirk-Othmer *Encyclopedia of Chemical Technology*

National Fire Protection Association *Manual of Hazardous Reactions*

ACGIH Documentation of TLVs and BEIs (6th edition, 1986)

Gosselin, R. E.; et al. *Clinical Toxicology of Commerical Products* (5th edition) (Williams & Wilkins, Baltimore, 1984)

REFERENCES

1. Otsuka, H.; et al. *Jpn. Kokai Tokkyo Koho* (Japan, 1979)
2. Nagamatsu, K. *Eisei Shikensho Hokoku* 1976, **94**, 41-43
3. Manzo, L.; et al. *Clin. Toxicol.* 1981, **18** (11), 1345-53
4. Colosio, C.l et al. *Br. J. Ind. Med.* 1991, **48**(2), 136-139
5. Sax, N. Irving *Dangerous Properties of Industrial Materials* (7th edition)

150. Triphenyltin Hydroxide

TRIPHENYLTIN HYDROXIDE

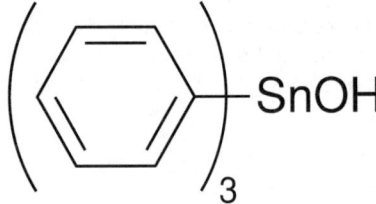

RISKS
Toxic by inhalation, in contact with skin and if swallowed (R23/24/25)

SAFETY PRECAUTIONS
Keep out of reach of children – Keep away from food, drink and animal feeding stuffs – If you feel unwell, seek medical advice (show label where possible) (S2, S13, S44)

IDENTIFIERS

SYNONYMS   Dowco 186; Du-Ter; ENT 28009; Erithane; Fenolovo; fentin hydroxide; hydroxytriphenylstannane; hydroxytriphenyltin; Haitin; K 19; stannane, hydroxytriphenyl-; Suzu H; triphenyltin oxide; Tenhide; TPTH; triphenylhydroxytin; triphenylstannol; Vancide KS

CHEMICAL ABSTRACTS No.	76-87-9
NIOSH No.	WH 8575000
HAZCHEM CODE	not available
UN No.	2786

THRESHOLD LIMIT VALUES

US TLV (TWA)	not available
US TLV (STEL)	not available

UK EXPOSURE LIMITS (OES)
Long-term (8 hr TWA value) .... not available
Short-term (10 min TWA value) .... not available

Germany
MAK .... not available

France
VME .... not available
VLE .... not available

Sweden
Short-term limit .... not available
Level limit .... not available

PHYSICAL PROPERTIES

Description   White powder.
Boiling point	not available
Melting point	122°C
Density	1.54 at 20°C
Vapour density	not available
Vapour pressure	3.572×10^{-7} mm Hg at 50°C
Flash point	not available
Explosive limits	not available
Autoignition temperature	not available

Solubility   Practically insoluble in water or alcohol. Soluble in ether or benzene.

PACKAGING AND TRANSPORTATION

Road transportation
hazard warning sign	2786 2786
Hazchem code	not available

Sea transportation
IMDG page No.	6221
class	6.1
label	marine pollutant
packaging group	III

Air transportation
ICAO/IATA code (UN No.)	2786
class	not available
label	not available
packaging group	not available
packing instructions	
cargo	not available
passenger	not available
passenger aircraft max. quantity	not available
cargo aircraft max. quantity	not available

150. Triphenyltin Hydroxide

MANUFACTURE

Triphenyltin hydroxide is prepared by alkaline hydrolysis of triphenyltin chloride.

USES

Triphenyltin hydroxide is used as a non-systemic fungicide. It is also used as an antifeeding compound for insect pest control, and as a marine antifouling paint.

CHEMICAL HAZARDS

Triphenyltin hydroxide reacts with many acidic compounds. It produces acrid smoke and toxic fumes of carbon dioxide, carbon monoxide, tin and tin oxides on decomposition.

BIOLOGICAL HAZARDS

Triphenyltin hydroxide is toxic to humans by ingestion and intraperitoneal administration. Rats whose diet included 400 ppm triphenyltin oxide died of starvation and haemorrhage after 7-34 days. Those fed 200 ppm showed reduced food consumption, but adapted over time and showed no abnormality of internal organs (1).

Vapour Inhalation

No information in available on the effects of inhalation.

Eye Contact

Triphenyltin hydroxide is a severe eye irritant in humans and rabbits.

Skin Contact

Triphenyltin hydroxide did not cause irritation to rabbits' skin (2), but caused skin irritation in experimental patch tests in man (3).

Swallowing

The oral LD_{50} in rats and mice is 46 mg/kg and 245 mg/kg respectively.

CARCINOGENICITY

An NCI Carcinogenesis Bioassay found no evidence that triphenyltin hydroxide is carcinogenic when fed to mice or rats (4).

MUTAGENICITY

A dominant lethal mutation study produced no evidence of mutagenicity in mice (5).

REPRODUCTIVE HAZARDS

Triphenyltin hydroxide has been classed by the EPA as teratogenic in rats (6). Experimental reproductive effects have been reported (7).

FIRST AID

Eyes Wash the eye with flowing water for 10 minutes.

Lungs Remove casualty from area of exposure. If unconscious, do not give anything to drink. Give artificial ventilation and chest compression or place in the recovery position as necessary. If conscious make the casualty lie or sit down quietly, give oxygen if available. Convulsions may occur and may cause unconsciousness.

Mouth Do not make the casualty vomit. Treat unconscious casualties as for lungs but if conscious give 1 pint of water to drink.

Skin Remove contaminated clothing immediately, drench the affected area with running water for at least 10 minutes.

In all cases of exposure, the patient should be transferred to hospital as soon as possible.

HANDLING AND STORAGE

Triphenyltin hydroxide should be handled wearing an approved respirator, chemical-resistant gloves, safety goggles and

150. Triphenyltin Hydroxide

other protective clothing. It should only be used in a chemical fume hood and handled and stored under nitrogen. It should be kept in a tightly closed container.

DISPOSAL

Eliminate all sources of ignition and ventilate the area. Wearing a laboratory coat or overalls, safety glasses and gloves, absorb the spill onto paper towels and allow to evaporate in a fume-cupboard. For large spills, absorb onto sand or vermiculite, and remove in buckets for atmospheric evaporation in a safe open area. Ideally, waste should be burned in an incinerator with afterburner.

FIRE PRECAUTIONS

Fires involving triphenyltin hydroxide should be extinguished using water spray, carbon dioxide, dry chemical powder, alcohol foam or polymer foam.

FURTHER READING

Bretherick, L. *Hazards in the Chemical Laboratory* (4th edition)

Bretherick, L. *Handbook of Reactive Chemical Hazards* (4th edition)

Sax, N. Irving *Dangerous Properties of Industrial Materials* (7th edition)

Encyclopedia of Occupational Safety & Health

Patty's *Industrial Hygiene and Toxicology*

Kirk-Othmer *Encyclopedia of Chemical Technology*

National Fire Protection Association *Manual of Hazardous Reactions*

ACGIH Documentation of TLVs and BEIs (6th edition, 1986)

The Merck Index (11th edition) (Merck & Co., New Jersey, 1989)

REFERENCES

1. Gaines, T.; et al. *Toxicol. Appl. Pharmacol.* 1968, **12**
2. *Toxicol. Appl. Pharmacol.* 1969, **14**, 637
3. Lisi, P.; et al. *Contact Dermatitis* 1987, **17**, 212-218
4. National Cancer Institute *Carcinogenesis Technical Report No. 139* (Bethesda, 1978)
5. *Toxicol. Appl. Pharmacol.* 1972, **23**, 288
6. *Fed. Reg.* 1985, **50**, 1107
7. Sax N. Irving *Dangerous Properties of Industrial Materials* (7th edition)

URANIUM
U

RISKS
Very toxic by inhalation and if swallowed – Danger of cumulative effects (R26/28, R33)

SAFETY PRECAUTIONS
When using do not eat, drink or smoke – In case of accident or if you feel unwell, seek medical advice immediately (show label where possible) (S20/21, S45)

IDENTIFIERS

SYNONYMS Uranium-238; Uranium I (^{238}U)
CHEMICAL ABSTRACTS No. 7440-61-1
NIOSH No. .... YR 3490000
HAZCHEM CODE .... not available
UN No. 2979

THRESHOLD LIMIT VALUES

US TLV (TWA) .... 0.2 mg/m^3
US TLV (STEL) .... 0.6 mg/m^3
UK EXPOSURE LIMITS (OES)
Long-term (8 hr TWA value) .... 0.2 mg/m^3
Short-term (10 min TWA value) .... 0.6 mg/m^3
Germany
MAK .... 0.25 mg/m^3
France
VME .... not available
VLE .... not available
Sweden
Short-term limit .... not available
Level limit .... not available

PHYSICAL PROPERTIES

Description A heavy, silvery-white, malleable, ductile, softer-than-steel, metallic element. Radioactive material. On vigorous shaking the metallic particles exhibit luminescence. A black powder when obtained by reduction.

Boiling point .... 3818°C
Melting point .... 1132°C
Density 18.95
Vapour density .... not available
Vapour pressure .... not available
Flash point .... not available
Explosive limits .... not available
Autoignition temperature .... not available
Solubility Insoluble in water, alkali and alcohol. Dissolved by acids.

PACKAGING AND TRANSPORTATION

Road transportation
hazard warning sign .... not available
Hazchem code .... not available

Sea transportation
IMDG page No. 7109-7111
class 7
label .... spontaneously combustible
packaging group .... not available

Air transportation
ICAO/IATA code (UN No.) 2979
class 7;4.2
label .... radioactive; spontaneous combustion
packaging group .... not available
packing instructions
cargo .... not available
passenger .... not available
passenger aircraft max. quantity ... not available
cargo aircraft max. quantity .... not available

151. Uranium

MANUFACTURE

Uranium is a naturally occurring element, widely distributed in nature, although only a few deposits are of commercial value. In the natural state it is a mixture of 3 isotopes: ^{238}U, which accounts for 99% of the element and ^{243}U and ^{235}U. ^{238}U has a half-life if 4.49×10^9 and decays by alpha emission. The end of the decay series is the stable ^{206}Pb isotope of lead. Beta- emission may occur from decay products, or from impurities such as radium. Uranium is extracted by concentrating and acid leaching of the ores, followed by separation of the solid and liquid phases, and the precipitation, filtration, purification and drying of the salts. The salts are calcined to the oxides then reduced and converted to UF_4. The fluoride is then reduced in a thermit type reaction to give the pure metal.

USES

Uranium is chiefly used in nuclear fuels. Depleted uranium is used in armour piercing shells, ship or aircraft ballast, and counter balances.

CHEMICAL HAZARDS

Uranium reacts with both air and water, and may form a pyrophoric surface when stored in the presence of moist air (1). The dust from the metal working is easily ignited (2), and the finely divided dust is pyrophoric (3). The metal incandesces in ammonia when heated to a dull red (4). It may ignite or explode during dissolution in bromine trifluoride, particularly when high concentrations of the hexafluoride are present (5). At high temperatures the metal ignites in carbon dioxide (3), and it also ignites in carbon dioxide/nitrogen mixtures (6). Use of a carbon tetrachloride extinguisher on a small uranium fire led to an explosion (7). Uranium powder ignites in fluorine at room temperature, in chlorine at 150-180°C, in bromine vapour at 210-240°C and in iodine vapour at 260°C (4). It reacts explosively with nitric acid (8). It ignites on heating with nitrogen oxide (4). It incandesces in nitryl fluoride (9). It incandesces in sulphur vapour, and with selenium also (4).

BIOLOGICAL HAZARDS

The biological hazards of uranium have been thoroughly reviewed in a 4 volume work based on the Manhatten project (10). A summary of the project is presented in an additional volume (11). Uranium is both chemically toxic and a radiation hazard. Soluble uranium compounds are rapidly absorbed and are particularly injurious to the kidneys. Insoluble uranium compounds are mainly injurious to the lungs, due to the deposition of radioactive particles which are only slowly cleared. Animal studies have shown that repeated exposure to uranium can build up a tolerance to toxicity, with cumulative doses being much greater that those required to produce acute toxicity in a single exposure. 15 and 25 year old assessments of the health of workers at one uranium plant failed to link exposure to uranium with any major kidney or blood disease (12,13).

Vapour Inhalation

Insoluble uranium is only slowly cleared from the lungs and chronic exposure may cause pulmonary fibrosis, pneumoconiosis, and changes to red and white blood cells. Central nervous system symptoms may also be observed (10,11).

Eye Contact

Insoluble uranium compounds were not toxic when placed in the eyes of rabbits (10,11).

Skin Contact

Insoluble uranium compounds caused no signs of poisoning when applied to rabbit skin (10,11).

Swallowing

Uranium is not well absorbed through the gastrointestinal tract, and most ingested material is eliminated in the faeces (10,11).

CARCINOGENICITY

Osteosarcoma and lung cancer may occur after inhalation of insoluble uranium, due to the effect of prolonged irradiation of the thorax. Rats injected with metallic uranium in the bone marrow and chest wall developed sarcomas (14).

MUTAGENICITY

Increases in sister chromatid exchanges and chromosome aberrations were found in workers from a nuclear fuel production and enrichment plant. The effect was believed to be due to chemical rather than radioactive toxicity. Dose dependent increases in chromosome aberrations were observed in the spermatozoa of mice injected with uranyl fluoride (16). Chromosome abberations were observed in V79 Chinese hamster cells treated with uranyl nitrate (17).

151. Uranium

REPRODUCTIVE HAZARDS

Chronic uranium poisoning may inhibit reproduction, and affect uterine and extrauterine development in experimental animals.

FIRST AID

Eyes Wash the eye with flowing water for 10 minutes.

Lungs Remove casualty from area of exposure. If unconscious, do not give anything to drink, give artificial ventilation and chest compression or place in the recovery position as necessary. If conscious make the casualty lie or sit down quietly, give oxygen if available. Lung congestion may occur – a conscious casualty with breathing difficulties should be placed in a sitting position.

Mouth Do not make the casualty vomit. Treat unconscious casualties as for lungs but if conscious give 1 pint of water to drink immediately; give repeated drinks of water (1 cupful every 10 minutes).

Skin Remove contaminated clothing immediately, drench the affected area with running water for at least 10 minutes.

In all cases of exposure, the patient should be transferred to hospital as soon as possible.

HANDLING AND STORAGE

Uranium must be handled in accordance with the Ionising Radiations Regulations where the main hazard is alpha emission, handling the material in a negative pressure glove box should prevent leakage into surrounding area. Rubber or plastic gloves, protective clothing and eye protection should be worn when handling uranium.

DISPOSAL

50

FIRE PRECAUTIONS

Fires involving uranium should be extinguished using graphite chips or asbestos blankets. Do not use carbon dioxide.

FURTHER READING

Bretherick, L. *Hazards in the Chemical Laboratory* (4th edition)

Bretherick, L. *Handbook of Reactive Chemical Hazards* (4th edition)

Sax, N. Irving *Dangerous Properties of Industrial Materials* (7th edition)

Encyclopaedia of Occupational Safety & Health

Patty's *Industrial Hygiene and Toxicology*

Kirk-Othmer *Encyclopedia of Chemical Technology*

National Fire Protection Association *Manual of Hazardous Reactions*

ACGIH Documentation of TLVs and BEIs (6th edition, 1986)

Viegtlin & Hodge *Pharmacology and Toxicology of Uranium compounds* (McGrawth, New York, 1953)

Laboratory Hazards Datasheet No. 103: Uranium and compounds (RSC, London, 1991)

REFERENCES

1. *MCA Case History* No. 1296
2. Bailar, J. C. *Comprehensive Inorganic Chemistry* 1973, 5, 40-42
3. Rieke, R. D.; et al. *J. Org. Chem.* 1979, **44**, 3446
4. Mellor, J. W. *A Comprehensive Treatise on Inorganic and Theoretical Chemistry* **12**, 31-32 (1942)
5. Johnson, K.; et al. *6th Nucl. Eng. Sci Conf.,* (New York, 1960), Reprint Paper No. 23
6. Rhein, R. A. *Rept. No. Cr-60125,* (Washington, NASA, 1964)
7. *NFPA 491M,* 1975, 435
8. *MCA Case History* No. 1104

151. Uranium

9. Aynsley, E. E.; et al. *J. Chem. Soc.* 1954, 1122
10. Voegtlin, C.; Hodge, H. C. (editors) *Pharmacology and Toxicology of Uranium Compounds* National Nuclear Energy Series, Div. IV Vol. 1 (in 4 parts) (McGraw-Hill, 1949)
11. Tannenbaum, A. (editor) *Toxicology of Uranium* National Nuclear Series, Div. IV, Vol. 23 (McGraw-Hill, 1951)
12. Mason, M. G.; et al *HASL-58, Symposium on Occupational Health Experience and Practices in the Uranium Industry* (New York City, Oct 1958)
13. Wing, J. F.; et al. *10th Ann. Meeting, Health Physics Soc.* (Los Angeles, CA, June 1963)
14. Hueper, W. C.; et al. *J. Natl. Cancer Inst.* 1952, **13**, 291
15. Mortin, F.; et al. *Br. J. Ind. Med.* 1991, **48**(2), 98-102
16. Hu, Q.; Zhu, S.; *Mutat. Res.* 1990, **244**(3), 209-214
17. Rosgaj, R.; et al. *Period. Biol.* 1983, **185**(4), 367-375

152. Vinyl Chloride

VINYL CHLORIDE
H$_2$C=CHCl

RISKS
May cause cancer – Extremely flammable liquefied gas (R45, R13)

SAFETY PRECAUTIONS
Avoid exposure-obtain special instructions before use – Keep container in a well ventilated place – Keep away from sources of ignition – No Smoking – If you feel unwell, seek medical advice (show label where possible) (S53, S9, S16, S44)

IDENTIFIERS

SYNONYMS   chloroethylene; chloroethene; ethene, chloro-; ethylene, chloro-; ethylene monochloride; monochloroethylene; Trovidur; vinyl chloride monomer; vinyl C monomer

CHEMICAL ABSTRACTS No.	75-01-4
NIOSH No.	KU 9625000
HAZCHEM CODE	2WE
UN No.	1086

THRESHOLD LIMIT VALUES

US TLV (TWA)	5 ppm (13 mg/m^3)
US TLV (STEL)	not available

UK EXPOSURE LIMITS (MEL)
Long-term (8 hr TWA value) .... 7 ppm
Short-term (10 min TWA value) .... not available

Germany
MAK .... not available

France
VME .... 1 ppm (plants after 1980); 3 ppm (plants before 1980)
VLE .... not available

Sweden
Short-term limit .... 5 ppm (13 mg/m^3)
Level limit .... 1 ppm (2.5 mg/m^3)

PHYSICAL PROPERTIES

Description   Colourless liquid or gas (when inhibited), faintly sweet odour. Polymerises in light or in presence of catalyst.

Boiling point	-13.9°C
Melting point	-160°C
Density	0.9195 at 15°C
Vapour density	2.15
Vapour pressure	2600 mm Hg at 25°C
Flash point	-78°C (open cup)
Explosive limits	4%-22%
Autoignition temperature	472°C

Solubility   Slightly soluble in water. Soluble in alcohol, benzene or carbon tetrachloride. Very soluble in ether.

PACKAGING AND TRANSPORTATION

Road transportation

hazard warning sign	1086 flammable gas
Hazchem code	2WE

Sea transportation

IMDG page No.	2186
class	2;2.1
label	flammable gas
packaging group	not available

Air transportation

ICAO/IATA code (UN No.)	1086
class	2;3
label	gas flammable
packaging group	not available
packing instructions cargo	200
passenger	forbidden
passenger aircraft max. quantity	forbidden
cargo aircraft max. quantity	150 kilograms

152. Vinyl Chloride

MANUFACTURE

Vinyl chloride is manufactured by the halogenation of ethylene. Ethylene is reacted with hydrogen chloride and oxygen to produce ethylene dichloride, which is subsequently cracked to produce vinyl chloride and hydrogen chloride.

USES

Vinyl chloride is used in the production of vinyl chloride hompolymer and copolymer resins, especially polyvinyl chloride and vinyl chloride-vinyl acetate copolymers. It is also used to produce methyl chloroform and vinylidene chloride.

CHEMICAL HAZARDS

Vinyl chloride gas is highly flammable, and forms explosive mixtures with air. Explosion can also occur on contact with oxides of nitrogen. Heat, light or prolonged exposure to air can cause the formation of peroxides, leading to explosive polymerisation. Vinyl chloride can react vigorously with oxidising agents. Being heavier than air, the gas may travel some distance along the ground to a source of ignition. Once burning it is very difficult to extinguish. Burning of vinyl chloride produces toxic and corrosive fumes of phosgene and hydrogen chloride (1), and carbon monoxide.

BIOLOGICAL HAZARDS

Plastics industry workers exposed to vinyl chloride have experienced a variety of toxic effects. Various animal studies have shown the liver to be the principal target organ, both via ingestion and inhalation exposure. Poisoning in man commonly produces hepatitis and congestive liver enlargement, sometimes accompanied by spleen enlargement. At high concentration, vinyl chloride gas acts as a general anaesthetic. Drowsiness and unconsciousness may be preceded by euphoria and gastrointestinal discomfort. Chronic exposure is associated with acroosteolysis (2,3), a syndrome involving reduced blood platelet count, enlarged spleen, liver damage, respiratory and circulatory obstruction, and changes in the bones. This is often accompanied by Raynaud's syndrome.

Vapour Inhalation

Inhalation is the commonest route of toxic exposure to vinyl chloride. An inhaled dose of 30 mg/m^3 caused poisoning in a 5 year-old child (4). Rats have shown poisoning after inhaling 1000 ppm for 6 hours or 500 ppm for 7 hours.

Eye Contact

Vinyl chloride is a severe eye irritant, causing eye redness and pain.

Skin Contact

Vinyl chloride is a severe skin irritant associated with scleroderma and dermatitis. Repeated exposure of the hands has reportedly caused lysis of the distal bones of the fingers. Contact with the liquid causes frostbite.

Swallowing

Vinyl chloride is moderately toxic by ingestion. The LD_{50} for rats is 500 mg/kg.

CARCINOGENICITY

Vinyl chloride has been classified as a human carcinogen by the International Agency for Research on Cancer (5). Cancer is most commonly produced in the liver, brain, lung and haemo-lymphopoietic system. The cancers reported in the literature have occurred in industrial workers exposed to high levels of vinyl chloride. In the USA, the TLV Committee of the ACGIH has adopted 5 ppm as the average exposure level below which there is no excess cancer risk.

MUTAGENICITY

Vinyl chloride was mutagenic in several strains of *Salmonella typhimurium* in the presence and absence of metabolic activation, with activation greatly increasing the mutagenic effect (6). Dominant lethal tests in mouse (7) and *Drosophila melanogaster* (8) were negative. It has been shown to cause chromosome alterations in laboratory animals (9,10) and exposed workers (11,12).

REPRODUCTIVE HAZARDS

One study has shown a significantly increased foetal mortality rate in the wives of workers exposed to vinyl chloride (13). Several studies in the USA have found excesses of various congenital deformities in children born in towns where there are vinyl chloride plants, though a definite association with vinyl chloride has not been proven (14,15).

152. Vinyl Chloride

FIRST AID

Eyes Wash the eye with flowing water for 10 minutes.

Lungs Remove casualty from area of exposure. If unconscious, do not give anything to drink. Give artificial ventilation and chest compression or place in the recovery position as necessary. If conscious make the casualty lie or sit down quietly, give oxygen if available. Lung congestion may occur – a conscious casualty with breathing difficulties should be placed in a sitting position. Shock may result – if so do not give any drinks, and if conscious, lie casualty flat with legs raised. Convulsions may occur and may cause unconsciousness.

Mouth Do not make the casualty vomit. Treat unconscious casualties as for lungs, but if conscious give 1 pint of water to drink. Convulsions may occur and may cause unconsciousness. Shock may result – do not give any drinks, and if conscious lie casualty flat with legs raised.

Skin Remove contaminated clothing immediately. Wash the affected area with soap and copious amounts of water. Absorption through the skin may cause symptoms similar to lung congestion.

In all cases of exposure, the patient should be transferred to hospital as soon as possible.

HANDLING AND STORAGE

Vinyl chloride should be handled using self-contained breathing apparatus, nitrile or Viton gloves (16), safety goggles and other protective clothing. It should only be used with local exhaust ventilation. It should be kept in a fireproof container, placed away from sources of ignition and oxidising materials and protected from physical damage. Use a cool, dark, well ventilated location, preferably outside.

DISPOSAL

Eliminate all sources of ignition, ventilate the area and wear a laboratory coat or overalls, rubber gloves, approved compressed air breathing apparatus and safety boots. Absorb small spills onto paper towels, remove to a safe open air site and allow to evaporate in a metal tray. Wash spillage site with detergent. For large spills, absorb onto vermiculite-sodium carbonate mixture (90-10) or sand-soda ash (90-10) and mix carefully. EITHER transport in dry buckets to a safe open air area for atmospheric evaporation OR shovel into paper boxes and incinerate.

FIRE PRECAUTIONS

Stop the flow of gas if possible. If this cannot be done, it may be safer to let the fire burn itself out, as extinguishing could cause an explosive build-up of gas (17). Keep containers cool by spraying with water. In other cases use a powder or carbon dioxide extinguisher.

FURTHER READING

Sax, N. Irving *Dangerous Properties of Industrial Materials* (7th edition)

Encyclopedia of Occupational Safety & Health

Patty's *Industrial Hygiene and Toxicology*

Kirk-Othmer *Encyclopedia of Chemical Technology*

National Fire Protection Association *Manual of Hazardous Reactions*

ACGIH *Documentation of TLVs and BEIs* (6th edition, 1986)

Toxicity Profile: Vinyl Chloride (BIBRA, Carshalton, 1991)

REFERENCES

1. Commission of the European Communities *International Chemical Safety Cards: Vinyl Chloride* (CEC, Luxembourg, 1990)
2. Wilson, R. H.; et al. *JAMA* 1967, **20**(8), 577
3. Dinman, B. D.; et al. *Arch. Env. Health* 1971, **22**, 61
4. *Gig. Tr. Prof. Zabol.* 1980, **24**, 80
5. *IARC Monographs on the evaluation of the carcinogenic risk of chemicals to humans* 1979, **19**, 377-438
6. Bartsch. H., et al. *Int. J. Cancer* 1975, **15** 429-437
7. Anderson, D.; et al. *Mutat. Res.* 1976, **40**, 359-370

152. Vinyl Chloride

8. Verburgt, F. G.; Vogel, E. *Mutat. Res.* 1977, **48**, 327-336
9. Solveig Walles, S. A.; et al. *Cancer Lett.* 1984, **25**(1), 13-18
10. Anderson, D.; et al. *Mutat. Res.* 1981, **90**(3), 261-272
11. Hansteen, I-L.; et al. *Mutat. Res* 1978, **51**, 271-278
12. Fucic, A.; et al. *Mutat. Res.* 1990, **242** (4), 265-270
13. Infante, P. F.; et al. *Lancet* 1976, **i**, 734-735, 1289-1290
14. Infante, P. F. *Ann. NY Acad. Sci.* 1976, **271** 49-57
15. Edmonds, L. D.; et al. *Teratology* 1978, **17**, 137-142
16. Forsberg, K.; Mansdorf, S. Z. *Quick selection guide to chemical protective clothing* (Van Nostrand Reinhold, New York, 1989)
17. *Fire Protection Guide on Hazardous Materials*(9th edition) (Natl. Fire Protection Assoc., Quincy, MA, USA, 1986)

153. Vinyl Cyclohexene Diepoxide

VINYL CYCLOHEXENE DIEPOXIDE

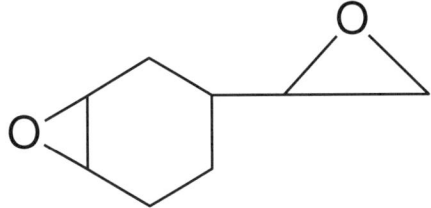

RISKS
Toxic by inhalation, in contact with skin and if swallowed – Possible risk of irreversible effects (R23/24/25, R40)

SAFETY PRECAUTIONS
Do not breathe vapour – Avoid contact with skin – If you feel unwell, seek medical advice (show label where possible) (S23, S24, S44)

IDENTIFIERS

SYNONYMS   Chissonox 206 monomer; 7-oxabicyclo [4.1.0] heptane, 3-oxyranyl-; 7-oxabicyclo [4.1.0] heptane, 3-(epoxyethyl)-; 4-vinylcyclohexene diepoxide; vinyl cyclohexene dioxide; 4- vinylcyclohexene dioxide

CHEMICAL ABSTRACTS No.	106-87-6
NIOSH No.	RN 8640000
HAZCHEM CODE	not available
UN No.	not available

THRESHOLD LIMIT VALUES

US TLV (TWA) .... 10 ppm (57 mg/m^3)
US TLV (STEL) .... not available
UK EXPOSURE LIMITS (OES)
Long-term (8 hr TWA value) .... 10 ppm (60 mg/m^3) under review
Short-term (10 min TWA value) .... not available

Germany
MAK .... not available

France
VME .... not available
VLE .... not available

Sweden
Short-term limit .... not available
Level limit .... not available

PHYSICAL PROPERTIES

Description   Colourless liquid.
Boiling point .... 227°C
Melting point .... Freezing point <55°C
Density .... 1.0986 at 20°C
Vapour density 4.07
Vapour pressure .... 0.1 mm Hg at 20°C
Flash point .... 110°C
Explosive limits .... not available
Autoignition temperature .... 393.3°C
Solubility   Soluble in water, dimethyl sulphoxide, ethanol or acetone.

PACKAGING AND TRANSPORTATION

Road transportation
hazard warning sign	not available
Hazchem code	not available

Sea transportation
IMDG page No.	not available
class	not available
label	not available
packaging group	not available

Air transportation
ICAO/IATA code	not available
class	not available
label	gas flammable
packaging group	not available
packing instructions cargo	not available
passenger	not available
passenger aircraft max. quantity	not available
cargo aircraft max. quantity	not available

153. Vinyl Cyclohexene Diepoxide

MANUFACTURE

Vinyl cyclohexene dioxide is produced by epoxidation of alkene with potassium hydrogen persulphide or peracetic acid.

USES

Vinyl cyclohexene dioxide is used in production of epoxy compounds, adhesives, in treating nylon and phenol-melamine resins and crease proofing cotton fabrics.

CHEMICAL HAZARDS

Vinyl cyclohexene dioxide is combustible on exposure to heat or flame.

BIOLOGICAL HAZARDS

Vinyl cyclohexene dioxide is a severe skin irritant and causes sensitisation. It is a central nervous system depressant. Carcinogenic, mutagenic and reproductive effects have been reported in animal studies.

Vapour Inhalation

The LC_{50} (4 hours) in rats is 800 ppm. Vasodilation and unsteady gait occurred during exposure; lung and liver congestion and occasional testicular atrophy were noted on autopsy.

Eye Contact

Vinyl cyclohexene dioxide is highly irritating to the eyes.

Skin Contact

Vinyl cyclohexene dioxide is a severe skin irritant. Skin sensitisation to vinyl cyclohexene dioxide has been reported in an electron microscopist, despite the use of latex or PVC gloves (1). Skin application caused acanthosis, parakeratosis and hyperkeratosis in mice and rats (2). The dermal LD_{50} in rabbits is 620 mg/kg.

Swallowing

The oral LD_{50} in rats is 2130 mg/kg.

CARCINOGENICITY

The IARC evaluation is that there is sufficient evidence for the carcinogenicity of vinyl cyclohexene dioxide (3). It was carcinogenic in mice on dermal application (4).

MUTAGENICITY

Vinyl cyclohexene dioxide is mutagenic in *Salmonella typhimurium* (5,6), but did not demonstrate dominant lethality in mice.

REPRODUCTIVE HAZARDS

Follicular atrophy of the ovary was reported in mice following oral administration of 50 and 100 mg for 13 weeks; major target organs were the forestomach, kidneys and testies (2). Vinyl cyclohexene diepoxides did not produce a significant number of anatomical defects in rat foetuses or chick embryos (7).

FIRST AID

Eyes Wash the eye with flowing water for 10 minutes.

Lungs Remove casualty from area of exposure. If unconscious, do not give anything to drink, give artificial ventilation and chest compression or place in the recovery position as necessary. If conscious make the casualty lie or sit down quietly, give oxygen if available. Lung congestion may occur – a conscious casualty with breathing difficulties should be placed in a sitting position.

Mouth Do not make the casualty vomit. Treat unconscious casualties as for lungs but if conscious give 1 pint of water to drink immediately; give repeated drinks of water (1 cupful every 10 minutes).

Skin Remove contaminated clothing immediately, drench the affected area with running water for at least 10 minutes.

In all cases of exposure, the patient should be transferred to hospital as soon as possible.

HANDLING AND STORAGE

Vinyl cyclohexane dioxide should be handled wearing an approved respirator, chemical-resistant gloves, safety goggles and other protective clothing. It should only be used in a chemical fume hood. It should be kept in a tightly closed container in a cool, dry place. The use of latex or PVC gloves did not protect an electron microscopist from absorption through the skin and subsequent allergic dermatitis (1).

DISPOSAL

Eliminate all sources of ignition and ventilate the area. Wearing a laboratory coat or overalls, safety glasses, gloves and self-contained breathing apparatus, absorb the spill onto paper towels and allow to evaporate in a fume-cupboard. For large spills, absorb onto sand or vermiculite, and remove in buckets for atmospheric evaporation in a safe open area. Ideally, waste should be burned in an incinerator with afterburner.

FIRE PRECAUTIONS

Fires involving vinyl cyclohexene dioxide should be extinguished using water spray, dry chemical powder, alcohol foam or polymer foam.

FURTHER READING

Sax, N. Irving *Dangerous Properties of Industrial Materials* (7th edition)

Encyclopaedia of Occupational Safety & Health

Patty's *Industrial Hygiene and Toxicology*

Kirk-Othmer *Encyclopedia of Chemical Technology*

National Fire Protection Association *Manual of Hazardous Reactions*

ACGIH Documentation of TLVs and BEIs (6th edition, 1986)

REFERENCES

1. Dannaker, C. J. *JOM, J. Occup. Med.* 1988, **30**(8), 641-643
2. Chhabra, R. S.; et al. *Fundam. Appl. Toxicol.* 1990, **14**(4), 745-751
3. IARC *Monographs on the evaluation of the carcinogenic risk of chemicals to humans* 1976, **11**, 141
4. Chhabra, R. S. *Fundam. Appl. Toxicol.* 1990, **14**(4), 752-763
5. Mortelmans, K.; et al. *Environ. Mol. Mutagen.* 1986, **8**(Suppl. 7), 1-119
6. Ringo, D. L.; et al. *J. Ultrastruct. Res.* 1982, **80**(3), 280-287
7. Roche, S. M.; et al. *Toxicol. Appl. Pharmacol.* 1968, **12**, 3

154. Xylenols

XYLENOLS

RISKS
Toxic in contact with skin and if swallowed – Causes burns (R24/25, R34)

SAFETY PRECAUTIONS
Keep out of reach of children – After contact with skin, wash immediately with plenty of water – If you feel unwell, seek medical advice (show label where possible) (S2, S28, S44)

IDENTIFIERS

SYNONYMS	dimethylphenol; phenol, dimethyl-
CHEMICAL ABSTRACTS No.	1300-71-6
NIOSH No.	ZE 5425000
HAZCHEM CODE	2X
UN No.	2261

THRESHOLD LIMIT VALUES

US TLV (TWA)	not available
US TLV (STEL)	not available
UK EXPOSURE LIMITS (OES)	
Long-term (8 hr TWA value)	not available
Short-term (10 min TWA value)	not available
Germany	
MAK	not available
France	
VME	not available
VLE	not available
Sweden	
Short-term limit	not available
Level limit	not available

PHYSICAL PROPERTIES

Description With the exception of 2,4-xylenol, which is often encountered as a yellow-brown liquid, the commoner xylenols are colourless crystalline solids.

Boiling point	203-226°C at 760 mm Hg
Melting point	48-75°C
Density	not available
Vapour density	not available
Vapour pressure	not available
Flash point	not available
Explosive limits	not available
Autoignition temperature	not available

Solubility Xylenols are slightly soluble in water, very soluble in alcohol, chloroform, ether or benzene and soluble in sodium hydroxide.

PACKAGING AND TRANSPORTATION

Road transportation

hazard warning sign	2261 toxic substance
Hazchem code	2X

Sea transportation

IMDG page No.	6280
class	6.1
label	poison; marine pollutant
packaging group	II

Air transportation

ICAO/IATA code (UN No.)	2261
class	6.1
label	poison
packaging group	II
packing instructions	
cargo	615
passenger	613
passenger aircraft max. quantity	25 kilograms
cargo aircraft max. quantity	100 kilograms

154. Xylenols

MANUFACTURE

Xylenols may be synthesised by oxidation of cymenes, oxidation decarboxylation of toluic acids and alkylation of phenol (1).

USES

Xylenols are used for the preparation of coal tar disinfectants and synthetic resins, in solvents, pesticides, pharmaceuticals, plasticisers, wetting agents, antioxidants and dyestuffs (2). The uses of xylenols have been reviewed (3).

CHEMICAL HAZARDS

Xylenols may emit acrid smoke and irritating fumes when heated to decomposition.

BIOLOGICAL HAZARDS

Signs and symptoms of exposure include dizziness, stomach pains, nausea, vomiting, dyspnoea and weakness. They are skin and eye irritants. Chronic exposure may cause liver and kidney damage, nervous system and digestive disorders (4,5).

Vapour Inhalation

No information is available concerning the effects of inhalation.

Eye Contact

Xylenols may cause severe eye irritation (4).

Skin Contact

Xylenols may cause severe skin irritation (4). They may be absorbed through the skin; the dermal LD_{50} in mice has been reported as 1040 mg/kg and 920 mg/kg for the 2,4- and 2,6-isomers respectively.

Swallowing

The oral LD_{50} in rats ranges from 296 mg/kg to 3200 mg/kg for the 2,6-and 2,4-isomers respectively. In mice values range from 383 mg/kg for the 2,5-isomer to 980 mg/kg for the 2,6-isomer.

CARCINOGENICITY

Tested on the skin of mice, 2,4-, 2,6-, 3,4- and 3,5-xylenol (the only isomers tested) caused increased incidence of papillomas (6).

MUTAGENICITY

No information is available concerning the mutagenicity of xylenols.

REPRODUCTIVE HAZARDS

No information is available concerning the reproductive hazards of xylenols.

FIRST AID

Eyes Wash the eye with flowing water for 10 minutes. Do NOT use Macrogol 300 in the eye.

Lungs Remove casualty from area of exposure. If unconscious, do not give anything to drink, give artificial ventilation and chest compression or place in the recovery position as necessary. If conscious make the casualty lie or sit down quietly, give oxygen if available. Lung congestion may occur – a conscious casualty with breathing difficulties should be placed in a sitting position. Convulsions may occur and may cause unconsciousness. Shock may result – do not give any drinks, and if conscious lie casualty flat with legs raised.

Mouth Do not make the casualty vomit. Treat unconscious casualties as for lungs but if conscious give 1 pint of water to drink immediately; give repeated drinks of water (1 cupful every 10 minutes). Convulsions may occur and may cause unconsciousness.

Skin Wearing protective gloves, remove contaminated clothing immediately, flush excess chemical off the skin with water, then wash with polyethylene glycol molecular weight 300 (Macrogol 300) for at least 30 minutes.

In all cases of exposure, the patient should be transferred to hospital as soon as possible.

154. Xylenols

HANDLING AND STORAGE

Xylenols should be handled wearing an approved respirator, rubber gloves and safety goggles (4). They should be kept in a cool, dry place or refrigerator (2).

DISPOSAL

Eliminate all sources of ignition and ventilate the area. Wearing a laboratory coat or overalls, safety glasses, gloves and self-contained breathing apparatus, absorb the spill onto paper towels and allow to evaporate in a fume-cupboard. For large spills, absorb onto sand or vermiculite, and remove in buckets for atmospheric evaporation in a safe open area. Ideally, waste should be burned in an incinerator with afterburner.

FIRE PRECAUTIONS

Fires involving xylenols should be extinguished using carbon dioxide, dry chemical powder, alcohol foam or polymer foam (4).

FURTHER READING

Sax, N. Irving *Dangerous Properties of Industrial Materials* (7th edition)

Encyclopaedia of Occupational Safety & Health

Patty's *Industrial Hygiene and Toxicology*

Kirk-Othmer *Encyclopedia of Chemical Technology*

National Fire Protection Association *Manual of Hazardous Reactions*

ACGIH Documentation of TLVs and BEIs (6th edition, 1986)

REFERENCES

1. Kharlampovich, G. D.; et al. *Khim. Prom.* 1968, **44**(1), 16-20
2. Keith, L. H.; Walters, D. B. (editors) *Compendium of Safety Data Sheets for Research and Industrial Chemicals* (VCH, Deerfield Park, 1987)
3. Kamegaya, E. *Chem. Econ. Eng. Rev.* 1970, **2**(2), 30-33
4. *Dangerous Prop. Ind. Mater. Rep.* 1987, **7**(3), 87-90
5. *Dangerous Prop. Ind. Mater. Rep.* 1984, **4**(1), 102-106
6. *Cancer Res.* 1959, **19**, 413-422

155. Xylidines

XYLIDINES

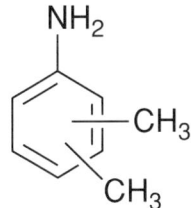

RISKS
Toxic by inhalation, in contact with skin and if swallowed – Danger of cumulative effects (R23/24/25, R33)

SAFETY PRECAUTIONS
After contact with skin, wash immediately with plenty of water – Wear protective clothing and gloves – If you feel unwell, seek medical advice (show label where possible) (S28, S36/37, S44)

IDENTIFIERS

SYNONYMS aminodimethylbenzene; Acid Leather Brown 2G; Acid Orange 24; benzenamine, *ar,ar*-dimethyl-; dimethylaniline; dimethylphenylamine

CHEMICAL ABSTRACTS No.	1300-73-8
NIOSH No.	ZE 8575000
HAZCHEM CODE	3X
UN No.	1711

THRESHOLD LIMIT VALUES

US TLV (TWA)	0.5 ppm (2.5 mg/m^3)
US TLV (STEL)	not available

UK EXPOSURE LIMITS (OES)
Long-term (8 hr TWA value) ..... 2 ppm (10 mg/m^3)
Short-term (10 min TWA value) ... 10 ppm (50 mg/m^3)

Germany
MAK .... 5 ppm (25 mg/m^3)

France
VME .... 2 ppm (10 mg/m^3)
VLE .... not available

Sweden
Short-term limit .... not available
Level limit .... not available

PHYSICAL PROPERTIES

Description Most of the common xylidines are red to dark-brown liquids (3,4-xylidine is pale brown crystals). Commercial xylidine is a mixture of isomers.

Boiling point	213-226°C
Melting point	not available
Density	0.97-0.99
Vapour density	4.17
Vapour pressure	<1 torr at 20°C
Flash point	96°C (closed cup)
Explosive limits	not available
Autoignition temperature	not available

Solubility Slightly soluble in water. Soluble in alcohol.

PACKAGING AND TRANSPORTATION

Road transportation
hazard warning sign	1711 toxic substance
Hazchem code	3X

Sea transportation
IMDG page No.	6280
class	6.1
label	poison
packaging group	II

Air transportation
ICAO/IATA code (UN No.)	1711
class	6.1
label	poison
packaging group	II
packing instructions cargo	611
passenger	609
passenger aircraft max. quantity	5 litres
cargo aircraft max. quantity	60 litres

155. Xylidines

MANUFACTURE

Xylidines are prepared by the reduction of the corresponding nitro compounds.

USES

Xylidines are used in dye manufacturing, pharmaceuticals and in organic synthesis (1). The 2,4-isomer is also used in the production of pigments and antioxidants.

CHEMICAL HAZARDS

Xylidines may form explosive chloramines in contact with hypochlorite bleaches, and ignite on contact with fuming nitric acid. Contact with strong oxidisers may be explosive. They may attack some plastics, rubber and coatings (1).

BIOLOGICAL HAZARDS

The toxicity of xylidines is similar to that of aniline; the onset is insidious, with headache, dizziness and cyanosis which may not be recognised as signs of overexposure (2). Xylidines do no have adequate warning properties. Results of studies in dogs and rats showed divergent metabolic pathways for the two species (3).

Vapour Inhalation

The LC_{50} (7 hours) in mice is 149 ppm. Repeated exposure (45 ppm 7 hours/day for 20-40 weeks) caused liver damage in experimental animals. Repeated exposure of cats to 138 ppm caused cyanosis, loss of coordination, prostration and death, with pulmonary oedema, pneumonia and liver and kidney damage found on autopsy (2).

Eye Contact

No information is available on the effects of xylidines on the eyes.

Skin Contact

Xylidines may be absorbed through the skin. In rabbits skin absorption causing cyanosis and death did not produce local effects (2). Their dermal toxicity in cats is similar to that of aniline.

Swallowing

The oral LD_{50} in rats ranges from 467 mg/kg to 1297 mg/kg for the 2,4- and 2,5-isomers respectively. In mice values range from 250 mg/kg to 1027 mg/kg for the 2,4-and 2,3-isomers respectively. In dogs xylidenes are less toxic and form less methaemoglobin than aniline (4).

CARCINOGENICITY

2,4-And 2,5-xylidine appear to be carcinogenic in animals. They differ in their tumorigenicity; the 2,5-isomer increased tumour incidence in male rats and male and female mice, while the 2,4-isomer caused lung tumours in female mice only (5). An increase in incidence of liver and subcutaneous tumours was reported in rats after oral administration of 2,4- or 2,5-xylidine, but it was the opinion of the IARC that the data were inadequately reported and did "not allow an evaluation of the carcinogenicity" of the isomers to be made (6). 2,6-Xylidine has been reported as carcinogenic in rats (7).

MUTAGENICITY

2,3-, 2,4-, 2,5-and 3,4-xylidine are weakly mutagenic in *Salmonella typhimurium* (5,8). Neither 2,4- nor 2, 5-xylidine induced DNA damage in V79 cells (5). 2,4-xylidine elicited positive DNA repair responses in primary cultured rat hepatocytes (9). 2,6-Xylidine induced sister chromatid exchanges (with or without metabolic activation) and chromosome aberrations in Chinese hamster ovary cells (10). 2,6-Xylidine had no effect on the frequency of micronuclei in mouse bone marrow cells *in vivo* (11), and did not induce unscheduled DNA synthesis in rodent hepatocytes (12).

REPRODUCTIVE HAZARDS

No information is available concerning the reproductive hazards of xylidines.

FIRST AID

Eyes Wash the eye with flowing water for 10 minutes.

Lungs Remove casualty from area of exposure. If unconscious, do not give anything to drink. Give artificial ventilation and chest compression or place in the recovery position as necessary. If conscious make the casualty lie or sit down

155. Xylidines

quietly, give oxygen if available. Convulsions may occur and may cause unconsciousness. Shock may result – if so do not give any drinks, and if conscious, lie casualty flat with legs raised.

Mouth Do not make the casualty vomit. Treat unconscious casualties as for lungs, but if conscious give 1 pint of water to drink.

Skin Remove contaminated clothing immediately, wash the affected area with soap and copious amounts of water. Absorption through the skin may cause symptoms similar to those of inhalation.

In all cases of exposure, the patient should be transferred to hospital as soon as possible.

HANDLING AND STORAGE

Xylidines should be handled wearing an approved respirator, long rubber or neoprene gloves, safety goggles and rubber apron. They should only be used in a chemical fume hood (13). They should be kept in a tightly closed container, in a cool, dry, well-ventilated place, away from oxidisers and protected against physical damage. Long term storage should be under an inert atmosphere (1).

DISPOSAL

Wear a laboratory coat, safety spectacles, butyl rubber gloves and suitable safety shoes, and have an approved self-contained breathing apparatus or canister respirator available. Absorb small liquid spills onto paper towels, evaporate in an iron pan in a fume cupboard, add crumpled paper and burn carefully. Brush small solid spills onto paper, put in an iron pan, cover with crumpled paper and burn carefully in a safe place outside. For large spills,
EITHER cover with sand and mix carefully. Shovel into containers, disperse in an excess solution of dilute hydrochloric acid, mix well and leave to stand for 24 hours stirring occasionally. Carefully decant acid extract into the drains, diluting with a large volume of cold tap water. Wash the sand thoroughly with cold water.
OR cover with a sand-soda ash mix (90-10), mix well, shovel into cardboard boxes, pack with crumpled paper and incinerate.

FIRE PRECAUTIONS

Fires involving xylidines should be extinguished using water spray, carbon dioxide, dry chemical powder, alcohol foam or polymer foam. Container explosion may occur under fire conditions.

FURTHER READING

Bretherick, L. *Hazards in the Chemical Laboratory* (4th edition)

Bretherick, L. *Handbook of Reactive Chemical Hazards* (4th edition)

Sax, N. Irving *Dangerous Properties of Industrial Materials* (7th edition)

Encyclopedia of Occupational Safety & Health

Patty's *Industrial Hygiene and Toxicology*

Kirk-Othmer *Encyclopedia of Chemical Technology*

National Fire Protection Association *Manual of Hazardous Reactions*

ACGIH Documentation of TLVs and BEIs (6th edition, 1986)

REFERENCES

1. Keith, L. H.; Walters, D. B. (editors) *Compendium of Safety Data Sheets for Research and Industrial Chemicals* (VCH, Deerfield Park, 1987)
2. Proctor, N. H.; Hughes J. P. *Chemical Hazards of the Workplace* (Lippincott Co., Philadelphia, 1978)
3. Short, C. R.; et al. *Toxicology* 1989, **57**(1), 45-58
4. Treon, J. F.; et al. *Arch. Ind. Hyg. Occup. Med.* 1950, **1**, 506
5. Zimmer, D.; et al. *Mutat. Res.* 1980, **77**, 317-326
6. IARC *Monographs on the evaluation of the carcinogenic risk of chemicals to humans* 1977, **16**, 367-385
7. *NTP Technical Report No. 278*
8. Kugler-Steigmeier, M. E.; et al. *Mutat. Res.* 1989, **211**(2), 279-289
9. Yoshimi, N.; et al. *Mutat. Res.* 1988, **206**(2), 183-191

155. Xylidines

10. Galloway, S. M.; et al. *Environ. Mol. Mutagen.* 1987, **10**(Suppl. 10), 1-175
11. Parton, J. W.; et al. *Mutat. Res.* 1988, **206**(2), 281-283
12. Mirsalis, J. C.; et al. *Environ. Mol. Mutagen.* 1989, **14**(3), 155-164
13. Lenga, R. E. *The Sigma-Aldrich Library of Chemical Safety Data* (2nd edition) (Sigma-Aldrich, Milwaukee, 1988)

156. Zinc Chromate

ZINC CHROMATE
ZnCrO$_4$

RISKS
May cause cancer – Harmful if swallowed – May cause sensitisation by skin contact (R45, R22, R43)

SAFETY PRECAUTIONS
Avoid exposure-obtain special instructions before use – If you feel unwell, seek medical advice (show label where possible) (S53, S44)

IDENTIFIERS

SYNONYMS buttercup yellow; chromic acid, zinc salt (ZnCrO$_4$); chromium zinc oxide (ZnCrO$_4$); chromic acid (H$_2$CrO$_4$), zinc salt(1:1); primrose yellow; zinc yellow; zinc chromate (ZnCrO$_4$); zinc chromium oxide (ZnCrO$_4$); zinc tetroxychromate; zinc tetraoxychromate

CHEMICAL ABSTRACTS No.	13530-65-9
NIOSH No.	GB 3290000
HAZCHEM CODE	not available
UN No.	not available

THRESHOLD LIMIT VALUES

US TLV (TWA)	0.01 mg/m^3 (as Cr)
US TLV (STEL)	not available

UK EXPOSURE LIMITS (OES)
Long-term (8 hr TWA value) .... 0.5 mg/m^3 (as Cr) (under review)
Short-term (10 min TWA value) .... not available

Germany
MAK .... not available

France
VME .... 0.05 mg/m^3 (as Cr)
VLE .... not available

Sweden
Short-term limit .... not available
Level limit .... 0.02 mg/m^3 (as Cr)

PHYSICAL PROPERTIES

Description	Yellow solid.
Boiling point	not available
Melting point	not available
Density	3.40
Vapour density	not available
Vapour pressure	not available
Flash point	not available
Explosive limits	not available
Autoignition temperature	not available
Solubility	Sparingly soluble in water.

PACKAGING AND TRANSPORTATION

Road transportation

hazard warning sign	not available
Hazchem code	not available

Sea transportation

IMDG page No.	not available
class	not available
label	not available
packaging group	not available

Air transportation

ICAO/IATA code	not available
class	not available
label	not available
packaging group	not available
packing instructions	
cargo	not available
passenger	not available
passenger aircraft max. quantity	not available
cargo aircraft max. quantity	not available

156. Zinc Chromate

MANUFACTURE

Zinc chromate is prepared by addition of chromic acid solution to a zinc oxide or zinc hydroxide slurry.

USES

Zinc chromate is used in surface coatings as a corrosion-resistant primer coating or in metal conditioners (wash primers). The German Technical Rule for Dangerous Substances TRG 602, concluded it was not yet possible to impose an overall ban on zinc chromate anticorrosive pigments until adequate substitutes are available (1).

CHEMICAL HAZARDS

No information is available concerning the chemical hazards of zinc chromate.

BIOLOGICAL HAZARDS

Zinc chromate is a human and animal carcinogen, and is mutagenic in bacterial assays and cultured mammalian cells.

Vapour Inhalation

Zinc chromate is a lung carcinogen.

Eye Contact

No information is available on the effects on the eye.

Skin Contact

Zinc chromate in primer paints can cause nasal ulceration and dermatitis in exposed workers (2).

Swallowing

No information is available on the effects of ingestion.

CARCINOGENICITY

The IARC evaluation states that there is "sufficient evidence for the carcinogenicity of ... zinc chromate in rats", and "sufficient evidence of respiratory carcinogenicity in men occupationally exposed during chromate production", although epidemiological data do not permit an evaluation of the relative contribution of the metal, chromium(III) and chromium(VI) compounds, or of soluble versus insoluble compounds (3-6).

The UK Industrial Injuries Advisory Council has extended the list of jobs covered for lung cancer compensation to include workers exposed to zinc chromate (7-9). Its carcinogenicity and has been reviewed (10,11). It has been identified by the EEC as a probable carcinogen (12). Moderate or heavy exposure to zinc chromate may constitute a risk of developing lung cancer (14). Zinc chromate induced bronchial carcinomas in 5/100 rats following intrabronchial pellet implantation (13).

MUTAGENICITY

Chromium(VI) compounds including zinc chromate were inactive or scarcely active in the *Salmonella*/microsome test when dissolved in water. Mutagenicity was increased when solubilised by 0.5N sodium hydroxide or nitrilotriacetic acid trisodium salt (NTA). It was directly clastogenic in the sister chromatid exchange assay in Chinese hamster ovary cells, NTA significantly increased chromosome damaging activity (14). Positive results have been reported in mammalian cell gene mutation and cell transformation assays (15). Chromate pigments containing zinc induced neoplastic transformation of Syrian hamster embryo cells (16). Its mutagenicity has been reviewed (10).

REPRODUCTIVE HAZARDS

Although there appears to be no data specific to zinc chromate, chromium(VI) trioxide is embryotoxic and teratogenic in animals (3).

FIRST AID

Eyes Wash the eye with plenty of water or normal saline for at least 20 minutes.

Lungs Wearing appropriate respiratory protection, remove casualty from area of exposure. If unconscious, do not give anything to drink, give artificial ventilation and chest compression or place in the recovery position as necessary. If conscious make the casualty lie or sit down quietly, give oxygen if available.

156. Zinc Chromate

Mouth Do not make the casualty vomit. Treat unconscious casualties as for lungs but if conscious give a glass or two of milk or water to drink immediately.

Skin Remove contaminated clothing immediately, wash the affected area with soap and copious amounts of water. Absorption through the skin may cause symptoms similar to those of inhalation.

In all cases of exposure, the patient should be transferred to hospital as soon as possible.

HANDLING AND STORAGE

Zinc chromate should be handled wearing an approved respirator, rubber gloves, boots, safety goggles and other protective clothing. All equipment should be fully decontaminated or disposed of after use. It should be kept in a tightly closed container, in a well ventilated area away from heat and water (2).

DISPOSAL

Wearing a laboratory coat or overalls, safety glasses (with a polycarbonate visor), rubber gloves, suitable safety boots and breathing apparatus if the spill is large and in a confined area, cover with a reducing agent, eg. sodium metabisulphite, sodium thiosulphate or a ferrous salt (do not use carbon or sulphur), mix and spray with water. Transfer the slurry to a large container of water, neutralise with soda ash, and run to waste with excess cold running water. For large spills absorb onto sand and arrange for removal by a licenced contractor.

FIRE PRECAUTIONS

Fires involving zinc chromate should be extinguished using water spray, carbon dioxide, dry chemical powder, alcohol foam or polymer foam.

FURTHER READING

Bretherick, L. *Hazards in the Chemical Laboratory* (4th edition)

Bretherick, L. *Handbook of Reactive Chemical Hazards* (4th edition)

Sax, N. Irving *Dangerous Properties of Industrial Materials* (7th edition)

Encyclopaedia of Occupational Safety & Health

Patty's *Industrial Hygiene and Toxicology*

Kirk-Othmer *Encyclopedia of Chemical Technology*

National Fire Protection Association *Manual of Hazardous Reactions*

ACGIH *Documentation of TLVs and BEIs* (6th edition, 1986)

Environmental Health Criteria 61: Chromium (WHO, Geneva, 1988)

REFERENCES

1. *Bundesarbeitsblatt* 1988, **5**, no pages given
2. *Laboratory Hazards Data Sheet no. 76: Zinc & zinc compounds* (RSC, London, 1988)
3. IARC *Monographs on the evaluation of the carcinogenic risk of chemicals to humans* 1980, **23** 205-323
4. Davies, J. M. *Br. J. Ind. Med.* 1984, **41**(2), 158-159
5. Frentzel Beyme, R. *J. Cancer Res. Clin. Oncol.* 1983, **105**(2), 183-188
6. *Monograph on exposure to chemicals in the workplace. Zinc chromate* (Natl. Cancer Inst., Bethesda, 1986)
7. *Occupational lung cancer* (Industrial Injuries Advisory Council, London, 1986)
8. *Health Saf. Inf. Bull.* 1988, **136**, 11-12
9. *Occup. Saf. Health* 1987, **17**(5), 4
10. *Monograph on human exposure to chemicals in the workplace* (NTIS, Springfield, 1986)
11. Santodonato, J. *NTIS Report* 1985, SRC-TR-1125 Order No. PB86-155165/GAR
12. *Promosafe* 1986, **13**(5), 414
13. Levy, L. S.; et al. *Br. J. Ind. Med.* 1986, **43**(4), 243-256
14. Venier, P.; et al. *Mutat. Res.* 1985, **156**(3), 219-228

156. Zinc Chromate

15. *Environmental Health Criteria 61: Chromium* (WHO, Genvea, 1988)
16. Elias, Z.; et al. *Carcinogenesis (London)* 1989, **10**(11), 2043-2052

ZINC PHOSPHIDE
Zn$_3$P$_2$

157. Zinc Phosphide

PHYSICAL PROPERTIES

Description Cubic, dark grey cystals or powder with a faint phosphorus odour.

Boiling point	1100°C
Melting point	420°C
Density	4.55 at 13°C
Vapour density	not available
Vapour pressure	not available
Flash point	not available
Explosive limits	not available
Autoignition temperature	not available

Solubility   Insoluble in water or alcohol. Soluble in benzene or carbon disulphide

RISKS
Very toxic if swallowed – Contact with acids liberates very toxic gas (R28, R32)

SAFETY PRECAUTIONS
Keep locked up and out of reach of children – When using do not eat, drink or smoke – Do not breathe dust – After contact with skin, wash immediately with plenty of water – In case of accident or if you feel unwell, seek medical advice immediately (show label where possible) (S1/2, S20/21, S22, S28, S45)

IDENTIFIERS

SYNONYMS   Blue – ox; Delusol; Phosvin; Rumetan; Stutox; trizinc diphosphide

CHEMICAL ABSTRACTS No.	1314-84-7
NIOSH No.	ZH 4900000
HAZCHEM CODE	not available
UN No.	1714

PACKAGING AND TRANSPORTATION

Road transportation

hazard warning sign ... 1714 substance which in contact with water emits flammable gas; toxic

Hazchem code   not available

Sea transportation

IMDG page No.	4372
class	4.3
label	dangerous when wet; poison; marine pollutant
packaging group	I

Air transportation

ICAO/IATA code (UN No.)	1714
class	4.3;6.1
label	danger if wet; poison
packaging group	I
packing instructions	
cargo	412
passenger	forbidden
passenger aircraft max. quantity	forbidden
cargo aircraft max. quantity	15 kilograms

THRESHOLD LIMIT VALUES

US TLV (TWA)	not available
US TLV (STEL)	not available
UK EXPOSURE LIMITS (OES)	
Long term (8 hr TWA value)	not available
Short-term (10 min TWA value)	not available
Germany	
MAK	not available
France	
VME	not available
VLE	not available
Sweden	
Short-term limit	not available
Level limit	not available

157. Zinc Phosphide

MANUFACTURE

Zinc phosphide can be prepared by reducing trizinc phosphate with hydrogen at 600°C, by passing phosphorus vapour over zinc at 400°C, or by direct reaction between phosphorus and zinc powder under pressure or heat.

USES

Zinc phosphide is used in rodenticide formulations.

CHEMICAL HAZARDS

Zinc phosphide is stable when dry, but decomposes slowly in moist air to produce toxic and flammable phosphine vapours. Contact with acids produces phosphine. It reacts violently with concentrated sulphuric and nitric acids and other oxidising agents. Use of perchloric acid in analysis of zinc phosphide causes an explosion (1). Contact of a drop of water with a zinc phosphide rodenticide preparation causes ignition (2).

BIOLOGICAL HAZARDS

Zinc phosphide is toxic to humans and produces nausea, vomiting and central nervous system depression. It can be absorbed into the tissues where it slowly hydrolyses to produce phosphine. The methods and mechanisms of poisoning have been reviewed (3).

Vapour Inhalation

Inhalation of zinc phosphide may result in hydrolysis and formation of phosphine. Zinc phosphide inhalation may cause lung irritation, pulmonary oedema, dilation of the heart and hyperaemia of the visceral organs. Chronic exposure may cause anaemia, bronchitis and gastrointestinal, visual, speech and motor disturbances (4). The LC_{50} in rats was reported as 19.6 mg/l for 10% zinc phosphide (5).

Eye Contact

Contact of zinc phosphide with the eyes may produce pain, photophobia, conjunctival oedema and corneal damage (4).

Skin Contact

Contact of the skin with zinc phosphide may produce burns that heal slowly with scar formation (4). Mild skin irritation was produced when moistened zinc phosphide powder was applied to the skin of rabbits (5). The dermal LD_{50} in rabbits was reported as 2000-5000 mg/kg body weight, indicating little dermal absorption of the compound (5).

Swallowing

Ingestion of zinc phosphide may cause nausea, vomiting, diarrhoea, abdominal pain, chest tightness, coughing, headaches and dizziness. In severe cases cardiovascular and respiratory failure, renal and hepatic damage may occur. Symptoms may be delayed and death can occur up to one week later. A review of 20 cases of suicidal ingestion of zinc phosphide formulations showed most lethal doses to be greater than 20 g (range 4.5-180 g) while most non-fatal ingestions were less than 20 g (range 0.5-50 g) (7). Reported LD_{50} values in animal studies are 40.5 mg/kg (8), 2.7 mg/kg (5) and 12 mg/kg (9) in rats; 40 mg/kg (10) in mice; LD_{50} of 250 mg/kg in cats and 40 mg/kg in rabbits (11). Female rats fed 0-500 mg/kg zinc phosphide for 13 weeks showed death in some animals at doses >200 mg/kg, dose dependent depilation and decreased red blood cell count at all doses, and organ weight gains at >200 mg/kg (12). In another study in rats, reduced weight gain and 2/6 deaths at 200 mg/kg and 6/6 deaths at 300 mg/kg were reported, with pathological symptoms of liver damage and lung haemorrhage (13).

CARCINOGENICITY

No information is available concerning the carcinogenicity of zinc phosphide.

MUTAGENICITY

No information is available concerning the mutagenicity of zinc phosphide.

REPRODUCTIVE HAZARDS

No information is available concerning the reproductive hazards of zinc phosphide.

FIRST AID

Eyes Wash the eye with flowing water for 10 minutes.

157. Zinc Phosphide

Lungs Remove casualty from area of exposure. If unconscious, do not give anything to drink, give artificial ventilation and chest compression or place in the recovery position as necessary. If conscious make the casualty lie or sit down quietly, give oxygen if available. Lung congestion may occur – a conscious casualty with breathing difficulties should be placed in a sitting position.

Mouth Do not make the casualty vomit. Treat unconscious casualties as for lungs but if conscious give 1 pint of water to drink. Lung congestion may occur – a conscious casualty with breathing difficulties should be placed in a sitting position.

Skin Remove contaminated clothing immediately, drench the affected area with running water for a least 10 minutes.

In all cases of exposure, the patient should be transferred to hospital as soon as possible.

HANDLING AND STORAGE

Zinc phosphide should be handled wearing an approved respirator, chemical-resistant gloves, safety goggles and other protective clothing. It should only be used in a chemical fume hood (14). It should be kept in a tightly closed container and protected from moisture. Store away from oxidising materials.

DISPOSAL

Wearing a laboratory coat or overalls, safety glasses, rubber gloves and suitable safety shoes, scoop up spill into a large beaker, carefully add water and stir until dissolved. Run the solution to waste with excess water.

For larger spills add soda ash at intervals. Decant liquid after 24 hours and neutralise with 6M hydrochloric acid. Discharge supernatant to drain with x1000 dilution of cold tap water. The sludge should be removed by a licenced contractor. For solutions cover and mix with dry soda ash, and shovel into buckets. Carefully add cold water, neutralise with 6M hydrochloric acid and wash down the drain with x1000 volume of cold water.

FIRE PRECAUTIONS

Fires involving zinc phosphide should be extinguished using dry chemical powder, alcohol foam or polymer foam. Do not use acid, carbon dioxide, halogenated agents or water. May produce irritating phosphorus oxides fumes in fires. Produces flammable phosphine when wet.

FURTHER READING

Bretherick, L. *Handbook of Reactive Chemical Hazards* (4th edition)

Sax, N. Irving *Dangerous Properties of Industrial Materials* (7th edition)

Environmental Health Criterial 73: Phosphine and selected Metal Phosphides (Geneva, WHO, 1988)

REFERENCES

1. Muir, G. D. *private communication* in Bretherick, L. *Handbook of Reactive Chemical Hazards* (4th edition) p.963
2. Fehse, W. *Feuerscheutz* 1938, **18**, 17
3. Casteel, S. W.; et al. *Vet. Hum. Toxicol.* 1986, **28**(2), 151-154
4. *Dangerous Properties of Industrial Materials Report* 1985, **5**, 103-106
5. US EPA *Fed. Reg.* 1983, **48**(37), 7714-7716
6. Rao *personal communication* in *Environmental Health Criteria 73: Phosphine and Selected Metal Phosphides* IPCS/WHO, 1988, p.70
7. Stephenson, J. B. P. *Arch. Environ. Health* 1967, **15**, 83-88.
8. Dieke, S. H.; Richter, C. P. *Publ. Health Rep.* 1946, **61**, 672-679
9. *Malaysian Agricultural J.* 1979, **52**(2), 166
10. *Yakkyoku Pharmacy* 1980, **31**, 1247
11. *J. Am. Pharm. Assoc.* 1952, **42**, 468
12. Bai, K. M.; et al. *Indian J. Exper. Biol.* 1980, **18**(8), 854-857
13. WHO/FAO *Data Sheet on Pesticides, No.24: zinc phosphide* (WHO, Geneva, 1976) (VBC/DS/77.24)
14. Lenga, R. E. *The Sigma-Aldrich Library of Chemical Safety Data* (2nd edition) (Sigma-Aldrich, Milwaukee, 1988)

INDEX OF NAMES AND SYNONYMS

The numbers in this index refer to the **NUMBERS** of the data sheets and **not** to page numbers.

Data sheets 1-79 appear in Volume 4a.

Data sheets 80-157 appear in Volume 4b.

A

Acco Fast Bordeaux GP Salt	103
acenaphthene, 5-nitro-	99
acenaphthylene, 1,2-dihydro-5-nitro-	99
acetatotriphenylstannane	149
acetic acid, bromo-	21
acetic acid, chloro-	38
acetic acid, mercapto-	141
ACETONE CYANOHYDRIN	1
acetonitrile, chloro-	39
acetonitrile, trichloro-	146
(acetyloxy)triphenylstannane	149
acetoxytriphenylstannane	149
acetoxytriphenyltin	149
acetylene tetrabromide	138
acetylene tetrachloride	139
Acid Leather Brown 2G	155
Acid Orange 24	155
Acquinite	148
ACRYLAMIDE	2
acrylic acid, 2,2-dimethyl-trimethylene ester	96
acrylic acid, 2,3-epoxy-propyl ester	66
acrylic acid, oxydiethylene ester	54
acrylic amide	2
AD 6	95
AD 6 (suspending agent)	95
Aero Liquid HCN	76
Aldifen	62
ALLYL ALCOHOL	3
allyl alcohol oxide	65
ALLYL CHLORIDE	5
ALLYLAMINE	4
allylic alcohol	3
Alzogur	47
Amacel Developed Navy SD	48
Amarthol Fast Bordeaux GP Base	103
Amarthol Fast Bordeaux GP Salt	103
Amarthol Fast Orange R Base	100
Amarthol Fast Scarlet G Base	109
Amarthol Fast Scarlet G Salt	109
amidocyanogen	47
2-amino-1-methylbenzene	144
4-amino-1-methylbenzene	145
1-amino-2-methylbenzene	144
1-amino-3,4-dichlorobenzene	49
4-amino-3-methoxyazobenzene	103
4-amino-3-methoxyazobenzene	103
3-aminoaniline	115
4-aminoaniline	116
m-aminoaniline	115
p-aminoaniline	116
aminoanisole	9
aminobenzene	8
p-aminochlorobenzene	40
p-aminodiethylaniline	55
p-aminodimethylaniline	58
aminodimethylbenzene	155
aminoethylene	15
2-aminonitrobenzene	101
4-aminonitrobenzene	102
m-aminonitrobenzene	100
o-aminonitrobenzene	101
p-aminonitrobenzene	102
aminophen	8
4-aminophenetole	113
p-aminophenetole	113
3-aminopropene	4
3-aminopropylene	4
2-aminotoluene	144
4-aminotoluene	145
o-aminotoluene	144
p-aminotoluene	145
AMMONIUM FLUORIDE	6
AMMONIUM FLUOROSILICATE	7
ammonium fluosilicate	7
ammonium hexafluorosilicate	7
ammonium silicofluoride	7
Amprolene	72
ANCA 1040	20
anhydrous hydrazine	75
ANILINE	8
aniline oil	8
aniline, 2,4,6-trichloro-	147
aniline, 4,4'-methylene-bis[2-chloro-	50
aniline, m-nitro-	100
aniline, N,N-diethyl-	53
aniline, N,N-dimethyl-	56
aniline, N-methyl-	90
aniline, o-nitro-	101
aniline, p-chloro-	40
aniline, p-nitro-	102
anilinobenzene	64
anilinoethane	67
anilinomethane	90
ANISIDINE	9
p-anisidine, 2-nitro-	103
Anprolene	72
Anproline	72
Antol	68
Anyvim	8
Aquachloral	36
ARSENIC	10
arsenic acid	11
arsenic acid anhydride	11
arsenic anhydride	11
arsenic black	10
arsenic chloride	12
arsenic(III) chloride	12
arsenic hydride	14
arsenic oxide	13
arsenic oxide	11
arsenic(III) oxide	13
arsenic(V) oxide	11
ARSENIC PENTOXIDE	11
arsenic sesquioxide	13
ARSENIC TRICHLORIDE	12
arsenic trihydride	14
ARSENIC TRIOXIDE	13
arsenic-75	10
arsenicals	10
arsenious acid	13
arsenious chloride	12

269

INDEX

arsenious oxide	13	
arsenious trichloride	12	
arsenious trioxide	13	
arseniuretted hydrogen	14	
arsenous acid	13	
arsenous acid anhydride	13	
arsenous anhydride	13	
arsenous chloride	12	
arsenous hydride	14	
arsenous oxide	13	
arsenous oxide anhydride	13	
ARSINE	14	
Artificial ant oil	74	
Atul Fast Bordeaux GP Base	103	
Axoene fast Bordeaux GP Salt	103	
azacyclopropane	15	
aziran	15	
AZIRIDINE	15	
aziridine, 2-methyl-	91	
Azoamine Red Zh	102	
Azobase MNA	100	
Azobase NAS	103	
Azoene Fast Bordeaux GP Base	103	
Azoene Fast Orange GR Base	101	
Azoene Fast Orange GR Salt	101	
Azoene Fast Scarlet GC Base	109	
Azoene Fast Scarlet GC Base	109	
Azoene Fast Scarlet GC Salt	109	
Azoene Fast Scarlet GC Salt	109	
Azofix Bordeaux GP	103	
Azofix Orange GR	101	
Azofix Scarlet G Salt	109	
Azogene Fast Blue B	48	
Azogene Fast Bordeaux G	103	
Azogene Fast Scarlet G	109	

B

Ba 35846	121
Bacillol	45
Baker's P & S liquid ointment	114
BASF Ursal D	116
Batasan	149
benzanenamine, 4-methyl-	145
benzenamine	8
benzenamine, 2,4,6-trichloro	147
benzenamine, 2-methyl	144
benzenamine, 2-methyl-5-nitro-	109
benzenamine, 2-nitro-	101
benzenamine, 3-nitro-	100
benzenamine, 4,4'-methylene-bis[2-chloro-	50
benzenamine, 4-chloro-	40
benzenamine, 4-ethoxy-	113
benzenamine, 4-methoxy-2-nitro-	103
benzenamine, 4-nitro-	102
benzenamine, ar,ar-dimethyl-	155
Benzenamine, N,N,4-trimethyl	60
benzenamine, N,N-diethyl-	53
benzenamine, N,N-dimethyl-	56
benzenamine, N-methyl-	90
benzenamine, N-phenyl-	64
benzene, (epoxyethyl)-	118
benzene, 1-chloro-4-nitro-	44
benzene, 1-methyl-2,4-dinitro-	63
benzene, 1-methyl-2-nitro-	107
benzene, 1-methyl-4-nitro-	108
benzene-2,4-diisocyanato-1-methyl-	143
benzeneamine	8
1,3-benzenediamine	115
1,4-benzenediamine	116
m-benzenediamine	115
p-benzenediamine	116
1,4-benzenediamine, N,N-diethyl-	55
1,4-benzenediamine, N,N-dimethyl-	58
1,3-benzenediol	127
m-benzenediol	127
benzidine, 3,3'-dimethoxy-	48
benzidine, 3,3'-dimethyl-	142
Benzofur D	116
1,4-benzoquine	16
1,4-benzoquinone	16
p-BENZOQUINONE	16
Betaprone	124
Bi 3411	36
bichloride of mercury	84
Big dipper	64
bioxirane	24
2,2-bioxirane	24
[1,1'-biphenyl]-4,4'-diamine, 3,3'-dimethoxy-	48
[1,1'-biphenyl]-4,4'diamine, 3,3'-dimethyl-	142
BIS(2-CHLOROETHYL) ETHER	17
bis(chloro-2-ethyl)oxide	17
bis(hydroxymethyl)acetylene	26
bisulphite	137
1,1'-bi[ethylene oxide]	24
Blue - ox	157
Blue Base Irga B	48
Blue Bass NB	48
Blue BN Base	48
Blue Oil	8
Bonoform	139
Bordeaux Base Ciba IV	103
Bordeaux Base Irga IV	103
Bordeaux Base NGP	103
Bordeaux GP Base	103
Bordeaux GP Salt	103
Bordeaux GPS Salt	103
Bordeaux Salt Ciba IV	103
Bordeaux Salt NGP	103
boron bromide	18
boron chloride	19
boron fluoride	20
BORON TRIBROMIDE	18
BORON TRICHLORIDE	19
BORON TRIFLUORIDE	20
BPL	124
Brentamine Fast Bordeaux GP Base	103
Brentamine Fast Orange GR Base	101
Brentamine Fast Orange GR Salt	101
Brestan	149
Brestan 60	149
Bromo-o-gas	22
bromoacetate ion	21
2-bromoacetic acid	21
α-bromoacetic acid	21
BROMOACETIC ACID	21
bromoacetic acid, ethyl ester	68
Bromofume	71
BROMOMETHANE	22
1-BROMOPROPANE	23
1,3-butadiene diepoxide	24
BUTADIENE DIEPOXIDE	24
butadiene dioxide	24
butane diepoxide	24
butane, 1,2;3,4-diepoxy-	24
butanenitrile	27
n-butanenitrile	27
2-butenal	46
butoxycarbonyl chloride	25
buttercup yellow	156
butyl chlorocarbonate	25
BUTYL CHLOROFORMATE	25
2-butyne-1,4-diol	26
butynediol	26
2-butynediol	26
1,4-BUTYNEDIOL	26
butyric acid nitriie	27

INDEX

n-butyronitrile	27	
BUTYRONITRILE	27	
C		
C.I. 10355	64	
C.I. 37025	101	
C.I. 37030	100	
C.I. 37035	102	
C.I. 37105	109	
C.I. 37107	145	
C.I. 37135	103	
C.I. 37230	142	
C.I. 76000	8	
C.I. 76060	116	
C.I. 76075	58	
C.I. 76505	127	
CI 77180	28	
C.I. 77760	88	
C.I. 77805	128	
C.I. Azoic Coupling Component 107	145	
C.I. Azoic Diazo Component 1	103	
C.I. Azoic Diazo Component 6	101	
C.I. Azoic Diazo Component 7	100	
C.I. Azoic Diazo Component 12	109	
C.I. Azoic Diazo Component 87	102	
C.I. Azoic Diazo Component 113	142	
C.I. Developer 11	115	
C.I. Developer 13	116	
C.I. Developer 17	102	
C.I. Developer 4	127	
CI Disperse Black 6	48	
C.I. Oxidation base 1	8	
C.I. Oxidation Base 10	116	
C.I. Oxidation Base 31	127	
C.I. Pigment Yellow 32	136	
p-CA	40	
caddy	29	
CADMIUM	28	
CADMIUM CHLORIDE	29	
CADMIUM CYANIDE	30	
cadmium dichloride	29	
cadmium dicyanide	30	
cadmium difluoride	31	
cadmium dinitrate	32	
CADMIUM FLUORIDE	31	
cadmium fume	33	
cadmium monosulphate	34	
cadmium monoxide	33	
CADMIUM NITRATE	32	
CADMIUM OXIDE	33	
cadmium oxide fume	33	
cadmium sulfate	34	
CADMIUM SULPHATE	34	
calochlor	84	
carbamonitrile	47	
carbimide	47	
carbolic acid	114	
carbon dichloride oxide	119	
CARBON MONOXIDE	35	
carbon oxide (CO)	35	
carbon oxychloride	119	
carbonchloridic acid, propyl ester	126	
carbonic dichloride	119	
carbonic oxide	35	
carbonochloridic acid ethyl ester	70	
carbonochloridic acid, butyl ester	25	
carbonochloridic acid, methyl ester	92	
carbonyl	119	
cathyl chloride	70	
CD F2	31	
Cellitazol B	48	
Cellon	139	
Celmide	71	
CG	119	
Chem-tol	112	
Chemox PE	62	
chinone	16	
Chissonox 206 monomer	153	
α-chloracetic acid	38	
chloracetonitrile	39	
CHLORAL HYDRATE	36	
Chloraldurat	36	
chlorallylene	5	
chlorex	17	
chloride, carbonyl dichloride	119	
CHLORINE	37	
chlorine mol.	37	
3-chloro-1-propylene	5	
1-CHLORO-2,4-DINITRO-BENZENE	41	
1-chloro-2-propene	5	
Chloroacetic acid	38	
CHLOROACETIC ACID	38	
chloroacetic acid ethyl ester	69	
2-chloroacetonitrile	39	
α-chloroacetonitrile	39	
CHLOROACETONITRILE	39	
chloroallylene	5	
4-chloroaniline	40	
p-CHLOROANILINE	40	
4-chlorobenzenamine	40	
chlorocarbonic acid ethyl ester	70	
chlorocarbonic acid, methyl ester	92	
1-chloro-2-(β-chloroethoxy)ethane	17	
chlorodimethyl ether	43	
chloroethanoic acid	38	
chloroethanol	42	
β-chloroethanol	42	
δ-chloroethanol	42	
2-CHLOROETHANOL	42	
chloroethene	152	
2-chloroethyl alcohol	42	
β-chloroethyl alcohol	42	
2-chloroethyl ether	17	
chloroethylene	152	
chloroformic acid	70	
chloroformic acid, butyl ester	25	
chloroformic acid, methyl ester	92	
chloroformyl chloride	119	
chloromethoxymethane	43	
chloromethyl cyanide	39	
CHLOROMETHYL METHYL ETHER	43	
4-chloronitrobenzene	44	
p-CHLORONITROBENZENE	44	
Chlorophen	112	
4-chlorophenylamine	40	
p-chlorophenylamine	40	
chloropicrin	148	
3-chloropropene	5	
1-chloropropene-2	5	
3-chloropropylene	5	
α-chloropropylene	5	
chromic acid (H_2CrO_4), strontium salt (1:1)	136	
chromic acid (H_2CrO_4), zinc salt (1:1)	156	
chromic acid, zinc salt ($ZnCrO_4$)	156	
chromium zinc oxide ($ZnCrO_4$)	156	
Cianurina	85	
Cibacete Diazo Navy Blue 2B	48	
Citrine ointment	87	
colloidal arsenic	10	
colloidal selenium	128	
corrosive sublimate	84	
CRESOLS	45	
cresylic acid	45	
CROTONALDEHYDE	46	
crude arsenic	13	
Cryptogil OL	112	
cryptohalite	7

INDEX

Term	Page
Curafume	22
CYANAMIDE	47
cyanoacetonitrile	81
cyanoamine	47
N-cyanoamine	47
cyanogen nitride	47
cyanogenamide	47
1-cyanopropane	27
2-cyanopropene	89
cyanotrichloromethane	146
1,4-cyclohexadiene dioxide	16
2,5-cyclohexadiene-1,4-dione	16
1,4-cyclohexadienedione	16
cyclohexane, 1,1'-methylene-bis[isocyanatodicyclohexyl-methane diisocyanate	52
Cyclon	76
Cyclone B	76

D

Term	Page
Dainichi Fast Scarlet G Base	109
Daito Bordeaux Base GP	103
Daito Bordeaux Salt GP	103
Daito Orange Base R	100
Daito Scarlet Base G	109
Dawson 100	22
DBE	71
DCA	49
3,4-DCA	49
DCEE	17
DEA	53
DEB	24
Deep Lemon Yellow	136
Delusol	157
Desmodur T80	143
Developer C	115
Developer H	115
Developer M	115
Developer O	127
Developer P	102
Developer PF	116
Developer R	127
Developer RS	127
Devol Bordeaux B	103
Devol Boreaux GP Salt	103
Devol Orange B	101
Devol Orange R	100
Devol Orange Salt B	101
Devol Red GG	102
Devol Scarlet B	109
Devol Scarlet G Salt	109
DFA	64
4,4'-di-o-toluidine	142
4,4'-di-o-toluidine	142
di-isocyanate de toluylene	143
Diabase Bordeaux GP	103
Diabase Scarlet G	109
Diacel Navy DC	48
Diamet Kh	50
diamide	75
diamine	75
4,4'-diamino-3,3'-dimethyl-biphenyl	142
4,4'-diamino-3,3'dimethyl(1,1'-biphenyl)	142
1,3-diaminobenzene	115
1,4-diaminobenzene	116
m-diaminobenzene	115
p-diaminobenzene	116
1,4-diaminobenzol	116
diaminoditolyl	142
diammonium fluosilicate	7
diammonium hexafluorosilicate	7
diammonium hexafluorosilicate(2-)	7
diammonium silicon hexafluoride	7
dianisidine	48
o-DIANISIDINE	48
diarsenic pentoxide	11
diarsenic trioxide	13
Diasalt Bordeaux GP	103
Diazo Fast Bordeaux GP	103
Diazo Fast Orange GR	101
Diazo Fast Orange R	100
Diazo Fast Scarlet G	109
1,2-dibromoethane	71
sym-dibromoethane	71
α,β-dibromoethane	71
dibromomercury	83
3,4-dichloranilin	49
3,4-dichloraniline	49
2,2'-dichlorethyl ether	17
sym-dichloroethyl ether	17
dichlorine	37
1,1-DICHLORO-1-NITRO-ETHANE	51
2,2-DICHLORO-4,4'-METHYLENE-DIANILINE (MOCA)	50
4,5-dichloroaniline	49
3,4-DICHLOROANILINE	49
3,4-dichlorobenzenamine	49
dichlorocadmium	29
2,2'-dichlorodiethyl ether	17
β,β'-dichlorodiethyl ether	17
$\alpha,\alpha,$-dichlorodimethyl ether	43
dichloroether	17
dichloroethyl ether	17
2,2'-dichloroethyl ether	17
β,β'-dichloroethyl ether	17
dichloroethyl oxide	17
di(β-chloroethyl)ether	17
dichloromercury	84
dicyanomercury	85
dicyanomethane	81
DICYCLOHEXYLMETHANE-4,4'-DIISOCYANATE	52
diepoxybutane	24
1,2,3,4-diepoxybutane	24
1,2;3,4-diepoxybutane	24
4-(diethylamino)aniline	55
p-(diethylamino)aniline	55
N,N-diethylaminobenzene	53
diethylaniline	53
N,N-DIETHYLANILINE	53
N,N-diethylbenzenamine	53
DIETHYLENE GLYCOL DIACRYLATE	54
diethylene glycol dichloride	17
diethylphenylamine	53
N,N-DIETHYLPHENYLENE-DIAMINE	55
dihydro-1H-azirine	15
1,2-dihydro-5-nitro-acenaphthylene	99
dihydroazirene	15
dihydrogen selenide	132
dihydrooxirene	72
m-dihydroxybenzene	127
diisocyanatotoluene	143
dimenthylopropane diacrylate	96
3,3'-dimethlbiphenyl-4,4'-diamine	142
3,3'-dimethoxybenzidine	48
dimethyl sulfate	59
DIMETHYL SULPHATE	59
3,3'-dimethyl-(1,1'-biphenyl)4,4'-diamine	142
3,3'-dimethyl-4,4'-biphenyl-diamine	142
3,3'-dimethyl-4-4'-diphenyl-diamine	142
dimethyl-p-phenylenediamine	58
dimethyl-p-toluidine	60
dimethylamine, N-nitroso-	57
4-(dimethylamino)aniline	58
p-(dimethylamino)aniline	58
(dimethylamino)benzene	56
p-(dimethylamino)toluene	60
p-dimethylaminophenylamine	58

INDEX

Term	Page
dimethylaniline	155
dimethylaniline	56
N,N-DIMETHYLANILINE	56
N,N-dimethylbenzenamine	56
3,3'-dimethylbenzidine	142
dimethylene oxide	72
dimethyleneimine	15
dimethylphenylamine	155
dimethylnitromethane isonitropropane	106
DIMETHYLNITROSAMINE	57
dimethylolpropane diacrylate	96
dimethylphenol	154
dimethylphenylamine	56
N,N-dimethylphenylamine	56
dimethylphenylaminium chloride	56
N,N-DIMETHYLPHENYLENEDIAMINE	58
N,N-DIMETHYLTOLUIDINE	60
2,2-dimethyltrimethylene acrylate	96
dinitrochlorobenzene	41
2,4-dinitrochlorobenzene	41
dinitrogen dioxide	61
dinitrogen tetraoxide	61
DINITROGEN TETROXIDE	61
α-dinitrophenol	62
2,4-DINITROPHENOL	62
2,4-dinitrophenyl chloride	41
DINITROTOLUENE	63
2,4-dinitrotoluol	63
Dinofan	62
1,4-dioxybenzene	16
dioxybutadiene	24
N,N-diphenylamine	64
DIPHENYLAMINE	64
diphosgene	119
dipotassium hexafluorosilicate	122
dipotassium hexafluorosilicate(2-)	122
Direct Brown BR	115
Direct Brown GG	115
diselenium dichloride (Se_2Cl_2)	133
DMN	57
DMNA	57
DMPD	58
DNC8	41
2,4-DNP	62
DNT	63
2,4-DNT	63
Dolochlor	148
Dormal	36
Dow Pentachlorophenol DP-2 Antimicrobial	112
Dowcide 7	112
Dowco 186	150
Dowfume	22
Dowfume EDB	71
Dowfume MC-2 soil fumigant	22
Dowfume W-100	71
Dowfume W-8	71
Dowfume W-85	71
Dowfume W-90	71
Dowicide 6	140
Dowicide EC-7	112
Dowicide G	112
DPA	64
DPD	55
Du-Ter	150
Durafur Black R	116
Durafur Developer G	127
Durgasol Bordeaux GP Salt	103
Durotox	112

E

Term	Page
E.O.	72
ECF	70
EDB	71
EDB-85	71
EDB-Bee	71
Edco	22
EI	15
Electronic E-2	132
Embafume	22
Emisan 6	84
ENT 25208	149
ENT 28009	150
ENT 4,504	17
ENT-15,345	71
ENT-26263	72
ENT-26592	24
ENT-50324	15
epihydrin alcohol	65
2,3-EPOXY-1-PROPANOL	65
epoxyethane	72
1,2-epoxyethane	72
(epoxyethyl)benzene	118
1,2-epoxyethylbenzene	118
2,3-epoxypropan-1-ol	65
epoxypropane	95
1,2-epoxypropane	95
2,3-EPOXYPROPYL ACRYLATE	66
epoxystyrene	118
α,β-epoxystyrene	118
Erithane	150
erythritol anhydride	24
essence of mirbane	104
essence of myrbane	104
ethane pentachloride	111
ethane, 1,1'-oxybis[2-chloro-]	17
ethane, 1,1,2,2-tetrachloro-	139
ethane, 1,2-epoxy-1-phenyl-	118
ethane, pentachloro-	111
ethane,1,1,2,2-tetrabromo-	138
ethanol, 2-chloro-	42
ethene chlorohydrin	42
ethene oxide	72
ethene, chloro-	152
ether, bis(2-chloroethyl)	17
ether, bis(chloroethyl)bis(β-chloroethyl)ether	17
ether, chloromethyl methyl	43
ethide	51
4-ethoxyaniline	113
p-ethoxyaniline	113
4-ethoxybenzenamine	113
ethoxycarbonylmethylbromide	68
ethyl 2-bromoacetate	68
ethyl 2-chloroacetate	69
ethyl bromacetate	68
ethyl α-bromoacetate	68
ETHYL BROMOACETATE	68
ethyl carbonochloridate	70
ethyl chloroacetate	69
ethyl α-chloroacetate	69
ETHYL CHLOROACETATE	69
ethyl chlorocarbonate	70
ethyl chloroethanoate	69
ETHYL CHLOROFORMATE	70
ethyl chloromethanoate	70
ethyl ester	70
ethyl monobromoacetate	68
ethyl monochloracetate	69
N-ethylaminobenzene	67
ethylaniline	67
N-ETHYLANILINE	67
N-ethylbenzenamine	67
ethylchlorohydrin	42
ethylchlorohydrine	42
ethylene bromide	71
ethylene carboxamide	2
ethylene chlorohydrin	42
1,2-ethylene dibromide	71
ETHYLENE DIBROMIDE	71
ethylene glycol, chlorohydrin	42
ethylene monochloride	152

INDEX

Term	Page
ETHYLENE OXIDE	72
ethylene, chloro-	152
ethyleneimine	15
ethylenimine	15
ethylimine	15
ethylphenylamine	67
ethylynyl methanol	125
ethynylcarbinol	125
ETO	72
exhaust gas	35
Exolit LPKN	120
Exolit VPK-n	120
Exolite	120
Exolite 405	120
Exsel	130

F

Term	Page
Fast Blue B Base	48
Fast Blue Base B	48
Fast Blue DSC Base	48
Fast Bordeaux 3NA Base	103
Fast Bordeaux Base GP	103
Fast Bordeaux Base J	103
Fast Bordeaux GDN	103
Fast Bordeaux GND base	103
Fast Bordeaux GP	103
Fast Bordeaux GP Base	103
Fast Bordeaux GP Salt	103
Fast Bordeaux GP-T Base	103
Fast Bordeaux Salt GP	103
Fast Bordeaux Salt J	103
Fast Dark Blue Base R	142
Fast GPN	103
Fast Orange Base JR	101
Fast Orange Base R	100
Fast Orange M Base	100
Fast Orange MM Base	100
Fast Orange O Base	101
Fast Orange O Salt	101
Fast Orange R Base	100
Fast Orange R Salt	100
Fast Orange Salt JR	101
Fast Red 2G Base	102
Fast Red Base 2J	102
Fast Red Base GG	102
Fast Red GG Base	102
Fast Red MP Base	102
Fast Red P Base	102
Fast Red SG Base	109
Fast Scarlet Base J	109
Fast Scarlet G	109
Fast Scarlet G Base	109
Fast Scarlet G Salt	109
Fast Scarlet GC Base	109
Fast Scarlet J Salt	109
Fast Scarlet M 4NT	109
Fast Scarlet T Base	109
Felsules	36
FEMA No. 2433	72
Fenolovo	150
Fenoloyo acetate	149
Fenoxyl Carbon N	62
fentin acetate	149
fentin hydroxide	150
fermenticide liquid	137
flue gas	35
FLUORINE	73
Flux Maag	98
formic acid, chloro-, butyl ester	25
formic acid, chloro-, methyl ester	92
formic acid, chloro-, propyl ester	126
2-formylfuran	74
Fouramine D	116
Fouramine RS	127
Fourrine 79	127
Fourrine EW	127
Fourrine I	116
Fourrine L	116
Fumo-gas	71
Fur Black 41867	116
Fur Brown 41866	116
Fur Yellow	116
fural	74
furaldehyde	74
2-FURALDEHYDE	74
furale	74
2-furanaldehyde	74
furancarbonal	74
2-furancarbonal	74
2-furancarboxaldehyde	74
furfural	74
2-furfural	74
furfuraldehyde	74
2-furfuraldehyde	74
furfurole	74
furfurylaldehyde	74
furole	74
α-furole	74
Furro D	116
2-furylaldehyde	74
2-furylcarboxaldehyde	74
Futramine D	116

G

Term	Page
G 25	148
GC 6936	149
Glazel Renta	112
glycide	65
glycidol	65
glycidyl acrylate	66
glycidyl alcohol	65
glycidyl propenoate	66
Glycol bromode	71
glycol chlorohydrin	42
Glycol dibromide	71
glycol monochlorohydrin	42
grey arsenic	10
guanidine, 1-methyl-3-nitro-1-nitroso-	94
guanidine, N-methyl-N'-nitro-N-nitroso-	94

H

Term	Page
Haitin	150
Halon 1001	22
Haltox	22
Hansol Bordeaux GP Salt	103
Hiltonil Fast Blue B Base	48
Hiltonil Fast Bordeaux GP Base	103
Hiltonil Fast Orange GR Base	101
Hiltonil Fast Orange R Base	100
Hiltonil Fast Scarlet G Base	109
Hiltonil Fast Scarlet G Salt	109
Hiltonil Fast Scarlet GC Base	109
Hiltosal Fast Bordeaux GP Salt	103
Hiltosal Fast Orange GR Salt	101
Hindasol Bordeaux GP Salt	103
Hindasol Orange GR Salt	101
hydracrylic acid β-lactone	124
Hydral	36
HYDRAZINE	75
hydrazine base	75
hydrazine, anhydrous	75
hydrazine, aqueous solution	75
hydrazine, phenyl	117
hydrazine-benzene	117
hydrazinobenzene	117
hydrocyanic acid, liquefied	76
hydrogen arsenide	14
hydrogen cyanamide	47
hydrogen cyanide	76
HYDROGEN CYANIDE	76
hydrogen selenide	132
hydrogen selenide (H_2Se)	132

INDEX

HYDROGEN SULPHIDE	77
hydrogen sulphuric acid	77
m-hydroquinone	127
3-hydroxy-1,2-epoxypropanol	65
3-hydroxy-1-propane sulphonic acid γ-sultone	123
3-hydroxy-1-propane sulphonic acid sulphone	123
3-hydroxy-1-propane sulphonic acid sultone	123
1-hydroxy-2,4-dinitrobenzene	62
hydroxybenzene	114
1,3-hydroxybenzene	127
2-hydroxyethyl chloride	42
β-hydroxyethyl chloride	42
2-hydroxyisobutyronitrile	1
α-hydroxyisobutyronitrile	1
2,3-(hydroxymethyl)oxirane	65
m-hydroxyphenol	127
3-hydroxypropene	3
3-hydroxypropionic acid lacton	124
3-hydroxypropylene oxide	65
hydroxytoluene	45
hydroxytriphenylstannane	150
hydroxytriphenyltin	150
Hylene TM	143
3-hyroxyphenol	127

I

2-iodoacetic acid	78
IODOACETIC ACID	78
IODOMETHANE	79
Iscobrome	22
Iscobrome D	71
isocyanatomethane	93
isocyanic acid, methyl ester	93
isocyanic acid, 4-methyl-m-phenylene ester	143
isocyanic acid, methylphenylene ester	143
isopropene cyanide	89
isopropenylnitrile	89

K

K 19	150
Kako Bordeaux GP Base	103
Kako Bordeaux GP Salt	103
Kayafume	22
Kayaku Blue B Base	48
Kayaku Fast Bordeaux GP Base	103
Kayaku Fast Bordeaux Salt GP	103
Kayaku Scarlet G Base	109
Kessodrate	36
Kopfume	71
Kyanol	8

L

lactonitrile, 2-methyl-	1
Lake Blue B Base	48
Lake Maroon B base	103
Lake Scarlet G Base	109
Larvacide	148
Lauxtol A	112
Levoxine	75
Liromatin	149
Lirostanol	149
Lithosol Orange R Base	109
Lorinal	36

M

MAGNESIUM PHOSPHIDE	80
malonic acid dinitrile	81
malonic dinitrile	81
malonodinitrile	81
MALONONITRILE	81
mannitol mustard	75
Maroxol-50	62
MB	22
MBX	22
MCA	38
MEBR	22
Mercaptoacetate	141
mercaptoacetic acid	141
2-mercaptoacetic acid	141
2-mercaptoethonoic acid	141
α-mercatoacetic acid	141
mercuric bichloride	84
mercuric bromide	83
mercuric chloride	84
mercuric cyanide	85
mercuric dibromide	83
mercuric diiodide	86
mercuric iodide	86
mercuric nitrate	87
mercuric oxide	88
mercuric oxide (HgO)	88
MERCURY	82
mercury bichloride	84
mercury biiodide	86
MERCURY(II) BROMIDE	83
mercury bromide (HgBr$_2$)	83
MERCURY(II) CHLORIDE	84
mercury chloride (HgCl$_2$)	84
MERCURY(II) CYANIDE	85
mercury cyanide (Hg(CN)$_2$)	85
mercury dibromide	83
mercury dichloride	84
mercury dicyanide	85
mercury diiodide	86
mercury dinitrate	87
MERCURY(II) IODIDE	86
mercury iodide (HgI$_2$)	86
mercury monoxide	88
MERCURY(II) NITRATE	87
mercury nitrate (Hg(NO)$_3$)	87
MERCURY(II) OXIDE	88
mercury oxide (HgO)	88
mercury perchloride	84
mercury(2+) nitrate	87
mercury(2+) oxide	88
Merpol	72
Metafume	22
metallic arsenic	10
methacrylnitrile	89
methacrylonitrile	89
α-methacrylonitrile	89
METHACRYLONITRILE	89
methanamine, N-methyl-N-nitroso-	57
methane, bromo-	22
methane, chloromethoxy-	43
methane, iodo-	79
methane, isocyanato-	93
methane, trichloronitro-	148
Methogas	22
4-methoxy-2-nitroaniline	103
4-methoxy-2-nitroaniline	103
methoxyaniline	9
methoxybenzenamine	9
methoxycarbonyl chloride	92
methoxychloromethane	43
methoxymethyl chloride	43
methyl bromide	22
methyl chlorocarbonate	92
METHYL CHLOROFORMATE	92
methyl chloromethyl ether	43
methyl ethylene oxide	95
methyl iodide	79
METHYL ISOCYANATE	93
methyl lactonitrile	1
methyl sulfate	59
1-methyl-1-nitroso-3-nitro-guanidine	94
1-METHYL-3-NITRO-1-NITROSO-GUANIDINE	94

INDEX

N-methyl-N-nitroso-N'-nitro-guanidine	94	
4-methyl-phenylene diisocyante	143	
4-methyl-phenylene isocyanate	143	
2-methylacrylonitrile	89	
α-methylacrylonitrile	89	
(methylamino)benzene	90	
methylaniline	90	
2-methylaniline	144	
4-methylaniline	145	
o-methylaniline	144	
p-methylaniline	145	
N-METHYLANILINE	90	
2-methylazacyclopropane	91	
2-METHYLAZIRIDINE	91	
2-methylbenzenamine	144	
4-methylbenzenamine	145	
N-methylbenzenamine	90	
o-methylbenzenamine	144	
p-methylbenzenamine	145	
methylene cyanide	81	
methylenedinitrile	81	
2-methylethylenimine	91	
2-methyllactonitrile	1	
2-methylnitrobenzene	107	
4-methylnitrobenzene	108	
o-methylnitrobenzene	107	
p-methylnitrobenzene	108	
methylnitronitrosoguanidine	94	
methylnitrosoguanidine	94	
METHYLOXIRANE	95	
methylphenol	45	
methylphenylamine	90	
N-methylphenylamine	90	
2-methylpropenenitrile	89	
4-methylpyridine	121	
p-methylpyridine	121	
γ-methylpyridine	121	
3-(N-methylpyrollidino)pyridine	98	
MIA	78	
MIC	93	
Microlysin	148	
Millionate M	50	
mirbane oil	104	
Mitsui Blue B Base	48	
Mitsui Bordeaux GP Base	103	
Mitsui Bordeaux GP Salt	103	
Mitsui Scarlet G Base	109	
MKhUK	38	
MNA	100	
MNG	94	
MNNG	94	
MOCA	50	
molecular chlorine	37	
Mondur TDS	143	
monoallylamine	4	
monobromacetic acid	21	
monobromomethane	22	
monochloracetic acid	38	
monochloroacetic acid	38	
monochloroacetonitrile	39	
monochlorodimethyl ether	43	
monochloroethanoic acid	38	
monochloroethanoic acid ethyl ester	69	
2-monochloroethanol	42	
monochloroethylene	152	
monochloromethyl cyanide	39	
monochloromethyl methyl ether	43	
monohydroxybenzene	114	
monoiodoacetic acid	78	
monoiodomethane	79	
N-monomethylaniline	90	
monophenylhydrazine	117	
Muthmann's liquid	138	
N		
Nacconate IOO	143	
Nako H	116	
Nako TGG	127	
5-NAN	99	
naphthalene, 2-nitro-	105	
Naphthanil Blue B Base	48	
Naphthanil Bordeaux GP Base	103	
Naphthanil Diazo Bordeaux GP	103	
Naphthanil Scarlet G Base	109	
Naphthol AS-KG	145	
Naphthol AS-KGLL	145	
Naphthosol Fast Bordeaux GP Salt	103	
Naphtoelan Fast Bordeaux GP Base	103	
Naphtoelan Fast Bordeaux GP Salt	103	
Naphtoelan Fast Scarlet G Base	109	
Naphtoelan Fast Scarlet G Salt	109	
Naphtoelan Orange R Base	100	
Napthtoelan Red GG Base	102	
Natasol Bordeaux GP Salt	103	
Natasol Fast Orange GR	101	
NCI-C01865	63	
NCI-C01967	99	
NCI-C02039	40	
NCI-C02551	33	
NCI-C03736	8	
NCI-C50033	134	
NCI-C50088	72	
NCI-C50099	95	
NCI-C50124	114	
NCI-C50135	42	
NCI-C50533	143	
NCI-C53894	111	
NCI-C54933	112	
NCI-C55378	112	
NCI-C55845	16	
NCI-C56177	74	
NCI-C56655	112	
NCI-C60231	38	
NCI-C60399	82	
NCI-C60537	108	
NCI-CO3554	139	
NCI-CO4615	5	
NDMA	57	
Nefis	71	
neopentanediol diacrylate	96	
NEOPENTYL GLYCOL DIACRYLATE	96	
neutral ammonium fluoride	6	
Niax TDI	143	
nickel carbonyl	97	
NICKEL TETRACARBONYL	97	
nicotin	98	
NICOTINE	98	
L-nicotine	98	
(-)-nicotine	98	
S-nicotine	98	
Nitranilin	100	
p-nitraniline	102	
nitric acid, cadmium salt	32	
nitric acid, mercury(2+) salt	87	
N'-nitro-N-nitroso-N-methyl-guanidine	94	
5-NITRO-o-TOLUIDINE	109	
2-NITRO-p-ANISIDINE	103	
5-NITROACENAPHTHENE	99	
m-nitroaminobenzene	100	
2-nitroaniline	101	
3-nitroaniline	100	
4-nitroaniline	102	
m-NITROANILINE	100	
o-NITROANILINE	101	
p-NITROANILINE	102	
2-nitrobenzenamine	101	
3-nitrobenzenamine	100	
4-nitrobenzenamine	102	
NITROBENZENE	104	

INDEX

Term	Page
nitrobenzol	104
4-nitrochlorobenzene	44
p-nitrochlorobenzene	44
nitrochloroform	148
nitrogen oxide	61
nitrogen tetraoxide	61
nitrogen tetraoxide	61
Nitrokleenup	62
β-nitronaphthalene	105
2-NITRONAPHTHALENE	105
5-nitronaphthalene ethylene	99
Nitrophen	62
Nitrophene	62
p-nitrophenyl chloride	44
4-nitrophenylamine	102
m-nitrophenylamine	100
2-NITROPROPANE	106
N-nitroso-N-methylnitroguanidine	94
N-nitrosodimethylamine	57
nitrosoguanidine	94
2-nitrotoluene	107
4-nitrotoluene	108
o-NITROTOLUENE	107
p-NITROTOLUENE	108
4-nitrotoluol	108
nitrotrichloromethane	148
Nitrozol CF extra	102
NK ester A 2G	54
NK Ester A-NPG	96
No scald	64
Noctec	36
none	80
Nortec	36
NSC 1532	62
NSC 9369	94
Nycoton	36
Nycton	36

O

Term	Page
oil of mirbane	104
oil of myrbane	104
ONA	101
Orange Base Ciba II	101
Orange Base Irga	101
Orange Base Irga I	100
Orange Base Salt	101
Orange Ciba II	101
Orange Salt Irga II	101
Orsin	116
orvinyl carbinol	3
osmic acid	110
osmium oxide	110
OSMIUM TETROXIDE	110
7-oxabicyclo[4.1.0]heptane, 3-(epoxyethyl)-	153
7-oxabicyclo[4.1.0]heptane, 3-oxyranyl-	153
oxacyclopropane	72
oxane	72
1,2-oxathiolane, 2,2-dioxide	123
2-oxetanone	124
oxidoethane	72
α,β-oxidoethane	72
oxirane	72
oxirane, methyl-	95
oxirane, phenyl	118
oxiranemethanol	65
oxybenzene	114
oxydiethylene acrylate	54
oxydiethylene diacrylate	54
Oxyfume	72
Oxyfume 12	72
Oxytreat 35	75

P

Term	Page
Pelagol D	116
Pelagol DR	116
Pelagol Grey D	116
Pelagol Grey RS	127
Pelagol RS	127
Peltol D	116
Penta	112
Penta-Kil-Pentasol	112
PENTACHLOROETHANE	111
pentachlorophenate	112
2,3,4,5,6-pentachlorophenol	112
PENTACHLOROPHENOL	112
pentachlorophenol, Dowicide EC-7	112
pentachlorophenol, DP-2	112
pentachlorophenol, technical	112
Pentacon	112
pentalin	111
Penwar	112
Peratox	112
Permacide	112
Permagard	112
Permatox DP-2	112
Permatox Penta	112
Permite	112
Pestmaster	71
Pestmaster EDB-85	71
Phaldrone	36
Pharmasol Bordeaux GP	103
Pharmazoid Bordeaux GP	103
phenethylene oxide	118
p-phenetidin	113
4-phenetidine	113
p-PHENETIDINE	113
phenic acid	114
PHENOL	114
phenol, 2,3,4,6-tetrachloro-	140
phenol, 2,4-dinitro-	62
phenol, dimethyl-	154
phenol, methyl-	45
phentin acetate	149
phenyl hydrate	114
phenyl hydroxide	114
PHENYL OXIRANE	118
phenylamine	8
N-phenylaniline	64
N-phenylbenzenamine	64
1,4-phenylendeiamine	116
m-PHENYLENEDIAMINE	115
p-PHENYLENEDIAMINE	116
p-phenylenediamine, N,N-diethyl-	55
p-phenylenediamine, N,N-dimethyl-	58
1,3-phenylenediaminte	115
phenylethylene oxide	118
PHENYLHYDRAZINE	117
phenylic acid	114
phenylic alcohol	114
N-phenylmethylamine	90
2-phenyloxirane	118
phosgen	119
PHOSGENE	119
PHOSPHORUS, WHITE (YELLOW)	120
phosphorus-31	120
Phosvin	157
Picfume	148
p-picoline	121
γ-picoline	121
4-PICOLINE	121
Picride	148
PNA	102
PNCB	44
PNOT	109
PNT	108
POTASSIUM FLUORO-SILICATE	122
potassium fluorosilicate (K$_2$SiF$_6$)	122
potassium hexafluorosilicate	122

INDEX

Term	Page
potassium silicafluoride	122
potassium silicofluoride (K_2SiF_6)	122
Priltox	112
primrose yellow	156
propane sultone	123
γ-propane sultone	123
propane, 1,2-epoxy-	95
propane, 1-bromo-	23
propane, 2-nitro	106
propanedinitrile	81
propanenitrile, 2-hydroxy-2-methyl-	1
1-propanesulphonic acid-3-hydroxy-γ-sultone	123
1,3-PROPANESULTONE	123
3-propanolide	124
3-PROPANOLIDE	124
PROPARGYL ALCOHOL	125
2-propen-1-amine	4
2-propen-1-ol	3
2-propen-1-ol	3
propen-1-ol-3	3
1-propen-3-ol	3
propenamide	2
2-propenamide	2
2-propenamine	4
propene oxide	95
propene, 3-chloro	5
1-propene, 3-chloro-	5
2-propenenitrile, 2-methyl-	89
2-propenoic acid, 2,2-dimethyl-1,3-propanediyl ester	96
2-propenoic acid, oxiranylmethyl ester	66
2-propenoic acid, oxydi-2,1-ethanediyl ester	54
propenol	3
2-propenol	3
propenyl alcohol	3
2-propenyl alcohol	3
2-propenyl chloride	5
2-propenylamine	4
propiolactone	124
1,3-propiolactone	124
3-propiolactone	124
β-propiolactone	124
β-propionolactone	124
propyl bromide	23
n-propyl bromide	23
propyl chlorocarbonate	126
n-propyl chloroformate	126
PROPYL CHLOROFORMATE	126
propyl cyanide	27
propylene epoxide	95
propylene oxide	95
1,2-propylene oxide	95
propyleneimine	91
propylenimine	91
1,2-propylenimine	91
2-propyn-1-ol	125
1-propyne-3-ol	125
2-propynol	125
3-propynol	125
propynyl alcohol	125
2-propynyl alcohol	125
Prussic acid	76
PS	148
pyridine, 4-methyl-	121
pyridine,3-(1-methyl-2-pyrrolidinyl)-, s-	98
pyromucic aldehyde	74

Q

Term	Page
quicksilver	82
quinone	16
p-quinone	16
Quodorole	50

R

Term	Page
R 40B1	22
RCRA Waste No. P005	3
RCRA Waste No. P048	62
RCRA Waste No. P054	15
RCRA Waste No. P056	73
RCRA Waste No. P063	76
RCRA Waste No. P064	93
RCRA Waste No. P067	91
RCRA Waste No. P082	57
RCRA Waste No. P087	110
RCRA Waste No. P102	125
RCRA Waste No. PO24	40
RCRA Waste No. PO69	1
RCRA Waste No. U007	2
RCRA Waste No. U025	17
RCRA Waste No. U029	22
RCRA Waste No. U052	45
RCRA Waste No. U085	24
RCRA Waste No. U095	142
RCRA Waste No. U105	63
RCRA Waste No. U115	72
RCRA Waste No. U125	74
RCRA Waste No. U135	77
RCRA Waste No. U138	79
RCRA Waste No. U149	81
RCRA Waste No. U151	82
RCRA Waste No. U163	94
RCRA Waste No. U184	111
RCRA Waste No. U197	16
RCRA Waste No. U204	129
RCRA Waste No. U205	130
RCRA Waste No. U209	139
RCRA Waste No. U223	143
RCRA Waste No. U242	112
Rectules	36
Red 2G Base	102
red mercuric iodide	86
red mercuric oxide	88
Renal PF	116
resorcin	127
RESORCINOL	127
Rotox	22
Rubuinte TDI 80/20	143
Rumetan	157

S

Term	Page
Santar	88
Santar M	88
Santobrite	112
Sanyo Fast Bordeaux GP Base	103
Sanyo Fast Bordeaux Salt GP	103
Scaldip	64
Scarlet Base Ciba II	109
Scarlet Base Irga II	109
Scarlet Base NSP	109
Scarlet G Base	109
selane	132
SELENIUM	128
selenium chloride ($SeCl_4$)	135
selenium dihydride	132
SELENIUM DIOXIDE	129
SELENIUM DISULPHIDE	130
selenium fluoride (SeF_6)	131
SELENIUM HEXAFLUORIDE	131
SELENIUM HYDRIDE	132
SELENIUM MONOCHLORIDE	133
SELENIUM MONOSULPHIDE	134
selenium oxide (SeO_2)	129
selenium sulphide	134
selenium sulphide (SeS_2)	130
SELENIUM TETRACHLORIDE	135
Selenuim sulfide	134
Selsun Blue	130
Setacyl Diazo Navy R	48
Shell unkrauttod A	3
Shinnippan Fast Red GG Base	102

INDEX

Entry	Page
Shinnippon Fast Bordeaux GP Base	103
silicate (2-1), hexafluoro, dipotassium	122
silicate(2-), hexafluoro-, diammonium	7
Sinituho	112
SK-chloral hydrate	36
Soilbrom	71
Soilbrom-100	71
Soilbrom-40	71
Soilbrom-85	71
Soilbrom-90	71
Soilbrome-85	71
Soilfume	71
Solfo Black 2B Supra	62
Solfo Black B	62
Solfo Black BB	62
Solfo Black G	62
Solfo Black SB	62
Somnos	36
Somui sed	36
Sontec	36
SR 247	96
stannane, (acetyloxy) triphenyl-	149
stannane, acetoxytriphenyl-	149
stannane, hydroxytriphenyl-	150
Steara PBQ	16
sterilizing gas ethylene oxide 100%	72
stink damp	77
stromtium chromate (VI)	136
STRONTIUM CHROMATE	136
Strontium chromate (1:1)	136
strontium chromate (SrCrO$_4$)	136
strontium chromate 12170	136
strontium chromate A	136
strontium chromate X 2396	136
strontium yellow	136
Stutox	157
styrene 7,8-oxide	118
styrene epoxide	118
styrene oxide	118
styryl oxide	118
sublimate	84
Sugai Fast Bordeaux GP Base	103
Sugai Fast Scarlet G Base	109
Sulem	84
sulfur selenide	134
SULPHUR DIOXIDE	137
sulphur hydride	77
sulphur oxide (SO$_2$)	137
sulphurated hydrogen	77
sulphuric acid, cadmium salt	34
sulphuric acid, dimethyl ester	59
sulphurous acid anhydride	137
sulphurous anhydride	137
sulphurous oxide	137
Suzu	149
Suzu H	150
Symulon Bordeaux GP Base	103
Symulon Scarlet G Base	109

T

Entry	Page
T-Gas	72
TBE	138
TCP	140
2,4-TDI	143
Tekresol	45
Tenhide	150
Terabol	22
Term-I-Trol	112
Terr-o-gas 100	22
Tertrol D	116
Tertrosulphur Black PB	62
Tertrosulphur PBR	62
tetrabomoacetylene	138
1,1,2,2-TETRABROMOETHANE	138
1,1,2,2-tetrabromoethylene	138
tetracarbonylnickel	97
1,1,2,2-tetrachlorethane	139
tetrachloroethane	139
s-tetrachloroethane	139
1,1,2,2-TETRACHLOROETHANE	139
2,4,5,6-tetrachlorophenol	140
2,3,4,6-TETRACHLOROPHENOL	140
tetrachloroselenium	135
sym-tetrochloroethane	139
TGA2	54
2-thioglycolic acid	141
THIOGLYCOLIC ACID	141
thiovanic acid	141
Thompson's Wood Fix	112
tin triphenyl acetate	149
Tinestan	149
TL 1450	93
TL 337	15
TL 423	70
To NTu	21
o-tolidin	142
tolidine	142
2-tolidine	142
3,3'-tolidine	142
o,o'-tolidine	142
o-TOLIDINE	142
2,4-toluene diisocyanate	143
toluene, 2,4-dinitro-	63
toluene, o-nitro-	107
toluene, p-nitro-	108
TOLUENE-2,4-DIISOCYANATE	143
2,4-toluenediisocyanate	143
ar-toluenol	45
2-toluidine	144
4-toluidine	145
o-TOLUIDINE	144
p-TOLUIDINE	145
o-toluidine, 5-nitro-	109
p-toluidine, N,N,-dimethyl-	60
toluylene-2,4-diisocyanate	143
o-tolylamine	144
p-tolylamine	145
2,4-tolylenediisocyanate	143
Tosyl	36
TPTA	149
TPTH	150
Trawotoz	36
tribromoborane	18
tribromoboron	18
trichloroacetaldehyde hydrate	36
trichloroacetaldehyde monohydrate	36
TRICHLOROACETONITRILE	146
2,4,6-trichloroaniline	147
sym-trichloroaniline	147
TRICHLOROANILINE	147
trichloroarsine	12
2,4,6-trichlorobenzenamine	147
trichloroborane	19
trichloroboron	19
trichloromethyl cyanide	146
TRICHLORONITROMETHANE	148
tricresol	45
trifluoroborane	20
trifluoroboron	20
N,N,4-trimethylaniline	60
p-N,N-trimethylaniline	60
triphenylaceto stannane	149
triphenylhydroxytin	150
triphenylstannol	150
TRIPHENYLTIN ACETATE	149
TRIPHENYLTIN HYDROXIDE	150
triphenyltin oxide	150
tritox	146
trizinc diphosphide	157

INDEX

trona	18	Vandex	128		
Trovidur	152	Versneller NL 63/10	56	**X**	
TsAKS	47	VI-cad	29	XL all insecticide	98
Tubotin	149	vinyl C monomer	152	XYLENOLS	154
Tulabase Fast Bordeaux GP	103	vinyl carbinol	3	XYLIDINES	155
		VINYL CHLORIDE	152		
U		vinyl chloride monomer	152	**Y**	
Unifume	71	VINYL CYCLOHEXENE DIEPOXIDE	153	yellow mercury oxide	88
URANIUM	151	vinyl cyclohexene dioxide	153	**Z**	
Uranium I (^{238}U)	151	4-vinylcyclohexene diepoxide	153	Zaclon discoids	76
Uranium-238	151	4-vinylcyclohexene dioxide	153	ZINC CHROMATE	156
Ursol D	116	VP 19-40	149	zinc chromate ($ZnCrO_4$)	156
USAF A-4600	81			zinc chromium oxide ($ZnCrO_4$)	156
USAF EK-1995	47	**W**		ZINC PHOSPHIDE	157
USAF KF-5	39	weed drench	3	zinc tetraoxychromate	156
USAF P-220	16	Weedone	112	zinc tetroxychromate	156
USAF RH-8	1	Westron	139	zinc yellow	156
		white arsenic	13	Zoba Black D	116
V		WLN	31	Zytox	22
Vancide KS	150				

CAS REGISTRY NUMBER INDEX

Data sheets 1-79 appear in Volume 4a.

Data sheets 80-157 appear in Volume 4b.

51-28-5	2,4-Dinitrophenol	62
54-11-5	Nicotine	98
57-57-8	3-Propanolide	124
58-90-2	2,3,4,6-Tetrachlorophenol	140
62-53-3	Aniline	8
62-75-9	Dimethylnitrosamine	57
64-69-7	Iodoacetic acid	78
68-11-1	Thioglycolic acid	141
70-25-7	1-Methyl-3-nitro-1-nitrosoguanidine	94
74-83-9	Bromomethane	22
74-88-4	Iodomethane	79
74-90-8	Hydrogen cyanide	76
75-01-4	Vinyl chloride	152
75-21-8	Ethylene oxide	72
75-44-5	Phosgene	119
75-55-8	2-Methylaziridine	91
75-56-9	Methyloxirane	95
75-86-5	Acetone cyanohydrin	1
76-01-7	Pentachloroethane	111
76-06-2	Trichloronitromethane	148
76-87-9	Triphenyltin hydroxide	150
77-78-1	Dimethyl sulphate	59
79-06-1	Acrylamide	2
79-08-3	Bromoacetic acid	21
79-11-8	Chloroacetic acid	38
79-22-1	Methyl chloroformate	92
79-27-6	1,1,2,2-Tetrabromoethane	138
79-34-5	1,1,2,2-Tetrachloroethane	139
79-46-9	2-Nitropropane	106
87-86-5	Pentachlorophenol	112
88-72-2	o-Nitrotoluene	107
88-74-4	o-Nitroaniline	101
90-04-0	Anisidine	9
91-66-7	N,N-Diethylaniline	53
93-05-0	N,N-Diethylphenylenediamine	55
95-53-4	o-Toluidine	144
96-09-3	Phenyl oxirane	118
96-96-8	2-Nitro-p-anisidine	103
97-00-7	1-Chloro-2,4-dinitrobenzene	41
98-01-1	2-Furaldehyde	74
98-95-3	Nitrobenzene	104
99-09-2	m-Nitroaniline	100
99-55-8	5-Nitro-o-toluidine	109
99-98-9	N,N-Dimethylphenylenediamine	58
99-99-0	p-Nitrotoluene	108
100-00-5	p-Chloronitrobenzene	44
100-01-6	p-Nitroaniline	102
100-61-8	N-Methylaniline	90
100-63-0	Phenylhydrazine	117
101-14-4	2,2-Dichloro-4,4'-methylenedianiline (MOCA)	50
103-69-5	N-Ethylaniline	67
104-94-9	Anisidine	9
105-36-2	Ethyl bromoacetate	68
105-39-5	Ethyl chloroacetate	69
106-47-8	p-Chloroaniline	40
106-49-0	p-Toluidine	145
106-50-3	p-Phenylenediamine	116
106-51-4	p-Benzoquinone	16
106-87-6	Vinyl cyclohexene diepoxide	153
106-90-1	2,3-Epoxypropyl acrylate	66
106-93-4	Ethylene dibromide	71
106-94-5	1-Bromopropane	23
107-05-1	Allyl chloride	5
107-07-3	2-Chloroethanol	42
107-11-9	Allylamine	4
107-14-2	Chloroacetonitrile	39
107-18-6	Allyl alcohol	3
107-19-7	Propargyl alcohol	125
107-30-2	Chloromethyl methyl ether	43
108-45-2	m-Phenylenediamine	115
108-46-3	Resorcinol	127
108-89-4	4-Picoline	121
108-95-2	Phenol	114
109-61-5	Propyl chloroformate	126
109-74-0	Butyronitrile	27
109-77-3	Malononitrile	81
110-65-6	1,4-Butynediol	26
111-44-4	Bis(2-chloroethyl) ether	17
119-90-4	o-Dianisidine	48
119-93-7	o-Tolidine	142
121-14-2	Dinitrotoluene	63
121-69-7	N,N-Dimethylaniline	56
122-39-4	Diphenylamine	64
123-73-9	Crotonaldehyde	46

INDEX

CAS	Name	Page
126-98-7	Methacrylonitrile	89
151-56-4	Aziridine	15
156-43-4	p-Phenetidine	113
302-01-2	Hydrazine	75
302-17-0	Chloral hydrate	36
420-04-2	Cyanamide	47
541-41-3	Ethyl chloroformate	70
542-83-6	Cadmium cyanide	30
545-06-2	Trichloroacetonitrile	146
556-52-5	2,3-Epoxy-1-propanol	65
581-89-5	2-Nitronaphthalene	105
584-84-9	Toluene-2,4-diisocyanate	143
592-04-1	Mercury(II) cyanide	85
592-34-7	Butyl chloroformate	25
594-72-9	1,1-Dichloro-1-nitroethane	51
602-87-9	5-Nitroacenaphthene	99
624-83-9	Methyl isocyanate	93
630-08-0	Carbon monoxide	35
634-93-5	Trichloroaniline	147
900-95-8	Triphenyltin acetate	149
1120-71-4	1,3-Propanesultone	123
1300-71-6	Xylenols	154
1300-73-8	Xylidines	155
1303-28-2	Arsenic pentoxide	11
1306-19-0	Cadmium oxide	33
1314-84-7	Zinc phosphide	157
1319-77-3	Cresols	45
1327-53-3	Arsenic trioxide	13
1464-53-5	Butadiene diepoxide	24
2223-82-7	Neopentyl glycol diacrylate	96
4074-88-8	Diethylene glycol diacrylate	54
7439-97-6	Mercury	82
7440-38-2	Arsenic	10
7440-43-9	Cadmium	28
7440-61-1	Uranium	151
7446-08-4	Selenium dioxide	129
7446-09-5	Sulphur dioxide	137
7446-34-6	Selenium monosulphide	134
7487-94-7	Mercury(II) chloride	84
7488-56-4	Selenium disulphide	130
7637-07-2	Boron trifluoride	20
7723-14-0	Phosphorus, white (yellow)	120
7774-29-0	Mercury(II) iodide	86
7782-41-4	Fluorine	73
7782-49-2	Selenium	128
7782-50-5	Chlorine	37
7783-06-4	Hydrogen sulphide	77
7783-07-5	Selenium hydride	132
7783-79-1	Selenium hexafluoride	131
7784-34-1	Arsenic trichloride	12
7784-42-1	Arsine	14
7789-06-2	Strontium chromate	136
7789-47-1	Mercury(II) bromide	83
7790-79-6	Cadmium fluoride	31
10025-68-0	Selenium monochloride	133
10026-03-6	Selenium tetrachloride	135
10045-94-0	Mercury(II) nitrate	87
10102-44-0	Dinitrogen tetroxide	61
10108-64-2	Cadmium chloride	29
10124-36-4	Cadmium sulphate	34
10294-33-4	Boron tribromide	18
10294-34-5	Boron trichloride	19
10325-94-7	Cadmium nitrate	32
12057-74-8	Magnesium phosphide	80
12125-01-8	Ammonium fluoride	6
13463-39-3	Nickel tetracarbonyl	97
13530-65-9	Zinc chromate	156
16871-90-2	Potassium fluorosilicate	122
16919-19-0	Ammonium fluorosilicate	7
20816-12-0	Osmium tetroxide	110
21908-53-2	Mercury(II) oxide	88
27134-27-6	3,4-Dichloroaniline	49
28605-81-4	Dicyclohexylmethane-4,4'-diisocyanate	52
29256-93-7	N,N-Dimethyltoluidine	60